Plant Adaptation to Global Climate Change

Plant Adaptation to Global Climate Change

Editor

Amit Kumar Mishra

MDPI • Basel • Beijing • Wuhan • Barcelona • Belgrade • Manchester • Tokyo • Cluj • Tianjin

Editor
Amit Kumar Mishra
Department of Botany
Mizoram University
Aizawl
India

Editorial Office
MDPI
St. Alban-Anlage 66
4052 Basel, Switzerland

This is a reprint of articles from the Special Issue published online in the open access journal *Atmosphere* (ISSN 2073-4433) (available at: www.mdpi.com/journal/atmosphere/special_issues/Plant_Climate_Change).

For citation purposes, cite each article independently as indicated on the article page online and as indicated below:

LastName, A.A.; LastName, B.B.; LastName, C.C. Article Title. *Journal Name* **Year**, *Volume Number*, Page Range.

ISBN 978-3-0365-1528-1 (Hbk)
ISBN 978-3-0365-1527-4 (PDF)

Contents

About the Editor

Amit Kumar Mishra

Dr. Amit Kumar Mishra is currently working as an Assistant Professor at the Department of Botany, Mizoram University, Aizawl, India. Prior to joining Mizoram University, he worked as a post-doctoral research associate at Montana State University, USA; Texas A&M University, USA; and Ben-Gurion University of the Negev, Israel. He attained his PhD degree in botany from the Institute of Science, Banaras Hindu University, India. He utilizes integrated phenotypical, physiological, biochemical, and molecular approaches for unraveling the mysteries of plant survival under abiotic stresses. He has published papers in reputed peer-reviewed international journals. Dr. Mishra has received several awards of national and international repute. He is a life member of the Plant Sciences Section of the Indian Science Congress and an expert member of the Agro-ecosystem Specialist's group, IUCN-CEM (2017-2021). His area of expertise is plant stress physiology.

Preface to "Plant Adaptation to Global Climate Change"

The issue of climate change is inevitably accompanied by climate variabilities, such as high temperatures, varying patterns of rainfall, and other environmental factors (including biotic factors), and causes an undesirable impact on plant growth and global food security. The effect of climate change on vegetation may stem from the cellular to molecular levels. Consequently, the current literature on the effect of different environmental factors on vegetation is varied. In view of the future impacts of climate change, understanding the response of plants becomes critical in developing strategies to cope with the threats to plant growth and development and in advancing our existing knowledge about the influence of climate change on vegetation.

This book emerged from the Special Issue "Plant Adaptation to Global Climate Change", published in the journal *Atmosphere*. The articles presented in this book highlight important aspects concerning the impact of global climate change on vegetation. These studies are of interest to the environmental science research community, including those interested in assessing climate change impacts on vegetation and researchers working on simulation modeling.

The editor thanks the authors who generously contributed their time and expertise to ensuring the high quality of this work. The editor especially thanks Prof. Shashi Bhushan Agrawal and Prof. Madhoolika Agrawal, Laboratory of Air Pollution and Global Climate Change, Department of Botany, Institute of Science, Banaras Hindu University, India, and family members for constant inspiration. The editor also expresses his gratitude to the editors of the journal *Atmosphere*, to the reviewers, and to the production team members for their invaluable support and teamwork in the publication of the book. I sincerely hope that this work promotes further approaches to increasing the current understanding of the impact of climate change on vegetation.

Amit Kumar Mishra
Editor

atmosphere

MDPI

Editorial
Plant Adaptation to Global Climate Change

Amit Kumar Mishra

Department of Plant Sciences and Plant Pathology, Montana State University, Bozeman, MT 59717, USA; amit.mishra@montana.edu or amit.bhu.bot@gmail.com; Tel.: +1-830-486-3036

Citation: Mishra, A.K. Plant Adaptation to Global Climate Change. *Atmosphere* **2021**, *12*, 451. https://doi.org/10.3390/atmos12040451

Received: 29 March 2021
Accepted: 31 March 2021
Published: 31 March 2021

The problem of climate change is unavoidably accompanied by climate variabilities, such as high temperature, varying patterns of rainfall, and other environmental factors (including biotic factors), and causes an adverse impact on plant development and global food security. The effect of climate change on vegetation may be from cellular to the molecular level. Consequently, the existing literature on the plant's response to different environmental factors is varied. In view of the future impacts of climate change, understanding the response of plants becomes critical in developing strategies to cope with the threats to plant growth and development. To advance our current knowledge on the impact of climate change on vegetation, articles focusing on the urban, regional, and global levels as well as modeling studies were collected in this Special Issue. The *Atmosphere* Special Issue entitled "Plant Adaptation to Global Climate Change" comprises 13 original papers.

The impact of climate change on the harvested area, yield, and production of sugarcane has been studied in Thailand [1]. The study concluded a projected decrease in future sugarcane yield, harvested area, and production by 23.9–33.2%, 1.3–2.5%, and 24.9–34.9%, respectively, using the spatial regression using the instrumental variable. Highlighting the well-being of the sugarcane growers and instability of the sugar price under future global climate change is the important feature of the study.

Bakku et al. [2] demonstrated differentially expressed genes in rice (*Oryza sativa* cv. Koshihikari) seeds under high-temperature stress using the transcriptomics approach in Japan. The study showed up- and downregulation of more than 100 genes in grade 2 rice (Y2) and grade 3 rice (Y3) seeds, respectively. This study is among the first that suggests that high temperature during the seed filling and maturation in rice damages yield as well as kernel quality.

Analysis of tree rings provided a comprehensive understanding of growth dynamics and their adaptation to climate change using Chinese Torreya (*Torreya grandis* cv Merrillii) as a model system [3]. The analysis was performed using six stem sections from trees having ages between 60–90 years and local climate data. The results revealed that the accumulated radial growth enhanced linearly with time. The study suggested that the gradual growth, drought resistance, and several stems in a single tree could help the trees acclimate to different climate conditions.

Quantification of the isoprenoids between soil with litter and atmosphere in a Mediterranean *Pinus pinea* was performed in order to study the ground level isoprenoid exchanges [4]. The study showed that isoprenoid emissions were high, variable, and can be assessed by the dry weight of litter around the trunk. The findings recommend pervasive spatio-temporal analysis of ground-level isoprenoids' exchanges in different types of ecosystem. Gandia et al. [5] highlighted recognizing the response to environmental change of weed species by analyzing their distribution. The analysis of species led to the categorization of weeds as generalist, regional, or local species, corresponding to latitude and related temperature ranges. Three weed species, *Linaria micrantha* (Cav) Hoffmanns & Link, *Sonchus oleraceous* L., and *Sysimbrium irium* L., were categorized as generalist and *Stellaria media* (L.) Vill. was identified as a local species. The approach in the study can be used to designate weed distribution as a marker of changing climatic conditions.

To study the effects of temperature on the physiological and ecological characteristics of plants, two high-resolution thermal cameras were used to monitor the canopy leaf temperature distribution in a primary tropical rain forest in southwest China [6]. The study included 28 different tree species and the results suggest that both stomatal conductance and size of the leaves determined the difference in the mean leaf-to-air temperature. The findings indicate species-specific functional traits required to investigate and model the interactions of entities for developing the knowledge and prediction of impacts of climate change on vegetation.

The following seven papers in this Special Issue conducted studies using different models or approaches under the future climate change scenario. Simulation of differential impact on winter wheat (*Triticum aestivum* L.) by future projections of climate change (2025 and 2050), especially under increasing temperature was done using CSM-CERES-Wheat model coupled with different Representative Concentration Pathways (RCPs) and two Global Circulation Models (GCMs) in China [7]. The study indicated that the production of wheat in Guanzhong plain will increase (positive) under future climate change using crop simulation modeling. However, the negative impact will depend upon the climate change projections as GCMs showed both increase and decrease in the grain yield. The study also emphasized proper use of irrigation management as rainfed wheat is very sensitive to climate change. In a study, a scaling approach was used to measure the variation of scaling factors and their correlation at large scales in the estimation of actual transpiration of three boreal species in a forest [8]. The authors demonstrated that the scaled canopy transpiration signified a considerable fraction of forest evapotranspiration (>70%) and recommend the approach for the proper estimation of actual transpiration in the areas having low tree diversity. Mendoza et al. [9] emphasized the use of the Climate Data Science (CDS) Toolbox Species Distribution Model (SDM) in evaluating the appropriate areas of grapevine (*Vitis vinifera* L.) under the present and future climate conditions in France. The study proved different possible effects of future climate change on the spatial distribution of proper areas for grapevine crops. The maximum entropy modeling approach was utilized to foresee future habitat distribution of the susceptible *Prunus Africana* under the effect of climate change in Tanzania [10]. The results showed reductions in appropriate habitats for *P. Africana* under all imminent representative concentration pathways' scenarios as compared to present distributions. Various statistical methods were used to study the variations in the seasonality of Ethiopian highlands' climate, consequences for crop development, assessment of variations in the annual cycle, and long-term trends. [11] Coupled Model Intercomparison Project (CMIP5) Hadley2 data assimilated by the Inter-Sectoral Impact Model Intercomparison Project (ISIMIP) hydrological models used in the study provided understandings on the unimodal annual cycle of soil moisture in past and future eras. The study concluded that evaporation is increasing and might put stress on different land and water resources due to seasonal variations. An empirical hazard model was used to get the pattern of the global spread of Black Sigatoka Leaf Disease (*Mycosphaerella fijiensis*), an important pathogen on banana [12]. The results showed that agricultural trade might play a significant role in spreading the disease across countries and highlights the threat and prospective cost of relying on just a few varieties with genetic similarity to produce a particular crop globally. Climate change is negatively affecting the health of populations around the world, especially in low-income countries like East Africa. A Wet Bulb Globe Temperature (WBGT) approach, a common index, was used to evaluate the heat stress in occupational health in East Africa [13]. The results showed that heat stress is already influencing the areas of East Africa. The analysis of two terms of the agricultural calendar suggests that Kenya and Tanzania face substantial portions of their national landmass influenced by high WBGT values; a neighboring country (Uganda) is comparatively less affected.

The goal of this Special Issue is to present research with a broad perspective to understand the effects of climate change on vegetation, involving applied research and studies with different types of modeling approaches, and the 13 papers in this Special

Issue achieve this goal. I thank the authors for their significant contributions and hope that this issue triggers some ideas and collaboration or serve as a resource to move ahead in a rapidly changing climate.

Funding: This research received no external funding.

Institutional Review Board Statement: Not applicable.

Informed Consent Statement: Not applicable.

Acknowledgments: The editor would like to thank the authors from the United States of America, India, Japan, Canada, France, Sweden, China, Tanzania, Switzerland, Trinidad and Tobago, South Africa, Spain, and Thailand for their important contributions to this Special Issue, and the reviewers for their constructive comments to improve the quality of the manuscripts. The editor is grateful to Alicia Wang for her kind support in processing and publishing this Special Issue.

Conflicts of Interest: The authors declare no conflict of interest.

References

1. Pipitpukdee, S.; Attavanich, W.; Bejranonda, S. Climate Change Impacts on Sugarcane Production in Thailand. *Atmosphere* **2020**, *11*, 408. [CrossRef]
2. Bakku, R.K.; Rakwal, R.; Shibato, J.; Cho, K.; Kikuchi, S.; Yonekura, M.; Sarkar, A.; Shioda, S.; Agrawal, G.K. Transcriptomics of Mature Rice (Oryza Sativa L. Koshihikari) Seed under Hot Conditions by DNA Microarray Analyses. *Atmosphere* **2020**, *11*, 528. [CrossRef]
3. Chen, X. Historical Radial Growth of Chinese Torreya Trees and Adaptation to Climate Change. *Atmosphere* **2020**, *11*, 691. [CrossRef]
4. Mu, Z.; Llusià, J.; Peñuelas, J. Ground Level Isoprenoid Exchanges Associated with Pinus pinea Trees in A Mediterranean Turf. *Atmosphere* **2020**, *11*, 809. [CrossRef]
5. Gandía, M.L.; Casanova, C.; Sánchez, F.J.; Tenorio, J.L.; Santín-Montanyá, M.I. Arable Weed Patterns According to Temperature and Latitude Gradient in Central and Southern Spain. *Atmosphere* **2020**, *11*, 853. [CrossRef]
6. Song, Q.; Sun, C.; Deng, Y.; Bai, H.; Zhang, Y.; Yu, H.; Zhang, J.; Sha, L.; Zhou, W.; Liu, Y. Tree Surface Temperature in a Primary Tropical Rain Forest. *Atmosphere* **2020**, *11*, 798. [CrossRef]
7. Zheng, Z.; Cai, H.; Wang, Z.; Wang, X. Simulation of Climate Change Impacts on Phenology and Production of Winter Wheat in Northwestern China Using CERES-Wheat Model. *Atmosphere* **2020**, *11*, 681. [CrossRef]
8. Quiñonez-Piñón, M.R.; Valeo, C. Modelling Canopy Actual Transpiration in the Boreal Forest with Reduced Error Propagation. *Atmosphere* **2020**, *11*, 1158. [CrossRef]
9. Hinojos Mendoza, G.; Gutierrez Ramos, C.A.; Heredia Corral, D.M.; Soto Cruz, R.; Garbolino, E. Assessing Suitable Areas of Common Grapevine (*Vitis vinifera* L.) for Current and Future Climate Situations: The CDS Toolbox SDM. *Atmosphere* **2020**, *11*, 1201. [CrossRef]
10. Giliba, R.A.; Yengoh, G.T. Predicting Suitable Habitats of the African Cherry (Prunus africana) under Climate Change in Tanzania. *Atmosphere* **2020**, *11*, 988. [CrossRef]
11. Tarkegn, G.B.; Jury, M.R. Changes in the Seasonality of Ethiopian Highlands Climate and Implications for Crop Growth. *Atmosphere* **2020**, *11*, 892. [CrossRef]
12. Strobl, E.; Mohan, P. Climate and the Global Spread and Impact of Bananas' Black Leaf Sigatoka Disease. *Atmosphere* **2020**, *11*, 947. [CrossRef]
13. Yengoh, G.T.; Ardö, J. Climate Change and the Future Heat Stress Challenges among Smallholder Farmers in East Africa. *Atmosphere* **2020**, *11*, 753. [CrossRef]

atmosphere

MDPI

Article

Climate Change Impacts on Sugarcane Production in Thailand

Siwabhorn Pipitpukdee [1,2], Witsanu Attavanich [1,*] and Somskaow Bejranonda [1]

[1] Department of Economics, Kasetsart University, 50 Phahonyothin Rd., Chatuchak, Bangkok 10900, Thailand; siwabhorn.p@ku.th (S.P.); somskaow.b@ku.th (S.B.)

[2] Center for Advanced Studies for Agriculture and Food, Kasetsart University Institute for Advanced Studies, Kasetsart University, Bangkok 10900, Thailand

* Correspondence: witsanu.a@ku.ac.th

Received: 29 February 2020; Accepted: 17 April 2020; Published: 19 April 2020

check for updates

Abstract: This study investigated the impact of climate change on yield, harvested area, and production of sugarcane in Thailand using spatial regression together with an instrumental variable approach to address the possible selection bias. The data were comprised of new fine-scale weather outcomes merged together with a provincial-level panel of crops that spanned all provinces in Thailand from 1989–2016. We found that in general climate variables, both mean and variability, statistically determined the yield and harvested area of sugarcane. Increased population density reduced the harvested area for non-agricultural use. Considering simultaneous changes in climate and demand of land for non-agricultural development, we reveal that the future sugarcane yield, harvested area, and production are projected to decrease by 23.95–33.26%, 1.29–2.49%, and 24.94–34.93% during 2046–2055 from the baseline, respectively. Sugarcane production is projected to have the largest drop in the eastern and lower section of the central regions. Given the role of Thailand as a global exporter of sugar and the importance of sugarcane production in Thai agriculture, the projected declines in the production could adversely affect the well-being of one million sugarcane growers and the stability of sugar price in the world market.

Keywords: climate change impacts; sugarcane; yield; harvested area; production; Thai agriculture

1. Introduction

Sugar is a low-cost energy source that can alleviate malnutrition problems in the case of energy deficiency [1]. About 80% of the global sugar produced from sugarcane [2,3] are cultivated in 120 countries with approximately 27 million ha and an average production is 1.8–2 billion tons per year [4]. In addition to sugar, sugarcane can be used to produce several products such as falernum, molasses, rum, bagasse, and ethanol, creating economic benefits along the supply chain [2].

Among sugarcane producing countries, Thailand ranked fourth for sugar production, accounting for 8.10% of the world's total sugar production [5] and ranked second for sugar export contributing to 16.95% of global export quantity with an export value of 2.97 billion USD in 2019 [5,6]. At the national level, sugarcane production plays an increasing role in Thai agriculture. With support from government policies aiming to reduce rice production and promote alternative energy, the harvested area of sugarcane has steadily increased 44.61% in the last decade from 1.35 million ha in the 2010/2011 production year to 1.96 million ha in 2018/2019 [7] with approximately 1 million farmers in 2019 [8]. In 2018/2019, the harvested area of sugarcane ranked third among major economic crops in Thailand following rice (11 million ha) and natural rubber (3.66 million ha). It accounted for 12% of total land use for 11 major economic crops. Cassava and maize ranked fourth and fifth with harvested areas of 1.39 and 1.10 million ha, respectively.

Over the last several decades, it has become increasingly clear that human activities, especially burning fossil fuels and deforestation, are changing the world's climate conditions, through increases in temperatures, extreme temperatures, droughts, and rainfall intensity [9]. Agriculture is the most vulnerable economic sector through such changes and for the past 30 years numerous studies have attempted to estimate the effect of changing climate on crop yields and their production [10–15].

Climate change can directly affect crops through rising temperature and changing rainfall patterns, or indirectly affect crops through soil, nutrient, and increasing pests interference [16]. Studies revealed that crop yields have been affected by the variability of temperature, rainfall, and the interaction between them and climate change impacts will be different across locations, types of crop, scenarios, and farmer adaptation [17–21]. Although the world may be able to cope with food insecurity at the macro level, the problem may also exist at the micro level with the shortage of food in developing countries compensated by developed countries receiving the benefits from climate change [13]. Previous studies also revealed that climate change is projected to negatively affect the global food system and food supply may not be available to meet demand in the future [21–23].

For sugarcane, all previous studies only assessed the impact of climate change on yield. Overall, studies showed mixed findings regarding changes in sugarcane yield from climate change. Singels et al. [24] employed the Canegro model and revealed that future sugarcane yields with constant CO_2 concentration set at 360 ppm were expected to decline in two sites, ranging from 4.15% for rainfed crops at Piracicaba (Brazil) and 4.65% for irrigated crops at Ayr (Australia) from the 1980–2010 baseline period. On the other hand, sugarcane yield was predicted to increase 2.58% for La Mercy (South Africa). By adding CO_2 fertilization effect, Marin et al. [25] found that the sugarcane yield would increase 24% for rainfed sugarcane in the 2050s in São Paulo, Brazil. Moreover, Silva et al. [26] found that rainfall was positively correlated with sugarcane production, whereas the temperature negatively influenced production in municipalities within Paraiba, Brazil. They also found that the mesoregion of Mata Paraibana has a higher probability of producing sugarcane than other mesoregions.

The positive impact of climate change on sugarcane yield was also found in Mexico [27] and southern China [28]. In Mexico, Baez-Gonzalez et al. [27] developed the Agricultural Land Management Alternatives model and revealed the positive impacts of future climate change on sugarcane yields with increases of 1%–13% under the A2 scenario from the baseline. In southern China, Ruan et al. [28] used the Agricultural Production Systems Simulator (APSIM)-Sugarcane model and found that the largest percentage change in sugarcane yields occurred at high latitude locations (e.g., Hezhou), with increased mean values of 44.2% and 23.5% for Representative Concentration Pathway (RCP)4.5 and RCP8.5 in the 2060s, respectively. On the other hand, in Africa, Adhikari, Nejadhashemi, and Woznicki [29] reviewed studies projecting the climate change impacts on sugarcane production and revealed that sugarcane will be resilient to temperature rise, but it will be vulnerable to rainfall variability. Yield of sugarcane is projected to decline less than 5% in East Africa by 2030 as compared to 1998–2002 [30].

In Thailand, Yoshida et al. [31] present the only research study to explore the relationship between climate and sugarcane yield in the Northeastern region of Thailand. Their study revealed that sugarcane yield had a significant positive relationship with four months of accumulated rainfall. This finding could imply that sugarcane yield is likely increased where the rainfalls are projected to increase under climate change scenarios. Unfortunately, their study did not analyze this relationship at the national level and did not differentiate the heterogenous effect of climate change on sugarcane yield among regions of Thailand. To our knowledge, there is no study that projects the future change in yield, harvested area, and production of sugarcane in Thailand under climate change scenarios.

Therefore, this study aims to estimate the effect of climate change on yield, harvested area, and production of sugarcane in Thailand using the provincial-level panel data analysis. Then, we project future changes in yield, harvested area, and production of sugarcane under climate change using climate projections from the Fifth Assessment Report (AR5) of the Intergovernmental Panel on Climate Change (IPCC) [9].

Our study provides several contributions to climate change related to sugarcane production. First, our study is a pioneer in simultaneously investigating the effect of climate change including yield, harvested area, and production, and analyzing climate change impacts for a whole country at the provincial level. Second, we add the prices of output and input in the model and address the issue of endogeneity bias in economics using spatial econometrics and the instrumental variable approach as suggested by Miao and colleagues [14]. Third, unlike other studies done in Thailand, we put additional effort to estimate the weighted average of climate data for each province using weighted least square regression, as first introduced by Mendelsohn, Nordhaus, and Shaw [12]. Fourth, we include variables capturing climate variability and extreme events in the model and use the recent AR5 downscaled projections of precipitation at the watershed level to deeply understand the variation of future precipitation at the local level. Finally, we include and project the population density as a variable capturing the change in socio-economic condition that could affect harvested areas of sugarcane.

This article is organized as follows: Section 2 presents details of materials and methods used for the analysis; Section 3 provides results and discusses the findings; and Section 4 presents the conclusions and policy implications that were drawn from the findings.

2. Materials and Methods

2.1. Model Estimation Approach

To quantify the effect of climate change on the production of sugarcane in Thailand, we constructed models by including factors that determine yield and harvested area of sugarcane following Miao and colleagues [14]. The province-specific sugarcane yield model and harvested area are shown below in Equations (1) and (2), respectively:

$$Y_{jt} = \beta_o + \boldsymbol{\beta}_1 \boldsymbol{Climate}_{jt} + \boldsymbol{\beta}_2 \boldsymbol{Price}_{jt} + \beta_3 PctIrrig_{jt} + \beta_4 T_{jt} + \beta_5 T_{jt}^2 + u_j + \epsilon_{jt} \quad (1)$$

$$H_{jt} = \alpha_o + \alpha_1 \boldsymbol{Climate}_{jt} + \boldsymbol{\alpha}_2 \boldsymbol{Price}_{jt} + \alpha_3 PctIrrig_{jt} + \alpha_4 Popden_{jt} + \alpha_5 T_{jt} + \alpha_6 T_{jt}^2 + v_j + e_{jt} \quad (2)$$

where j and t are indexed for province and year, respectively. Y_{jt} is yield of sugarcane in province j at time t. For brevity, we will omit explanations for the subscripts. H is the harvested area of sugarcane. $\boldsymbol{\beta}$ and $\boldsymbol{\alpha}$ are vectors of parameters to be estimated. ***Climate*** is the vector of climate variables including growing season average temperature, extreme maximum temperature, total rainfall, maximum rainfall in 24 h, and the dummy variables capturing El Niño–Southern Oscillation (ENSO) phases including El Niño, La Niña, and neutral phases. ***Price*** is the vector capturing output and input prices (i.e., farm received price of sugarcane and wage rate of labor). *PctIrrig* is the percent of irrigated area to total area in the province and T and T^2 are time trend capturing technological progress. In the model of harvested area, we added a variable *Popden* capturing population density, which determines the demand for land and pressure of land on non-agricultural development use; u and v are region fixed effects and ϵ and e are error terms.

For estimation, this study uses spatial regression to address spatial bias because the climate conditions, input prices, and labor wage in a large region can be quite similar, and the provincial-level yield and harvested area of sugarcane may be correlated with those in neighboring provinces. We also address endogeneity bias from using sugarcane price and wage rate by employing the instrumental variable (IV) approach together with the generalized method of moment (GMM), following procedures suggested by Miao and colleagues [14]. By testing for the good IVs, this study uses one-year lagged variables of the Southern Oscillation Index (SOI), extreme maximum temperature, and total stock of sugar as IVs for the yield model. For the harvested area model, it uses one-year lagged variables of extreme maximum temperature, total stock of sugar, and total amount of rainfall as IVs. After obtaining estimated coefficients from the yield and harvested area models for sugarcane, we then obtain climate projections from IPCC AR5 to predict future yield and harvested area for sugarcane. Finally, we estimate the quantity of sugarcane production by multiplying yield to its corresponding harvested area.

2.2. Data

This study constructs a unique provincial-level panel dataset during 1989–2016—the longest period used compared to other studies done in Thailand—from several sources [31]. Yield and harvested area plus crop prices were obtained from the Office of Agricultural Economics, Ministry of Agriculture and Cooperatives. Irrigation area was obtained from the Royal Irrigation Department. We obtained the historical monthly climate data including average temperature, maximum temperature, and mean precipitation for all climate stations in Thailand from the Meteorological Department. Climate projections during 2046–2055 were obtained from the IPCC AR5. They are the average values of all general circulation models produced by the Royal Netherlands Meteorological Institute (KNMI) using IPCC AR5 report. We also collected population statistics and future population projections under the assumption of a moderate fertility rate at the provincial level from the Ministry of Interior and the National Economic and Social Development Council (NESDC), respectively. Lastly, we constructed dummy variables capturing three ENSO phases (i.e., El Niño, La Niña, and neutral) from the National Oceanic and Atmospheric Administration (NOAA).

Unlike other studies in Thailand, we linked the agricultural data organized by province and the climate data organized by station by conducting a spatial statistical analysis following Mendelsohn and co-workers [12]. While climatic variables examined in this study are measured frequently, there are some provinces with several weather stations and others with no stations. Furthermore, some provinces are large enough that there is variation in climate within the province. We therefore proceeded by constructing an average climate for each province using weighted least square regression by controlling for the distance from the centroid, latitude, longitude and height of climate stations. The weight is the inverse of the square root of a station's distance from the province center because closer stations usually contain more information about the climate of the center. We located the centroid of each province and drew a circle within the radius of 250 km by assuming that all the weather stations within this radius provide some useful climate information.

We estimated a separate regression for each province since the set of stations within 250 km and the weights (distances) are unique for each province. The regression fits a second-order polynomial over four climate variables, so that there were 20 final variables in the regression, plus a constant term. Four regressions for each of the 77 provinces and 36 years led to over 11,088 estimated regressions. Table A1 in the Appendix A shows examples of the estimated coefficients of the weighted least square regression for each climate variable in July 2016 in Nakon Sawan Province, the largest sugarcane producing province in Thailand. Overall, we observe that the models fit relatively well, especially for the average temperature variable. All predicted values of climate variables are statistically significant at 1% level. Table 1 provides a summary statistic of variables at the provincial level.

3. Results and Discussion

This section provides the estimated coefficients from sugarcane yield and its harvested area models, the projected changes in yield, harvested area, and production of sugarcane under climate change scenarios, as well as a discussion of the findings.

3.1. Estimated Results

The estimated coefficients from the sugarcane yield and its harvested area models are shown in Tables 2 and 3, respectively. Details are provided in Sections 3.1.1 and 3.1.2, respectively.

Table 1. Summary statistics of selected variables at the provincial level.

Variables	Mean	SD	Min	Max
Yield (kg/ha)	58,652.50	11,093.84	18,612.50	92,462.50
Harvested area (1000 ha)	22.89	29.09	0.03	161.41
Average temperature (°C)	27.59	0.67	25.57	29.10
Maximum rainfall in 24 h (mm/day)	33.69	3.93	22.98	47.28
Extreme maximum temperature (°C)	35.91	0.55	34.49	37.38
Total rainfall (mm)	1331.35	204.97	886.76	2007.98
Population density (person/km^2)	125.64	67.17	21.56	417.38
Lag received price (USD/ton)	25.01	4.54	13.27	42.90
Lag wage (USD)	6.47	1.26	4.88	9.91
%Irrigated area per province area	12.72	25.81	0	166.72
No. of observation	1242			

3.1.1. Determinants of Sugarcane Yields

All climate variables (excepting for the El Niño phase) statistically influenced sugarcane yield (Table 2). The inverted U-shape relationship between temperature and sugarcane yield was revealed and we found the U-shape relationship between rainfall and sugarcane yield. Moreover, an increase in extreme maximum temperature showed the harmful impact on sugarcane yield. On the other hand, the maximum rainfall within 24 h was positively correlated to sugarcane yield. This finding could be explained by the fact that a majority of land planting sugarcane in Thailand are dryland above the sea level. Therefore, an increase in rainfall intensity still improved sugarcane yield. Other studies reached a similar conclusion [24,28].

We also revealed that the period with extreme climatic events, especially the La Niña phase, had lower yield than the period with neutral phase. In addition to the climate conditions, increase in the percent of irrigated area to total land area significantly improved the yield of sugarcane. Farm price received and labor wage rate in the previous year are negatively correlated to sugarcane yield. An increase in expected price could lead to a change in rotation practice and expanding area under the crop to marginal, low quality acres [32], which could decrease yield per ha. Furthermore, the reduction in labor use was induced by an increase in wage rate. Finally, technological progress captured by the variable Time trend affected sugarcane yields with a U-shape relationship. We used the estimated coefficients of Time trend and its square term to calculate the rate of technological change to investigate the role of technological progress on sugarcane yield. Our results revealed that sugarcane yield increased 1.36% per year as a result of technological progress during 1992–2016 period.

3.1.2. Determinants of Harvested Area

We found that total rainfall non-linearly determined sugarcane harvested area with inverted U-shape relationship. Its harvested area in the La Niña phase was higher than that in the neutral phase. We also revealed that increases in the percent of irrigated area to total land area reduced sugarcane harvested area because sugarcane usually grows in rainfed areas. Sugarcane growers could obtain a higher yield or switch from sugarcane to other high-valued crops when they can access an irrigation system. Higher population density reduced the sugarcane harvested area as found in previous studies [14] due to higher demand of land for non-agricultural use. The one year-lagged labor wage rate positively correlated to sugarcane harvested area. Increase in expected wage rate could lead farmers to substitute land for labor and expand sugarcane acreage. Lastly, technological progress non-linearly affected the sugarcane harvested area with an inverted U-shape relationship as shown in Table 3. Similar to Section 3.1.1, we calculated the rate of technological change to investigate the role of technological progress on harvested area and found that harvested area slightly dropped 0.000008% per year during the same period implying that technological progress had little impact on the land use of sugarcane.

Table 2. Determinants of yield.

Variables	Coefficients	Standard Errors
Time trend	−1684.42 ***	278.09
Time trend_sq	127.56 ***	12.61
%Irrigated area per province area	100.52 ***	13.20
Average temperature	165,114.40 ***	22,821.98
Average temperature_sq	−2942.43 ***	416.03
Total rain	−37.08 ***	11.09
Total rain_sq	0.01 **	0.01
Maximum rain in 24 h	274.62 **	137.55
Extreme max. temperature	−8592.73 ***	1012.81
El Niño	−513.00	585.99
La Niña	−2244.31 ***	622.67
North	4057.12 ***	1438.25
Northeast	5618.21 ***	1462.12
Southeast	−12,246.34 ***	2241.63
East	−3348.69 ***	1279.85
Lag price	−645.31 ***	154.72
Lag wage	−8765.63 ***	640.63
Constant	-1.87×10^{6} ***	312,248.60
Observations	1242	
R-square_adj.	0.49	
Root mean square error (MSE)	6747.97	

Notes: *, **, and *** indicate significance at the 10%, 5%, and 1% level, respectively.

Table 3. Determinants of harvested area.

Variables	Coefficients	Standard Errors
Time trend	1.04 **	0.51
Time trend_sq	−0.05 **	0.02
Population density	−0.07 ***	0.02
%Irrigated area per province area	−0.09 **	0.05
Total rain	0.05 *	0.03
Total rain_sq	-2.20×10^{-5} **	9.45×10^{-6}
Maximum rain in 24 h	−0.44	0.43
Extreme max. temperature	−0.43	2.78
El Niño	−0.67	1.28
La Niña	7.65 ***	1.70
North	−16.46 ***	3.72
Northeast	−8.37 **	3.88
Southeast	−10.82	6.91
East	−19.34 ***	4.09
Lag price	0.23	0.37
Lag wage	10.78 ***	2.91
Constant	−36.04	95.16
Observations	1242	
R-square_adj.	0.11	
Root MSE	10.90	

Notes: *, **, and *** indicate significance at the 10%, 5%, and 1% level, respectively.

3.1.3. Improvement in Estimation

To check whether adding the new economic variables and our estimation method improved the fitness of the model, we compared models with and without prices and wage variables and also models with and without IVs and spatial regression. We revealed that our yield and harvested area models that included price and wage variables and used the IV approach plus spatial regression had higher

R^2 values and lower root mean square error (RMSE) values than the models without prices and wage variables (See Tables A2 and A3). While the ordinary least square (OLS) method provided the low value of the root mean square error (see model 3 in Table A3) in the harvested area model, it did not address the endogeneity problems from both spatial bias and omitted variables. We performed the Moran's I test and found the spatial autocorrelation in the model. These above evidences imply that the method used in the current study improves the estimation of the models. Future research should address the problem of endogeneity generated by spatial bias, simultaneity bias, and omitted variables before performing the estimation.

3.2. Simulation of Climate Change Impacts on Production of Sugarcane

To project the impact of climate change on yield, harvested area, and production of sugarcane during 2046–2055 from the baseline during 1992–2016, we obtained future climate projections including growing season temperature, total precipitation, extreme maximum temperature, and maximum precipitation within 24 h from IPCC AR5 [9]. Climate change scenarios RCP4.5 and RCP8.5 were selected to investigate the variation of projected results. RCP8.5 captures rising radiative forcing pathway leading to 8.5 W/m^2 in 2100, while RCP4.5 is stabilized without the overshoot pathway to 4.5 W/m^2 after 2100.

Figure 1 presents the regional projected changes in climate variables used in the model. Overall, we observed that the Northeastern region is projected to have the highest increase in growing season temperature and extreme maximum temperature from the baseline among other regions. Growing season temperatures of sugarcane (January to December) are projected to increase ranging from 1.08–1.22 °C and 1.48–1.68 °C under RCP4.5 and RCP8.5, respectively. Extreme maximum temperatures are also projected to rise ranging from 1.21–1.55 °C and 1.61–1.86 °C under RCP4.5 and RCP8.5, respectively. All regions are projected to have higher annual maximum precipitation within 24 h.

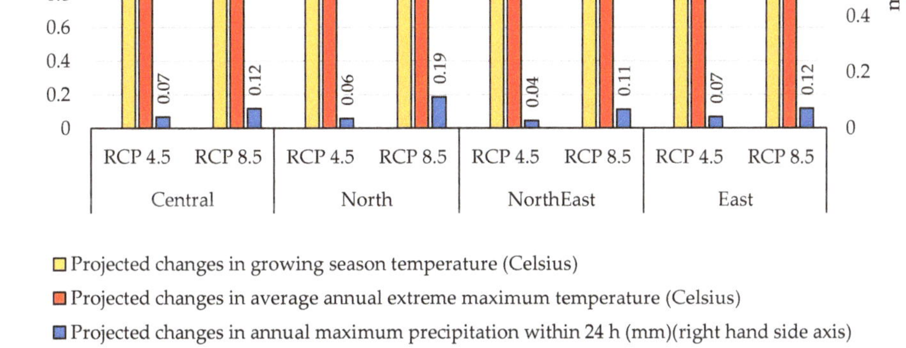

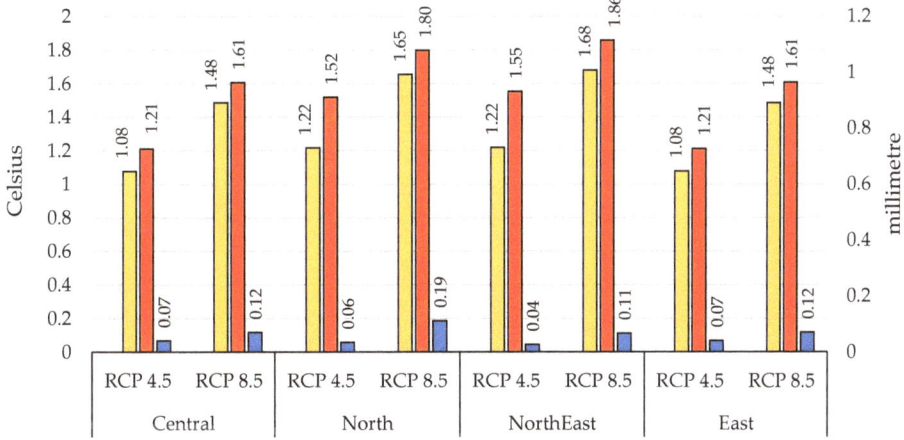

Figure 1. Projected changes in temperature (Celsius) during 2046–2055 under Representative Concentration Pathway (RCP)4.5 and RCP8.5 from the baseline during 1992–2016.

Since rainfall has high local variation, our study, unlike other studies in Thailand, used the latest IPCC AR5 downscaled projections of total annual rainfall at the watershed level provided by the Office of Natural Resources and Environmental Policy and Planning (ONEP). There are 25 watersheds in

Thailand and Figure 2 reveals that the total amount of rainfall under RCP8.5 will be higher than the total amount of rainfall under RCP4.5. Regions in the north, south, and upper section of northeast were projected to have higher future rainfall than the baseline, while the opposite was found in some provinces located in the lower-southern region. Unlike other studies, we obtained population statistics from Ministry of Interior and the National Economic and Social Development Council (NESDC), and then predicted future changes in population using the trend analysis with quadratic time trend and then quantified the projected population density to reflect changes in socio-economic conditions as shown in Figure 3. We observed that the population density was projected to increase in the central, eastern and southern regions, while it was forecasted to drop in the northeastern and northern regions.

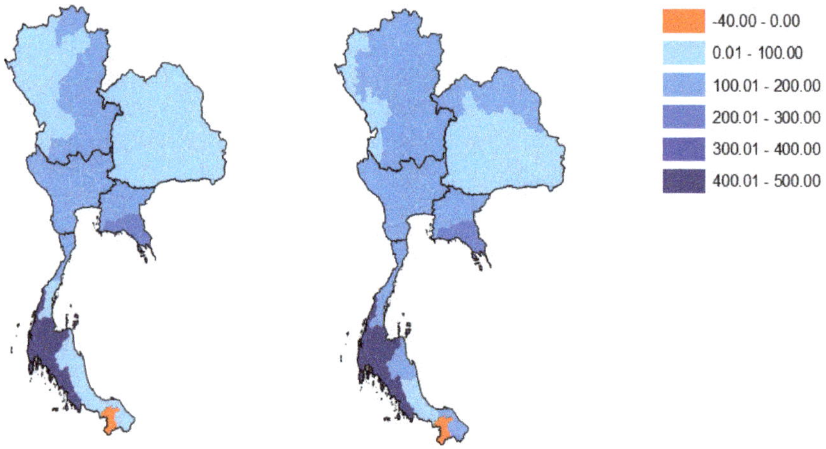

RCP 4.5 RCP 8.5

Figure 2. Projected changes in total annual rainfall (mm) during 2046–2055 under RCP 4.5 and RCP8.5 from baseline during 1992–2016.

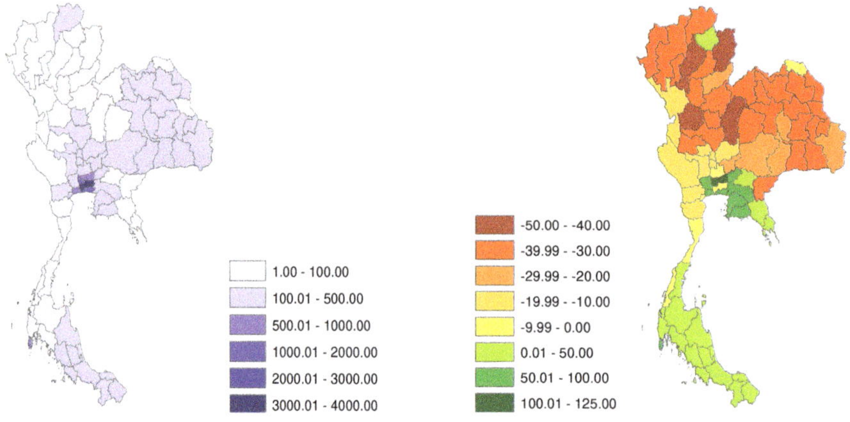

(**a**) (**b**)

Figure 3. Projected changes in population density (people/km^2) during 2046–2055 under scenario of moderate fertility rate. (**a**) Baseline of population density in 1992–2016 (people/km^2); (**b**) percent of change of population density in 2046–2055 from baseline.

After adding projections of climate and population density in estimated models from Tables 2 and 3, we found that future yields, harvested area, and production were projected to drop in all scenarios at the national level (Table 4). Future sugarcane yield was projected to drop 23.95% under RCP4.5 and 33.26% under RCP8.5 from the baseline. In other words, it was predicted to decline 0.59% and 0.87% per year during 1992–2016 period under RCP4.5 and RCP8.5, respectively. Although no study has investigated the impact of climate change on sugarcane yield in Thailand, our results were in line with findings in Brazil and Australia [24] and East Africa [30]. However, the magnitudes of the yield investigated in our study were higher than those in previous studies, which may come from the fact that a majority of sugarcane in Thailand has been grown in the rainfed area and the total precipitation in the Northeastern region was projected to increase less than other regions.

Table 4. National projected changes in yield, harvested area, and production of sugarcane under RCPs 4.5 and 8.5 during 2046–2055 from baseline 1992–2016.

Sugarcane	Baseline	Percent of Change under RCP4.5	Percent of Change under RCP8.5
Yield	61,360 (kg/ha)	−23.95	−33.26
Harvested area	1078 (1000 ha)	−1.29	−2.49
Production	66.17 (1000 MT)	−24.94	−34.93

By incorporating the role of changes in socio-economic condition captured by population density, we found that the harvested area of sugarcane was projected to slightly decline ranging from 1.29–2.49% from the baseline consistent with the findings of Miao and colleagues [14], or about 0.03–0.05% per year during 1992–2016. After multiplying projected sugarcane yield and its corresponding harvested area, this study reveals that sugarcane production is forecasted to decrease between 24.94–34.93% under two climate change scenarios from the baseline without CO_2 fertilization effect. As Thailand contributed 16.95% to the world's sugar export market, climate change could reduce the amount of sugar supplied to the world market.

Considering the distributional impacts of climate change at the provincial level, our findings revealed the reduction in future yield of sugarcane in all provinces ranges from 12.23–30.53% under RCP4.5 and 16.06–43.80% under RCP8.5 from the baseline, respectively as shown in Figure 4. The largest drop in yield was found in the lower section of the country. Prachuap Khiri Khan, Chachoengsao, Chon Buri, Rayong, and Nakhon Sawan were predicted to have the largest reduction. Mixed results were revealed for the harvested area of sugarcane as shown in Figure 5. A majority of provinces located in the northeastern and northern regions were projected to have an expansion of harvested area ranging from 2.78–19.45% under RCP4.5 and 0.35–16.79% under RCP8.5. On the other hand, some provinces located in the eastern and central regions were projected to face a reduction in harvested area with huge variations across provinces ranging from 0.03–93.07% under RCP4.5 and 0.37–98.45% under RCP8.5.

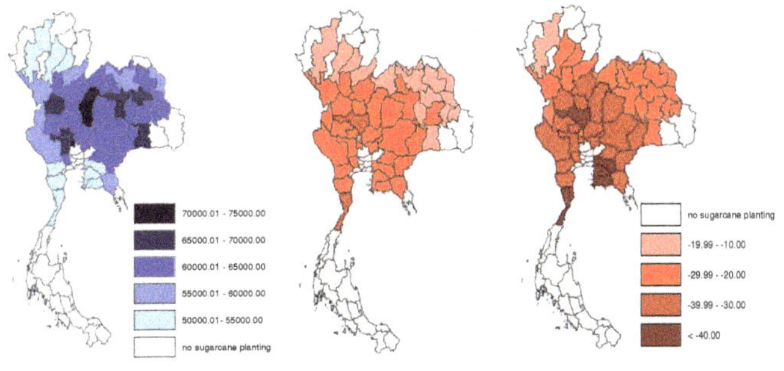

(a) (b) (c)

Figure 4. Projected percent changes in yield of sugarcane under climate change scenarios. (**a**) Baseline yield (kg/ha); (**b**) percent of change in yield under RCP4.5; (**c**) percent of change in yield under RCP8.5.

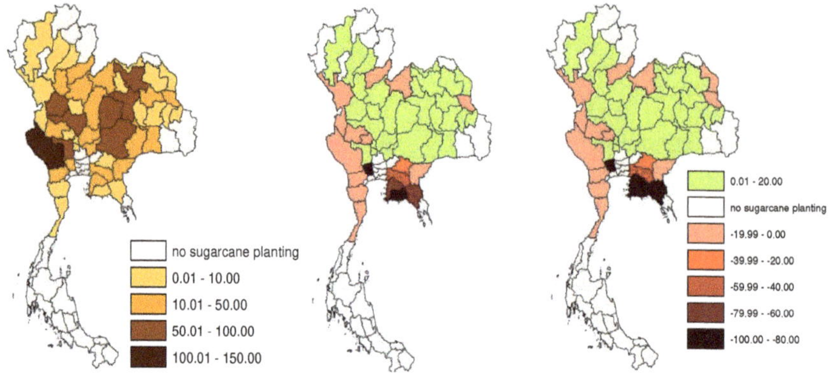

(**a**) (**b**) (**c**)

Figure 5. Projected percent changes in harvested area of sugarcane under climate change scenarios. (**a**) Baseline harvested area (1000 ha); (**b**) percent of change in harvested area under RCP4.5; (**c**) percent of change in harvested area under RCP8.5.

By multiplying yield and harvested area, we found that the sugarcane production was projected to decline at the national level (Table 4) approximately 24.94% under RCP4.5 and 34.93% under RCP8.5 from the baseline during 1992–2016, or equivalent to the declining of 0.62% and 0.92% per year under RCP4.5 and RCP8.5, respectively. Sugarcane production was also predicted to drop in all provinces implying that changes in yield dominated changes in harvested area as demonstrated in Figure 6. The largest drop was predicted in the eastern and lower section of the central regions. Production of the top five provinces (i.e., Kanchanaburi, Suphan Buri, Nakhon Sawan, Kamphaeng Phet, and Nakhon Ratchasima), accounting for 39.30% of total sugarcane production, was projected to decrease 20.13–26.65% under RCP4.5 and 30.35–38.09% under RCP8.5 from the baseline.

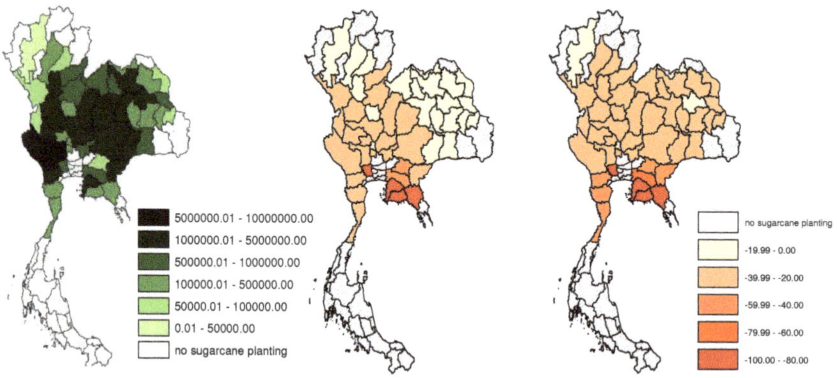

(**a**) (**b**) (**c**)

Figure 6. Projected percent changes in sugarcane production under climate change scenarios. (**a**) Baseline production (MT); (**b**) percent of change in production under RCP4.5; (**c**) percent of change in production under RCP8.5

Considering the role of technological progress in sugarcane production discussed in Sections 3.1.1 and 3.1.2, we may need to sustain the rate of technological progress on sugarcane production at least 0.62–0.92% per year in Thailand to address the future impact of climate change. A higher rate of technological progression on sugarcane production may be needed to fulfill the demand of sugarcane-related products given the rising population in the world, which is projected to reach 9.73 billion by 2050 [33].

In addition to Thailand, sugarcane producing countries should be aware of climate change impacts since previous studies also predicted a decline in sugarcane yield induced by climate change. For example, Singels et al. [24] projected the decline of sugarcane production in the rainfed area of Piracicaba (Brazil) and in the irrigated area of Ayr (Australia). Moreover, Adhikari, Nejadhashemi, and Woznicki [29] predicted a drop in sugarcane production in East Africa. Recent drought during the 2019/2020 season also caused a large fall of sugarcane production in India and Thailand [34]. Since India, Brazil, Thailand, and Australia are major sugarcane producing countries, climate change could also cause fluctuation in the world's markets of sugar, biofuel, and related sugarcane products. Importing and exporting countries plus traders of sugarcane-related products should consider the impact of climate change on sugarcane production in future planning.

4. Conclusions

The objectives of this study were to predict the impacts of climate change on yield, harvested area, and production of sugarcane in Thailand using spatial regression with the instrumental variable. A provincial-level panel dataset during 1989–2016 was constructed with downscaled climate projections under RCP4.5 and RCP8.5 from IPCC AR5 as well as projections of provincial-level future populations under a moderate fertility rate. Our results provide important implications on the well-being of almost one million sugarcane growers in Thailand and the vulnerability of sugar supplied in the world market as Thailand is ranked as the second largest exporter of sugar in the world market. The backward and forward linkage industries also could be affected by the vulnerability of sugarcane production. Several new contributions to climate change related sugarcane production were added.

For the determinants of crop yields, we found that in general climate variables, both mean and variability, statistically determined yields. In addition to climate variables, increased population density also reduced the harvested area for non-agricultural use. Technological progress also statistically determined yields with a non-linear effect. Input and output prices also affected production. Our simulated results demonstrate that sugarcane yield is projected to drop 23.95–33.26% from the baseline with the largest drop in the lower section of Thailand. The harvested area of sugarcane is projected to decline 1.29–2.49% from the baseline with expansion in the northeastern and northern regions and reduction in some provinces located in the eastern and central regions. Moreover, sugarcane production is forecasted to decrease 24.94–34.93% from the baseline with the largest drop in the Eastern and lower section of the central regions. As a result, the amount of sugar exported to the world could reduce approximately 2.49–3.49% and the standard of living of sugarcane growers could be diminished. To address the impact of climate change, the rate of technological progress on sugarcane production may need to increase at least 0.62–0.92% per year.

Several policy implications can be drawn from our findings. First, it is recommended that policy makers should raise awareness to farmers and private sectors on the serious effects of climate change on sugarcane production in predicted vulnerable areas, especially provinces in the eastern and central regions of Thailand. Second, to effectively reduce the impacts of climate change, the government should support the development of proper farm practices (e.g., moisture management, and soil and water conservation), crop insurance programs, and infrastructure (i.e., irrigation systems) to support the adaptation of farmers. Third, agricultural research and development should emphasize the development of heat-resistant species for sugarcane to sustainably adapt to the future warming world. Fourth, governments should promote research to quantify the impacts of climate change on sugarcane production at the finer scales (i.e., tambon and household level) to improve the accuracy of the projections and encourage researchers to analyze the climate change impacts on other crops, livestock, and fisheries. In addition, it is recommended to support the database development for climate change analysis in Thailand because one of the challenging problems of doing climate change research is the lack of a complete database. Last but not least, importing and exporting countries as well as traders of sugarcane-related products should consider the impact of climate change on sugarcane production in their future planning.

Author Contributions: Conceptualization, W.A. and S.B.; methodology, S.P., W.A., and S.B.; formal analysis, S.P. and W.A; investigation, W.A. and S.B.; writing—original draft preparation, S.P.; writing—review and editing, W.A.; visualization, S.P.; supervision, W.A. and S.B. All authors have read and agreed to the published version of the manuscript.

Funding: This research was partially funded by the Center for Advanced Studies for Agriculture and Food, the Institute for Advanced Studies, Kasetsart University under the Higher Education Research Promotion, and the National Research University Project of Thailand, Office of the Higher Education Commission, Ministry of Education, Thailand.

Conflicts of Interest: The authors declare no conflicts of interest.

Appendix A

Table A1. Weighted least square regression of climate variables of July 2016 in Nakon Sawan Province.

	Average Temperature		Total Rain		Maximum Rain in 24 h		Extreme Max. Temperature	
Latitude	14.5154	***	737.2921	*	−449.4323	***	−80.9482	***
	(2.1368)		(433.2796)		(127.9690)		(24.2753)	
Latitude_sq	0.0442	**	9.0523	**	8.9787	***	0.2565	
	(0.0174)		(3.5243)		(1.0409)		(0.1577)	
Longitude	19.5990	***	−9188.7530	***	−296.0857		−204.6332	***
	(6.4092)		(1299.5850)		(383.8320)		(39.7748)	
Longitude_sq	−0.0860	***	46.8822	***	1.4239		0.9611	***
	(0.0316)		(6.4126)		(1.8939)		(0.1910)	
Latitude * Longitude	−0.1616	***	−9.2311	**	1.9145	*	0.7447	***
	(0.0184)		(3.7253)		(1.1003)		(0.2413)	
Height	−0.3509	***	61.0837	***	16.6274	***	−0.6956	
	(0.0385)		(7.7988)		(2.3034)		(0.4141)	
Height_sq	0.0000		0.0021	***	0.0002	*	0.0000	*
	(0.0000)		(0.0006)		(0.0002)		(0.0000)	
Height_ * Latitude	0.0030	***	−0.7504	***	−0.2224	***	−0.0109	**
	(0.0003)		(0.0695)		(0.0205)		(0.0051)	
Height * Longitude	0.0030	***	−0.5090	***	−0.1332	***	0.0082	*
	(0.0003)		(0.0672)		(0.0198)		(0.0043)	
Constant	−1056.2490	***	451,024.7000	***	17,267.2400		10,929.9900	***
	(325.1456)		(65,928.9500)		(19,472.1000)		(2084.0220)	
R-squared	0.9140		0.5537		0.5219		0.6409	
Predicted value	28.29712	***	205.7859	***	53.06985	***	35.87393	***
	0.0141769		2.874609		0.8490152		0.131078	

Notes: *, **, and *** indicate significance at the 10%, 5%, and 1% level, respectively, and standard errors are reported in parentheses.

Table A2. Comparison of yield models with and without price and wage variables.

	1. Existing Model (IV and Spatial Regression with Price and Wage Variables)	2. IV and Spatial Regression without Price and Wage Variables	3. OLS without Price and Wage Variables
Variables	**Coefficients**	**Coefficients**	**Coefficients**
Time trend	−1684.42 ***	197.8	61.32
Time trend_sq	127.56 ***	26.60 ***	35.07 ***
%Irrigated area per province area	100.52 ***	110.9 ***	115.8 ***
Average temperature	165,114.40 ***	112,887 ***	81,758 ***
Average temperature_sq	−2942.43 ***	−2024 ***	−1422 ***
Total rain	−37.08 ***	-8.485	-6.193
Total rain_sq	0.01 **	0.00195	0.0016
Maximum rain in 24 h	274.62 **	−198.2	−366.5 ***
Extreme max. temperature	-8592.73 ***	-9363 ***	−11,825 ***
El Niño	-513.00	190.1	247.6
La Niña	−2244.31 ***	389.2	715.1 **
North	4057.12 ***	4386 ***	6610 ***
Northeast	5618.21 ***	6250 ***	8975 ***
Southeast	−12,246.34 ***	−10,585 ***	−13,390 ***
East	−3348.69 ***	−2834 **	−3860
Lag price	−645.31 ***	-	-
Lag wage	−8765.63 ***	-	-
Constant	-1.87×10^6 ***	-1.18×10^6 ***	−687,863 ***
Observations	1242	1242	1242
R-square_adj.	0.49	0.40	0.427
Root MSE	6747.97	7534.01	7562.85

Notes: *, **, and *** indicate significance at the 10%, 5%, and 1% level, respectively.

Table A3. Comparison of harvested area models with and without price and wage variables.

Harvested Area	1. Existing Model (IV and Spatial Regression with Price and Wage Variables)	2. IV and Spatial Regression without Price and Wage Variables	3. OLS without Price and Wage Variables
Variables	**Coefficients**	**Coefficients**	**Coefficients**
Time Trend	1.04 **	−0.528	0.249
Time Trend_sq	−0.05 **	0.0411 ***	0.0134 *
Population density	−0.07 ***	−0.0396 **	−0.278 ***
%Irrigated area per province area	−0.09 **	−0.144 ***	−0.038
Total rain	0.05 *	0.0172	−0.0347 **
Total rain_sq	-2.20×10^{-5} **	-1.30×10^{-5}	1.06×10^{-5} **
Maximum rain in 24 h	−0.44	0.338	0.204
Extreme max. temperature	−0.43	2.907 *	0.798
El Niño	−0.67	−0.551	1.407 ***
La Niña	7.65 ***	3.354 **	0.0332
North	−16.46 ***	−17.38 ***	−40.50 ***
Northeast	−8.37 **	−11.59 ***	−26.06 *
Southeast	−10.82	-7.51	-51.39 ***
East	−19.34 ***	−18.42 ***	−35.15 **
Lag price	0.23	-	-
Lag wage	10.78 ***	-	-
Constant	−36.04	−75.65	68.21 **
Observations	1242	1242	1242
R-square_adj.	0.11	0.09	0.0965
Root MSE	10.90	10.92	10.310

Notes: *, **, and *** indicate significance at the 10%, 5%, and 1% level, respectively.

References

1. FAO. Is sugar pure, white and deadly? In Proceedings of the Fiji/FAO Asia Pacific Sugar Conference, Suva, Fiji, 29–31 October 1997; pp. 41–44.
2. International Sugar Organization. The Sugar Market. 2019. Available online: https://www.isosugar.org/sugarsector/sugar (accessed on 23 August 2019).
3. Murphy, R. *Sugarcane: Production Systems, Uses and Economic Importance*; Nova Science: New York, NY, USA, 2017; p. 39.

4. FAOSTAT. Food and Agriculture Organization of the United Nations, Statistics Division. Forestry Production and Trade. Available online: http://www.fao.org/faostat/en/#data/FO (accessed on 4 April 2019).
5. USDA. Foreign Agricultural Service. Production, Supply and Distribution. 2020. Available online: https://apps.fas.usda.gov/psdonline/app/index.html#/app/advQuery (accessed on 28 February 2020).
6. UN Comtrade. World Cane Sugar Export's Statistics. 2019. Available online: https://comtrade.un.org/data/ https://comtrade.un.org/data (accessed on 25 August 2019).
7. Office of the Cane and Sugar Board. Production Report. 2019. Available online: http://www.ocsb.go.th/th/home/index.php (accessed on 20 February 2020).
8. Bangkok Post. Thai Sugar Body Seeks to Sweeten Environment and Health. 2019. Available online: https://www.bangkokpost.com/business/1758849/thai-sugar-body-seeks-to-sweeten-environment-and (accessed on 22 February 2020).
9. IPCC. 2013: Annex I: Atlas of Global and Regional Climate Projections. In *Climate Change 2013: The Physical Science Basis. Contribution of Working Group I to the Fifth Assessment Report of the Intergovermental Panel on Climate Change*; Cambridge University Press: Cambridge, UK; New York, NY, USA, 2013; pp. 1311–1393.
10. Adams, R.M.; Rosenzweig, C.; Peart, R.M.; Ritchie, J.T.; McCarl, B.A.; Glyer, J.D.; Curry, R.B.; Jones, J.W.; Boote, K.; Allen, L.H. Global climate change and US agriculture. *Nature* **1990**, *345*, 219–224. [CrossRef]
11. Attavanich, W.; McCarl, B.A. How is CO_2 affecting yields and technological progress? A statistical analysis. *Clim. Chang.* **2014**, *124*, 747–762. [CrossRef]
12. Mendelsohn, R.; Nordhaus, W.D.; Shaw, D. The impact of global warming on agriculture: A Ricardian analysis. *Am. Econ. Rev.* **1994**, *84*, 753–771.
13. Parry, M.L.; Rosenzweig, C.; Iglesias, A.; Livermore, M.; Fischer, G. Effects of climate change on global food production under SRES emissions and socio-economic scenarios. *Glob. Environ. Chang.* **2004**, *14*, 53–67. [CrossRef]
14. Miao, R.; Khanna, M.; Huang, H. Responsiveness of Crop Yield and Acreage to Prices and Climate. *Am. J. Agric. Econ.* **2015**, *98*, 191–211. [CrossRef]
15. Schlenker, W.; Roberts, M.J. Nonlinear temperature effects indicate severe damages to U.S. crop yields under climate change. *Proc. Natl. Acad. Sci. USA* **2009**, *106*, 15594–15598. [CrossRef]
16. Rosenzweig, C.; Iglesias, A.; Yang, X.; Epstein, P.R.; Chivian, E. Climate Change and Extreme Weather Events; Implications for Food Production, Plant Diseases, and Pests. *Glob. Chang. Hum. Health* **2001**, *2*, 90–104. [CrossRef]
17. Cammarano, D.; Ceccarelli, S.; Grando, S.; Romagosa, I.; Benbelkacem, A.; Akar, T.; Al-Yassin, A.; Pecchioni, N.; Francia, E.; Ronga, D. The impact of climate change on barley yield in the Mediterranean basin. *Eur. J. Agron.* **2019**, *106*, 1–11. [CrossRef]
18. Lobell, D.B.; Schlenker, W.; Costa-Roberts, J. Climate Trends and Global Crop Production since 1980. *Science* **2011**, *333*, 616–620. [CrossRef]
19. Raymundo, R.; Asseng, S.; Robertson, R.; Petsakos, A.; Hoogenboom, G.; Quiroz, R.; Hareau, G.; Wolf, J. Climate change impact on global potato production. *Eur. J. Agron.* **2018**, *100*, 87–98. [CrossRef]
20. Zhao, C.; Liu, B.; Piao, S.; Wang, X.; Lobell, D.B.; Huang, Y.; Huang, M.; Yao, Y.; Bassu, S.; Ciais, P.; et al. Temperature increase reduces global yields of major crops in four independent estimates. *Proc. Natl. Acad. Sci. USA* **2017**, *114*, 9326–9331. [CrossRef] [PubMed]
21. Attavanich, W.; McCarl, B.A.; Ahmedov, Z.; Fuller, S.W.; Vedenov, D.V. Effects of climate change on US grain transport. *Nat. Clim. Chang.* **2013**, *3*, 638–643. [CrossRef]
22. Brown, M.E.; Antle, J.; Backlund, P.; Carr, E.; Easterling, W.; Walsh, M.; Ammann, C.; Attavanich, W.; Barrett, C.; Bellemare, M.; et al. Climate Change, Global Food Security, and the U.S. Food System. Available online: http://www.usda.gov/oce/climate_change/FoodSecurity2015Assessment/FullAssessment.pdf (accessed on 10 May 2019).
23. Brown, M.E.; Carr, E.R.; Grace, K.L.; Wiebe, K.; Funk, C.C.; Attavanich, W.; Backlund, P.; Buja, L. Do markets and trade help or hurt the global food system adapt to climate change? *Food Policy* **2017**, *68*, 154–159. [CrossRef]
24. Singels, A.; Jones, M.; Marin, F.; Ruane, A.; Thorburn, P. Predicting Climate Change Impacts on Sugarcane Production at Sites in Australia, Brazil and South Africa Using the Canegro Model. *Sugar Tech.* **2013**, *16*, 347–355. [CrossRef]

25. Marin, F.R.; Jones, J.W.; Singels, A.; Royce, F.; Assad, E.; Pellegrino, G.Q.; Justino, F. Climate change impacts on sugarcane attainable yield in southern Brazil. *Clim. Chang.* **2012**, *117*, 227–239. [CrossRef]
26. Silva, W.K.D.M.; De Freitas, G.P.; Junior, L.M.C.; Pinto, P.A.L.D.A.; Abrahão, R. Effects of climate change on sugarcane production in the state of Paraíba (Brazil): A panel data approach (1990–2015). *Clim. Chang.* **2019**, *154*, 195–209. [CrossRef]
27. Baez-Gonzalez, A.D.; Kiniry, J.; Meki, M.N.; Williams, J.R.; Cilva, M.A.; Gonzalez, J.L.R.; Estala, A.M. Potential impact of future climate change on sugarcane under dryland conditions in Mexico. *J. Agron. Crop. Sci.* **2018**, *204*, 515–528. [CrossRef]
28. Ruan, H.; Feng, P.; Wang, B.; Xing, H.; O'Leary, G.J.; Huang, Z.; Guo, H.; Liu, D.L. Future climate change projects positive impacts on sugarcane productivity in southern China. *Eur. J. Agron.* **2018**, *96*, 108–119. [CrossRef]
29. Adhikari, U.; Nejadhashemi, A.P.; Woznicki, S.A. Climate change and eastern Africa: A review of impact on major crops. *Food Energy Secur.* **2015**, *4*, 110–132. [CrossRef]
30. Lobell, D.B.; Burke, M.; Tebaldi, C.; Mastrandrea, M.D.; Falcon, W.P.; Naylor, R.L. Prioritizing Climate Change Adaptation Needs for Food Security in 2030. *Science* **2008**, *319*, 607–610. [CrossRef]
31. Yoshida, K.; Srisutham, M.; Sritumboon, S.; Suanburi, D.; Janjirauttikul, N. Weather-induced economic damage to upland crops and the impact on farmer household income in Northeast Thailand. *Paddy Water Environ.* **2019**, *17*, 341–349. [CrossRef]
32. Feng, H.; Babcock, B.A. Impacts of Ethanol on Planted Acreage in Market Equilibrium. *Am. J. Agric. Econ.* **2010**, *92*, 789–802. [CrossRef]
33. FAO. *The Future of Food and Agriculture–Trends and Challenges*; Food and Agriculture Organization of the United Nations: Rome, Italy, 2017.
34. Reuters. Thailand's Sugar Output to Hit Nine-Year Low Due to Drought—Trade Body. 2020. Available online: https://www.reuters.com/article/thailand-sugar-output/thailands-sugar-output-to-hit-nine-year-low-due-to-drought-trade-body-idINKBN1ZU1Z3 (accessed on 13 April 2020).

atmosphere

MDPI

Article

Transcriptomics of Mature Rice (*Oryza Sativa* L. Koshihikari) Seed under Hot Conditions by DNA Microarray Analyses

Ranjith Kumar Bakku [1,2,†], Randeep Rakwal [3,4,5,6,*,†], Junko Shibato [5,†], Kyoungwon Cho [7,†], Soshi Kikuchi [8,†], Masami Yonekura [9,†], Abhijit Sarkar [10,†], Seiji Shioda [4,†] and Ganesh Kumar Agrawal [4,5,6,†]

1 Faculty of Engineering Information and Systems, Department of Computer Science, University of Tsukuba, 1-1-1 Tennodai, Tsukuba, Ibaraki 305-8572, Japan; ranjithkumar.bakku@gmail.com
2 Tsukuba Life Science Innovation Program (T-LSI), University of Tsukuba, 1-1-1 Tennodai, Tsukuba, Ibaraki 305-8572, Japan
3 Faculty of Health and Sport Sciences, University of Tsukuba, 1-1-1 Tennodai, Tsukuba, Ibaraki 305-8574, Japan
4 Global Research Center for Innovative Life Science, Peptide Drug Innovation, School of Pharmacy and Pharmaceutical Sciences, Hoshi University, 4-41 Ebara 2-chome, Shinagawa, Tokyo 142-8501, Japan; shioda@hoshi.ac.jp (S.S.); gkagrawal123@gmail.com (G.K.A.)
5 Research Laboratory for Biotechnology and Biochemistry (RLABB), PO Box 13265, Kathmandu 44600, Nepal; rjunko@nifty.com
6 Global Research Arch for Developing Education (GRADE) Academy Pvt. Ltd., Birgunj 44300, Nepal
7 Department of Biotechnology, College of Agriculture and Life Sciences, Chonnam National University, Gwangju 61186, Korea; kw.cho253@gmail.com
8 Genetic Resources Center, National Agriculture and Food Research Organization, Kannondai, Tsukuba, Ibaraki 305-8602, Japan; skikuchi@affrc.go.jp
9 Laboratory of Molecular Food Functionality, College of Agriculture, Ibaraki University, Ami, Ibaraki 300-0393, Japan; masami.yonekura.ogoori@vc.ibaraki.ac.jp
10 Laboratory of Applied Stress Biology, Department of Botany, University of Gour Banga, Malda 732103, West Bengal, India; abhijitbhu@gmail.com
* Correspondence: plantproteomics@gmail.com
† These authors contributed equally to this work.

check for updates

Received: 1 April 2020; Accepted: 17 May 2020; Published: 20 May 2020

Abstract: Higher temperature conditions during the final stages of rice seed development (seed filling and maturation) are known to cause damage to both rice yield and rice kernel quality. The western and central parts of Japan especially have seen record high temperatures during the past decade, resulting in the decrease of rice kernel quality. In this study, we looked at the rice harvested from a town in the central Kanto-plains (Japan) in 2010. The daytime temperatures were above the critical limits ranging from 34 to 38 °C at the final stages of seed development and maturity allowing us to investigate high-temperature effects in the actual field condition. Three sets of dry mature rice seeds (commercial), each with specific quality standards, were obtained from Japan Agriculture (JA Zen-Noh) branch in Ami-town of Ibaraki Prefecture in September 2010: grade 1 (top quality, labeled as Y1), grade 2 (medium quality, labeled as Y2), and grade 3 (out-of-grade or low quality, labeled as Y3). The research objective was to examine particular alterations in genome-wide gene expression in grade 2 (Y2) and grade 3 (Y3) seeds compared to grade 1 (Y1). We followed the high-temperature spike using a high-throughput omics-approach DNA microarray (Agilent 4 × 44 K rice oligo DNA chip) in conjunction with MapMan bioinformatics analysis. As expected, rice seed quality analysis revealed low quality in Y3 > Y2 over Y1 in taste, amylose, protein, and fatty acid degree, but not in water content. Differentially expressed gene (DEG) analysis from the transcriptomic profiling data revealed that there are more than one hundred upregulated (124 and 373) and downregulated (106 and 129) genes in Y2 (grade 2 rice seed) and Y3 (grade 3 rice seed), respectively. Bioinformatic analysis of DEGs selected as highly regulated differentially expressed (HRDE) genes revealed changes in function of genes related to metabolism, defense/stress response, fatty acid biosynthesis, and hormones. This research provides, for the first time, the seed transcriptome profile for the classified low grades (grade 2, and out-of-grade; i.e., grade 3) of rice under high-temperature stress condition.

Keywords: rice; heat stress; whole genome DNA microarray; yield loss; MapMan analysis; HRDE

---

1. Introduction

With the rise in mean global temperatures, the earth's biosphere is warming up gradually. According to the statistics of the North American Space Agency's (NASA) earth observatory data, the average global temperature increased above 1 °C since the year 1880 [1]. In the next 100 years, the surface temperatures are expected to rise between 2 °C to 6 °C if greenhouse gas emissions continue. This is a serious threat for our future generations as it directly influences the habitable conditions—both flora and fauna. A rise in temperature induces heat stress and seriously affects plants, which play a key role in providing food, oxygen, and shelter to several species of fauna including humans. Moreover, its effects on the productivity of food crops will result in a food crisis for the growing population [2]. Recently, Zhao et al. (2017) [3] estimated that for every one-degree rise in global temperature, yields of food crops like wheat, rice, maize, and soybean will be reduced by 6%, 3.2%, 7.4%, and 3.1%, respectively [3]. Therefore, there is a grave need to understand the effects of increasing temperatures and the biological responses induced in food crops, to address challenges in developing next-generation crops [2,3].

Heat stress damages both physiological and molecular level mechanisms in plants [4–7]. Major damages include scorching and abscission of leaves, shoot and stems, fruit/seed damage, reduced photosynthesis, seed germination, increase in reactive oxygen species (ROS), and osmotic stress [5,6,8–11]. Furthermore, the ROS accumulated during heat stress damages molecular components by inducing oxidative stress [12,13]. For example, the formation of hydroxyl radicle could induce irreversible DNA single-strand breaks, peroxidation of lipids, protein proteolysis, and damage of photosynthesis system II [14–16]. In addition, hydrogen peroxide (H_2O_2) formed during heat stress could alter the balance between starch biosynthesis and degradation mechanisms by upregulation of starch degradation and downregulation of the biosynthesis genes. Additionally, plants also develop heat-tolerance by responding to heat stress by altering their gene expression and synthesizing specific heat shock proteins (HSPs), transporters, and enzymes.

Rice (*Oryza sativa*), being one of the top three food crops produced and consumed by nearly 3.5 billion people around the world, has been well studied for its yield and quality with respect to changes in global temperatures. According to the Food and Agriculture Organization (FAO) of the United Nations, Asia is the largest producer of rice in the world (nearly 90%). The average rice production statistics between 1994–2018 show that China and India stand as the top two producers (193 and 140 million tons respectively), while Japan is ranked 10th (11 million tons) in the world. Japan is also the 13th largest consumer of rice in the world as of 2018. In Japan, as rice is a primary staple food crop with high socio-economic importance, any reduction in yield and damage to the quality of the grain is a serious issue. The decrease in rice productivity is generally due to reduced pollen germination and in turn, affects the spikelet fertility and yield [17–21]. Moreover, spikelet sterility occurs when air temperature reaches a threshold limit of 35 °C during flowering time [22]. For example, in 2007, the Kanto and Tokaido regions of Japan faced a marked rise in temperatures to 40 °C during the summer months. The samples collected in this region showed high rates of spikelet fertility damages [19]. In addition to high-temperature injury in grain productivity, it also affects the quality of rice. For example, if rice crops are exposed to high temperatures during the first two weeks after an early emergence, the rice grain turns into immature kernels with white portions [23–26]. A decade back, high proportions of first-grade rice kernel were produced in the warmer regions (west Japan) like Kyushu, while production was lower in northern regions of Japan (like Tohoku and Hokkaido), due to cold damage. However, in the past years, the rice crop quality from North to West Japan has been completely reversed due to climatic changes [18]. In the year 2010, the average temperature across Japan approached 28 °C to 29 °C, as compared to normal day temperatures (26 °C). That year recorded the hottest summer (June-September 2010) ever experienced with day temperatures ranging between 35 °C to 38 °C in East, South, Central, and West regions of Japan. On average, the daytime temperatures were nearly 1.8 and 2 °C above normal in August. An exposure to such anomalies in temperature for a couple of hours is enough to induce spikelet sterility and reduce crop productivity in rice. As such these conditions during 2010 resulted in producing milky white kernels in the first-grade rice kernel crop, and, therefore reduced their production throughout Japan (Figure 1). This grain chalkiness is due to the induced production of starch hydrolyzing enzymes (α-amylases) as a result of high temperatures [27–30].

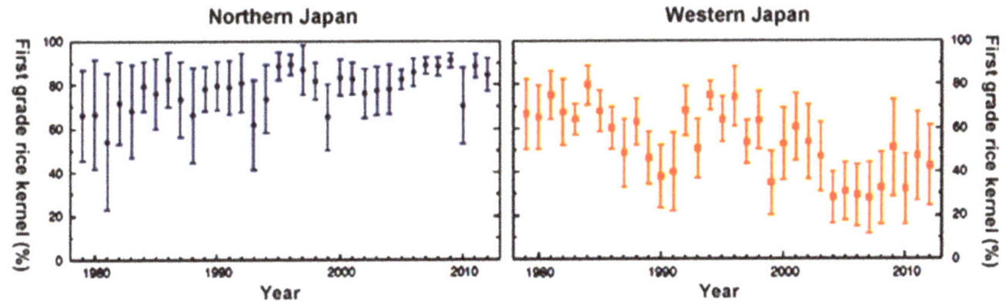

Figure 1. Temporal change in the proportion of first grade rice kernel in Northern Japan and Western Japan. This figure, created based on Ministry of Agriculture, Forestry and Fisheries data, has been obtained with permission from Prof. Shunji Ohta, Faculty of Human Sciences, Waseda University-Recent Impacts of Climatic Extreme Events on Everyday Food in Japan: The Need for an Adaptation Strategy for Climate Change [31].

It was observed that the productivity and quality of rice also depend on its variety along with the temperature. However, no studies were reported on how different varieties of the same cultivar respond to heat stress in the open field conditions. Especially in Japan, Koshihikari is a highly grown rice cultivar/variety for the past few decades, and also is the most widely affected variety due to climate changes in recent years. Understanding the importance of this cultivar in Japan, the current study focuses on exploring transcriptome level differences under heat stress between grades 1, 2, and 3 (labeled as Y1, Y2, and Y3 for the purpose of the experiment) of the Koshihikari rice variety. Each grade of rice is categorized by Japan's National Food Agency (NFA) based on the grain's physiological characteristics like weight per volume, moisture content, appearance, region in which it was grown, etc., which determine its quality [32]. For this open field sample, sampling was done from a region in East Japan (Kanto-plains) during the high-temperature season of the year 2010, and included three different grades of rice. The seed samples were analyzed for genome-wide gene expression (DNA microarray technique) and compared using bioinformatic techniques for identifying differentially expressed genes (DEG).

2. Results and Discussion

2.1. Koshihikari Rice Seed Quality in Grades 1 to 3

The seed quality was tested based on quality criteria like taste value, percentage of amylase, protein, water content, and fatty acid degree in the rice seed (Table 1). Grade 1 (Y1) Koshihikari rice is identified to have a high taste value of 86. Compared to Y1, grade 2 (Y2) and grade 3 (Y3) are relatively less tasty by 5 to 3 points as determined by professional analysis. In addition, the Y2 and Y3 grains have a high percentage of amylose and water content in comparison to Y1, indicating sticky and chalky rice. In general, lower amylose content means lower chalkiness of the grains, resulting in harder rice. Low water content indicates lower moisture levels which in turn produce less sticky rice after cooking [33,34]. Quality of rice also depends on its aroma and flavor. The surface lipids of grains form free fatty acids as they undergo hydrolysis during storage of the grain. The free fatty acids formed in this way are susceptible to oxidation and eventual formation of hydrocarbons such as aldehydes and ketones, which give a foul odor to the seeds [35,36]. Y1 was found to have a lesser degree (15.5) of unsaturated fatty acids in comparison to Y2 (17) and Y3 (20), indicating the presence of good aroma with grade 1. The overall grain quality analysis clearly indicated that under extreme temperatures, the grain quality is affected more (to negative values) in Y3 followed by Y2, and is least affected in Y1 (Grade 1).

Table 1. Rice grain (cv. Koshihikari * seed) quality analyses **.

Components	1st Grade	2nd Grade	Out of Grade (3rd)
Taste Value (point)	86 ***	81	83
Amylose (%)	17.9	18.5	18.2 ****
Protein (%)	6.6	7.5	7.6 *****
Water Content (%)	14.1	14.4	14.3
Fatty Acid Degree (mg/100 g)	15.5	17.0	20.0 ******

* Rice seeds, grades 1 to 3 were obtained from JA (Japan Agriculture). ** Analyses were done by rice analyzer (SATAKE, Japan). *** Above 85 is very good taste, as determined by a professional taster. **** Categorized as not so sticky. ***** Categorized as hard rice. ****** Higher the degree, increased oxidation of fatty acids.

2.2. *Investigation of the Koshihikari Rice Seed Transcriptomes in Grades 2 and 3*

This work looks at differences in transcripts accumulated in the dried endosperm after the maturation process. From the analysis of DEGs in Y3 and Y2 in comparison to Y1, a greater number of DEGs were observed in Y3 (502 genes) than Y2 (230 genes), as shown in Figure 2. Individually, a total of 373 upregulated and 129 downregulated genes were observed in Y3 while Y2 resulted in 124 and 106 up and downregulated genes, respectively. Among all the DEGs in grades Y3 and Y2, similar expression patterns were also observed for a few common genes present in both grades. In this category, there are nearly 59 upregulated genes and 33 downregulated genes.

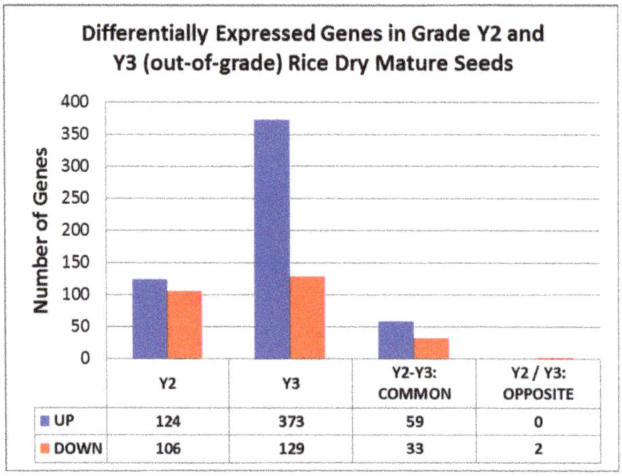

Figure 2. Figure indicating the differentially expressed genes (DEGs) in Y2 and Y3.

2.3. MapMan Analysis of Koshihikari Rice Seed Differentially Expressed Genes in Grades 2 and 3

In order to visualize the DEGs that are involved in key pathways, a total of 42,560 Rice Japonica genes were plotted against the pathway maps of the Mapman tool. Among all the genes, DEGs with high fold changes were selected for the analysis, as discussed in methods for Mapman analysis. From the fold change cut-off threshold, a total of 161 and 490 highly regulated differentially expressed (HRDE) genes for each Y2 and Y3 were identified. Among these, 92 genes are common for both grades. A total of 68 and 398 genes are unique for Y2 and Y3, respectively resulting in a total of 560 HRDEs for both. An overview of these HRDEs in Y1-Y2 and Y1-Y3, and the related 36 pathway bins, are shown in (Figure 3A,B). Results clearly indicate that the highest fraction of the gene regulation occurred in 16, 17, 20, 26, 27, 28, 29, 30, 33, and 34 Mapman bins. The number of genes associated with each bin, specific to each pathway, is listed in supplementary Table S1. These bins are related to secondary metabolism, hormone metabolism, stress response, miscellaneous enzyme families, RNA, DNA, protein, signaling, development, and transport. In addition, bin 35 has the most differentially expressed genes; however, these genes are without any functional annotation.

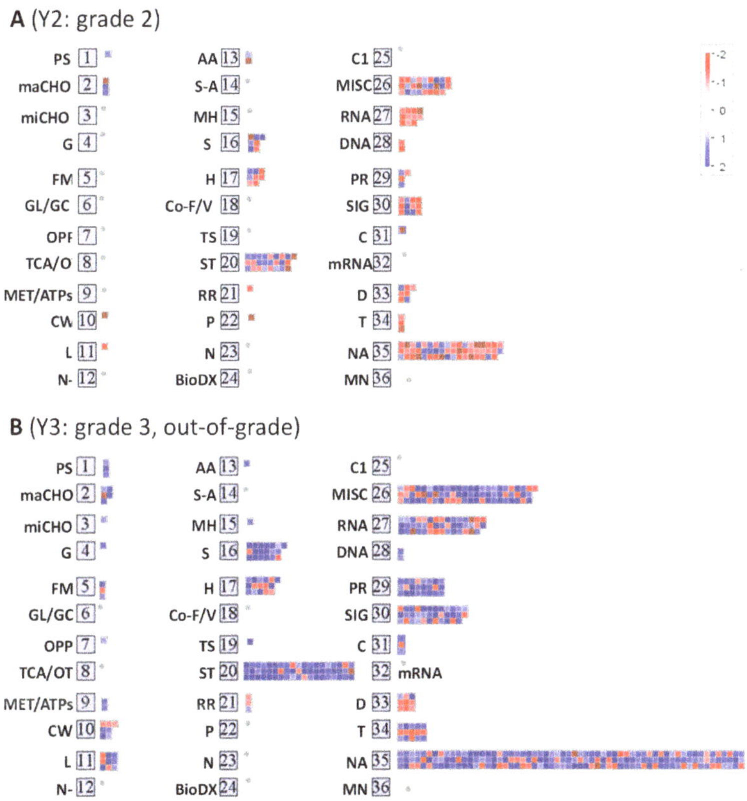

Figure 3. Mapman Bins for functional categorization of high temperature-responsive genes in (**A**) Y2: grade 2 Koshihikari rice seed and (**B**) Y3: grade 3 Koshihikari rice seed. Non-redundant 640 genes which expression are changed over 2-fold in the seeds of both grades 2 (Y2) and 3 (Y3) were functionally categorized into MapMan bins as described in Materials and Methods. The heat map with grid boxes shows the genes (blue, upregulated; red, downregulated) in each BIN for each grade seeds. Number of genes associated to each bin is listed in supplementary Table S1. The 36 BINS abbreviations: PS, photosynthesis; maCHO, major carbohydrate metabolism; miCHO, minor carbohydrate metabolism;

G, glycolysis; FM, fermentation; GL/GC, gluconeogenese/glyoxlate cycle; OPP, oxidative pentose phosphate; TCA/OT, tricarboxylic acid/organic acid transformations; MET/ATPs, mitochondrial electron transport/adenosine triphosphate; CW, cell wall; L, lipid metabolism; N-, nitrogen metabolism; AA, amino acid metabolism; S-A, sulfur assimilation; MH, metal handling; S, secondary metabolism; H, hormone metabolism; Co-F/V, co-factor and vitamin metabolism; TS, tetrapyrrole synthesis; ST, stress; RR, redox regulation; P, polyamine metabolism; N, nucleotide metabolism; BioDX, biodegradation of xenobiotics; C1, C1-metabolism; MISC, miscellaneous; RNA, ribonucleic acid; DNA, deoxyribonucleic acid; PR, protein; SIG, signaling; C, cell; mRNA, messenger RNA; D, development; T, transport; NA, not assigned; MN, mineral nutrition.

The overall expression profile indicates that stress response genes (bin 20) are strongly regulated in both Y2 and Y3. Y3 exhibited a larger number of strongly upregulated genes in comparison to Y2. Further, the differential gene expression patterns of Y2 and Y3 in three major processes, namely cell function, metabolism, and abiotic-biotic stress (Figures 4–6) were explored.

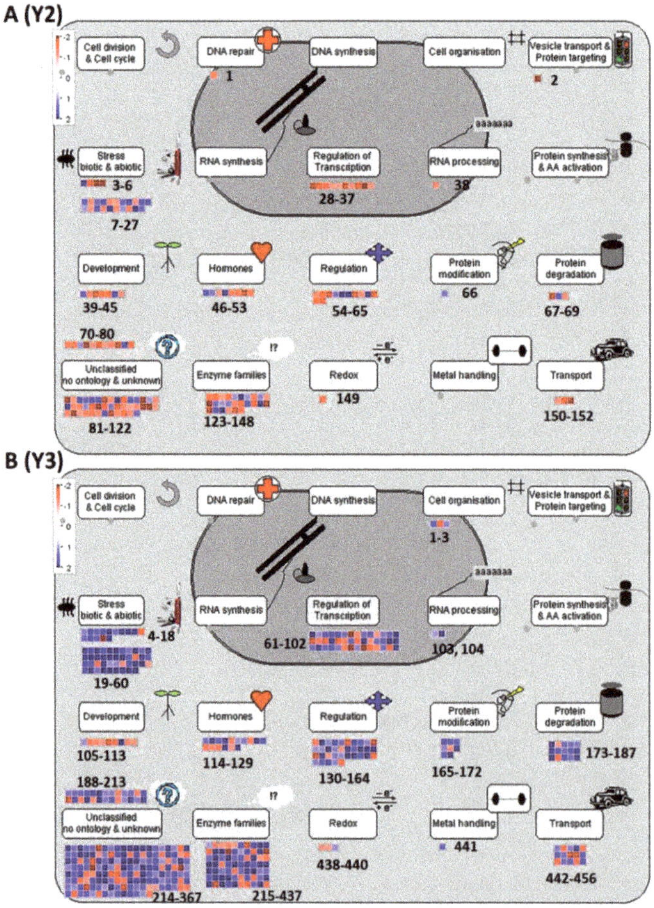

Figure 4. A cell function map of highly regulated differentially expressed (HRDE) genes that are categorized based on various functions, generated using Mapman. (**A**) Map for HRDE genes related to Koshihikari rice grade Y2. (**B**) Map for HRDE genes related to Koshihikari rice grade Y3. Blue and red colored data points indicated highly up or downregulated genes. Numbering for each data point on the figure was given in left to right order to identify relevant gene information from supplementary Table S2. These numbers represent serial numbers for the genes in the supplementary table accordingly.

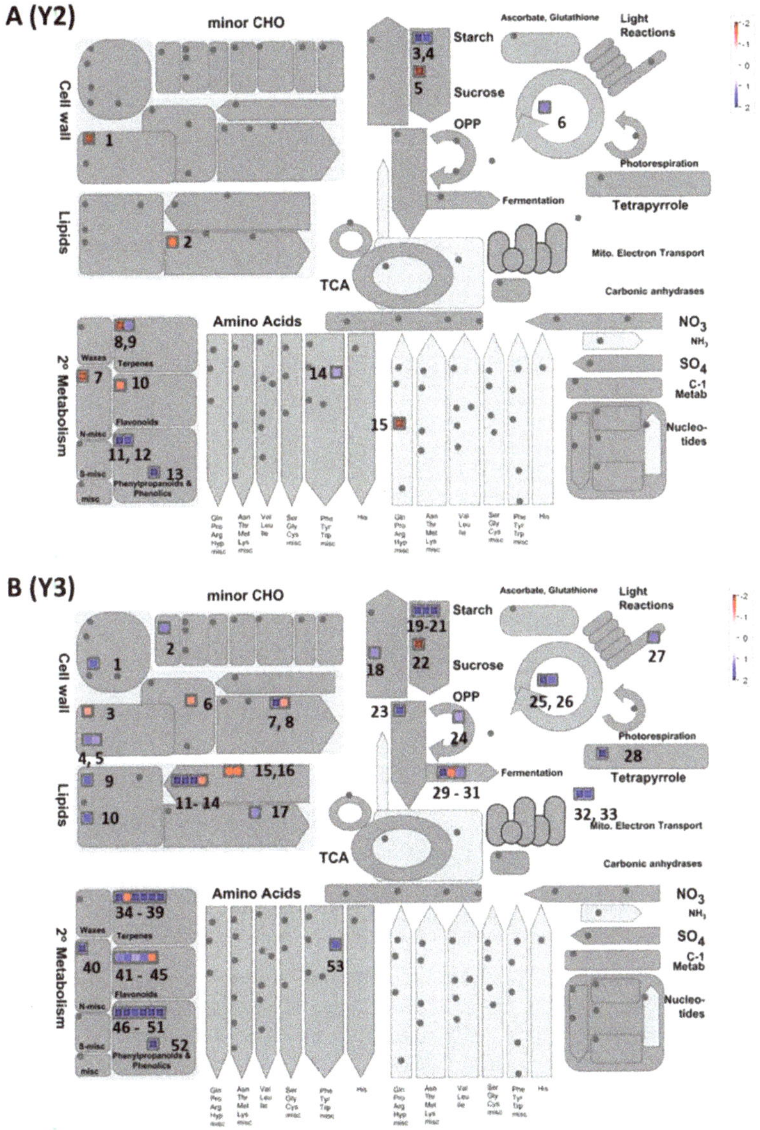

Figure 5. A metabolism overview map of highly regulated differentially expressed (HRDE) genes that are categorized into various metabolic pathways generated using Mapman. (**A**) Map for HRDE genes related to Koshihikari rice grade Y2. (**B**) Map for HRDE genes related to Koshihikari rice grade Y3. Numbering for each data point on the figure was given in left to right order to identify relevant gene information from supplementary Table S3. These numbers represent serial numbers for the genes in the supplementary table accordingly. Expression of specific genes of interest like PME, cellulose synthases, pectin lyases, XET, raffinose synthase, TAG lipases, DGK, Omega-6-desaturases, SPT2, amylase and Susy family genes can be observed in Y2(1, 3), Y3(6), Y3(7, 8), Y3(4), Y3(2), Y2(2), Y3(9), Y3(12, 13, 14), Y3 (10), respectively.

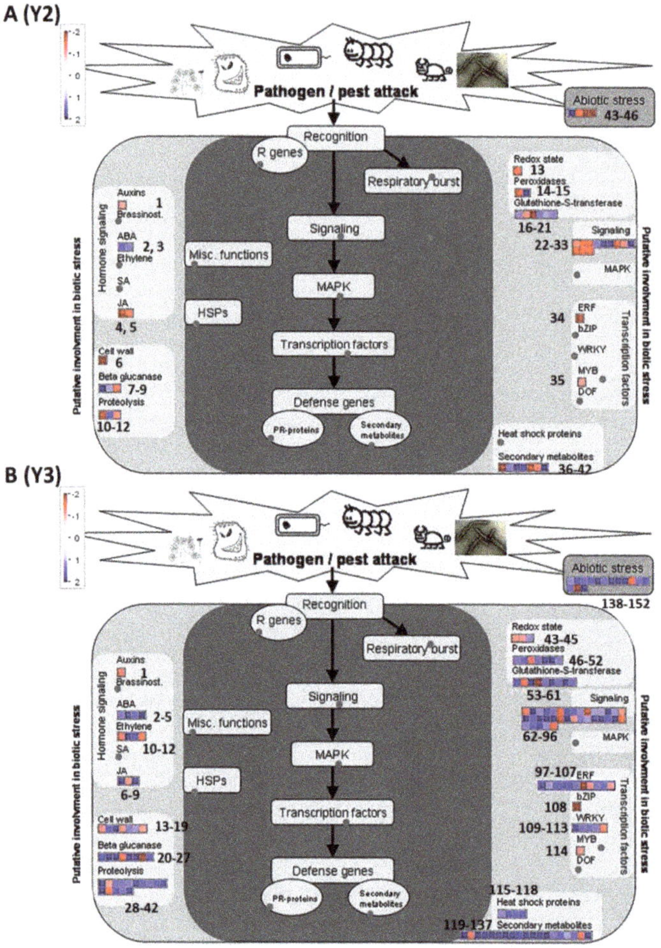

Figure 6. An abiotic-biotic map of differentially expressed genes that are categorized into abiotic and biotic stress-responsive pathways generated using Mapman. (**A**) Map for HRDE genes related to Koshihikari rice grade Y2. (**B**) Map for HRDE genes related to Koshihikari rice grade Y3. Numbering for each data point on the figure was given in left to right order to identify relevant gene information from supplementary Table S4. These numbers represent serial numbers for the genes in the supplementary table accordingly. Expression of specific genes of interest like sHsp, MYB4, DREB1/CBF, ERF, OsWRKY, GGPS, anthocyanins, and flavonoid biosynthesis, COMT, laccasses, OsSAUR, FIP1, lipoxygenase2 and OPR family genes can be observed in Y3(115/138), Y2/Y3(35/114), Y3(101, 102), Y3(98), Y3(110-112), Y2(41),Y3(133-135), Y3(178,179), Y2/Y3(47/186), Y2/Y3(1), Y2/Y3(2,3/2,3,4,5), Y2/Y3 (4/10) and Y2/Y2 (5/11,12) grid boxes, respectively. Y2/Y3 represents similar genes detected in both Y2 and Y3.

For each of the above pathway maps, a total of 15, 151 and 46 HRDE genes of Y2 and 53, 457 and 152 HRDE genes of Y3, were mapped as data points. All these genes are listed in supplementary Tables S2–S4, along with their bin-numbers and functions. For the cell function, metabolism-overview and biotic-abiotic stress pathway maps (Figures 4–6), the numbering for each data point was given in left to right order to identify relevant gene information from supplementary Tables S2–S4. Serial numbers for the genes in the supplementary tables were numbered accordingly. Overall mapping results indicate that around 650 genes related to various cellular functions were differentially regulated in both Y2 and Y3, respectively (Figure 4, supplementary Table S2). Y2 appears to respond specifically by down regulation of genes in biotic-abiotic stress, transcription, RNA processing, protein degradation, hormones and regulation. While Y3 exhibited quite opposite response by upregulation of genes in respective mechanisms, with few downregulated genes. Genes related to protein modification are downregulated in both Y2 and Y3.

Specifically, from the metabolic-overview map (Figure 5), results indicate that highly up-, or downregulated genes are mainly related to cell wall, lipid, starch, and secondary metabolic pathways. In Y2, very few genes were differentially regulated (7 downregulated and 8 upregulated). Among these, genes related to cell wall remodeling proteins (pectin methyl esterase's- PMEs), sucrose synthase, triacylglycerol (TAG) lipase, phosphotase synthase and a few secondary metabolism-related genes (tyrosine decarboxylate, terpene cyclase, isoflavonoid reductase) were highly downregulated. On the other hand, Y3 data showed upregulation of cell wall modification proteins like transglucosylases (XET), diacylglycerol kinases, omega-6 fatty acid desaturases, male sterility proteins, α-amylase isozymes, rubisco interacting proteins, mitochondrial electron-transport proteins (transposons) and many proteins related to secondary metabolites (terpenes, flavonoids, and phenolics) (Supplementary Table S3). Out of both Y2 and Y3, commonly upregulated metabolic pathways were related to starch, chitinase, phenylpropanoids and phenolics, and aromatic amino acid synthases. While the commonly downregulated genes are related to PME and sucrose synthase.

On the other hand, among the cellular functions, the biotic and abiotic stress responses are the major events during heat stress. Y2 exhibited downregulation of genes related to auxins, jasmonic acid (JA), lipoxygense, PME, proteases, peroxidases, protein kinases, ATP binding proteins, signaling G-proteins, calcium ion binding proteins and abiotic stress-related germin like proteins (Figure 6). On the other hand, Y3 showed more upregulated genes especially related to glucanase, proteolysis, peroxidases, glutathione-S-transferases, signaling, transcription factors (TFs), HSPs, secondary metabolites, and abiotic stress responses. Despite heat stress, in Y2 there were no genes related to ethylene, HSP, and TFs like b-zip that crossed the threshold set for detection of HRDE genes, in our two-color dye-swap DNA microarray strategy. There are genes with a 1-fold upregulation which are not considered as the threshold is set to 2. So, there is a minimal level of gene expression in Y2 indicating that it is not as sensitive as Y3 to heat stress.

2.4. *High Temperature-Triggered Regulatory Events in Koshihikari Rice Seeds of Grades 2 and 3*

Overall results indicated that the effect of high temperature on Koshihikari rice of grades 2 and 3 is quite different. Most of the genes were downregulated in Y2 type, while Y3 exhibited upregulation.

2.4.1. Cell Wall Damage Repair

Initial response to heat shock is observed in cell wall remodeling proteins like PMEs and XETs. PMEs are involved in regulating cell wall plasticity, porosity, and modulation of Ca^{2+} ion channels [37], while XETs are involved in secondary cell wall strengthening [38,39]. The regulation of XETs and PMEs under heat shock is previously observed in rice, *Arabidopsis*, and many plant varieties [38,40,41]. During heat stress, Y2 and Y3 exhibited down regulation of PMEs genes Os09g37360.1 and Os04g46740.1, respectively. In addition, Y3 alone shows downregulation of genes related to cellulose synthesis (Os02g09930.1) and pectin lyases (Os10g26940.1 and Os02g15690.1), while upregulation of XET gene (Os06g22919.2) indicating cell wall response in Y3 is more sensitive to heat shock. An upregulation of the raffinose synthase gene (Os08g38710.1) in Y3, which prevents plants from oxidative stress damage [42–44], was also observed. Besides, the pectinase activity upregulation of cellulases like endo-glucosidase, prominently in Y3 can be seen (supplementary Table S3, Metabolism overview; and Figure 5).

2.4.2. Lipid Remodelling

The lipid signaling is another key process involved in abiotic stress responses by the plants. In Y2, there is no more than one gene related to lipid mechanism which was identified as the HRDE gene. This gene (Os08g04800.1) codes for a TAG lipase protein and is related to lipid degradation. TAG lipase was found to be strongly downregulated (−1.63). TAG lipases de-esterify fatty acids from TAG (accumulated in lipid bodies) or plastglobular lipids and are generally accumulated during seed germination [45]. TAG lipases are observed to be accumulated under heat stress in *Arabidopsis* [45,46]; however, the current data shows the opposite result. It is to be noted that rice and *Arabidopsis* are different, and therefore, the same gene could have unique functions in different plants and tissues. On the other hand, in Y3 there were no genes related to TAG lipases identified as HRDEs. However, the same gene was found in Y3 with very minimal (three times less than Y2) downregulation (not HRDE). This indicated that in this case, strong downregulation of TAG lipases could be common in Koshihikari as this is a DEG analysis in comparison to Y1 also. So, the result in Y3 indicates that Y3 might have started the accumulation of this gene under heat stress.

Furthermore, the genes related to diacyl-glycerol kinases (DGK), phospholipid synthesis, sphingolipid transferases, and fatty acid omega 6 desaturases were strongly upregulated in Y3 (supplementary Table S3, Metabolism overview; and Figure 5). In general, DGKs are involved in phosphatidic acid (PA) synthesis as well as sphingo-lipid synthesis. The PA produced by the mediation of DGKs has a positive regulatory impact on sphingo-lipid kinases and a negative effect on abscisic acid synthesis pathways [47]. PA is also involved in the growth and development of plant root hair and also functions as membrane-localized signal to recruit specific proteins and change their activity [48–50]. Therefore, by upregulation of this DGK specific gene (Os04g54200.1), Y3 is preparing for abiotic stress due to increased temperatures.

The fatty acid desaturases, on the other hand, are involved in maintaining membrane fluidity and are usually upregulated under low-temperature stress. However, studies also identified that high temperatures also enhance desaturases involved in the eukaryotic pathway [51]. In the current work upregulation of omega-6-desturases (Os07g23410.2, Os07g23430.1, Os02g48560.5) FAD2 and FAD3 related genes was observed. This clearly indicates a remodeling of fatty acids in endoplasmic reticulum [52]. In addition, the FAD2 gene was also found to be downregulated. This could be possibly due to a stop in the FA 18:1 synthesis from plastids [53,54].

Sphingo-lipids, on the other hand, also play an essential role in preventing plants from heat stress. Therefore, the data were screened for any changes in Sphingo-lipid genes and resulted in the identification of an upregulated of serine palmitoyl transferase 2 (SPT2) (Os01g70370.1) in Y3. This enzyme is involved in de novo synthesis of sphiganine and di-hydro shpinganine to form sphingo-lipids and ceramides. The ceramides further induce at least one heat shock protein-like $\alpha\beta$-crystallin upon heat stress [55]. In accordance with this, an upregulation of "α-crystallin-Hsps IbpA," a small Hsp (sHsp) related gene (Os02g48140.1) was also observed (Supplementary Table S4, Aiotic-Biotic stress; Figure 6). The SPT2 enzyme is known to play a key role in the male gametophyte development in *Arabidopsis* [56]. Upregulation of SPT2 might be a heat shock recovery mechanism in order to prevent gametophyte damage.

2.4.3. Transcription Factor Activation

Besides differential expression of major genes involved in the metabolic process, several changes in the TF genes related to abiotic-stress responses were also observed. Specifically, in both Y2 and Y3 the MYB and bZIP TF's were downregulated while upregulation of AP2/ERF and OsWRKY was observed only in Y3 (Supplementary Table S4, Aiotic-Biotic stress; Figure 6). MYB4 TF-related gene (Os09g36730.1) was downregulated in both Y2 and Y3. Recent studies in *Arabidopsis* have indicated that upregulation of MYB4 TFs induce the biosynthesis of secondary metabolites (hydroxycinnamate esters) that in turn increase UV-B hypersensitivity in plants [57,58]. Downregulation of these MYB4 TFs in Y2 and Y3 indicate increased tolerance to UV-B radiation. AP2/ERF TFs were upregulated in Y3 alone. The AP2/ERF (APETALA2/ethylene response factor) TF's are one of the most important groups in plants that help in the stress defense mechanism [59]. These TF's induce a set of abiotic stress-related genes. In this study, AP2/ERF family proteins like dehydration-responsive element-binding protein

(DREB1/CBF) (Os09g35020.1 and Os09g35010.1) and ethylene-responsive element-binding factor (ERF) (Os04g46220) were upregulated in Y3. These genes were known to play a key role in ABA independent stress tolerance, especially related to osmotic stress [60,61]. The other major class of TFs identified in this study is the OsWRKY TFs. The WRKY TF's are well known for their roles in the regulation of plant growth, development and apoptosis, and responses to biotic and abiotic stresses [62]. There are 74 and 109 WRKY members in each of *Arabidopsis* and rice genome, respectively [63,64]. In the current analysis, especially in Y3, an upregulation of OsWRKY genes 24, 28, 45, and 49 (Os01g61080.1, Os06g44010.1, Os05g25770.1, Os05g49100.1), and downregulation of OsWRKY24 (Os07g02060.1) was observed. However, no such specific responses were observed in Y2. Among these, the OsWRKY28 and OsWRKY45 were well studied. OsWRKY28 known to be highly responsive to As(V) and regulation of As(V)/Pi uptake or tolerance in rice [65]. In addition, Cai et al. (2014) showed that OsWRKY28 rice mutants resulted in various disorders like downregulation of JA biosynthesis genes in the shoots, irregular spikelet development, altered flower closing, and anther dehiscence and eventually resulting in lower fertility [66]. In the current study, JA gene expression was observed to decrease in Y2 while OsWRKY28 is not differentially regulated. On the other hand, in Y3, upregulation of OsWRKY28 along with JA biosynthesis genes is seen. In addition, JA is also known to regulate OsWRKY45 gene expression [67]. The OsWRKY45 family is also known to participate in ABA and salt stress signaling [68]. Therefore, it is obvious from our study that the expression of OsWRKY28, JA, and OsWRKY45 are linked to each other. This clearly indicates that Y3 might be moving towards heat stress damage recovery by overexpressing these TFs and in turn producing stress-responsive hormones.

2.4.4. Secondary Metabolites

Secondary metabolites are produced by plants through several metabolic pathways to support their survival in the environment. Some of the most commonly studied secondary metabolites are alkaloids, flavonoids, and terpenoids. Most plants regulate synthesis of these compounds for their survival based on the environmental conditions. Among most environmental factors, temperature is known to influence plant secondary metabolite production significantly [69]. In the current study, upregulation of some genes related to ent-kaurene synthase, anthocyanins, flavonoid biosynthesis, lignin biosynthesis, and laccases was observed. Three genes related to ent-kaurene synthase were found to be strongly upregulated in Y3 (Os04909900.1, Os02936140, and Os02936210), while only one gene (Os01914630) was mildly upregulated in Y2. These enzymes are known for their role in gibberellin phytohormone biosynthesis and serve as intermediates in specific di-terpenoid metabolism [70]. Additionally, specific genes belonging to the anthocyanins and flavonoid biosynthesis pathway in Y3 were found to be upregulated. Some of these are chalcone synthase (Os11g32650.1), leucoanthocyanidin reductase (Os06g44170.1) and dihydroflavonol-4-reductase (DFR) (Os03g08624.1). On the contrary, in Y2, these genes were not found. However, the isoflavonone reductase related gene with moderate expression was found to be downregulated in both Y2 (Os01g01660.1) and Y3 (Os02g56460.1). This class of flavonoids was identified to exhibit antioxidative functions in plants like tomato [71]. These are also responsible for fruit and flower color in several plants. In *Arabidopsis*, the mutants with DFR deficient genes resulted in decreased levels of proanthocyanidin brown tannins on their seed coats [72,73]. Another important metabolite that is very essential for plants is lignin. It provides structural support for the plants to grow and also protect the plant from pathogens. Its accumulation could be considered as a repair mechanism to prevent cell wall damage during heat stress. Genes related (Os04g01470.1, Os04g09680.1, and Os11g42200.1) to lignin biosynthesis were also found to be upregulated in both Y2 and Y3. Os04g01470.1 and Os04g09680.1 are the caffeic acid 3-O-methyltransferase (COMT) genes. These are involved in converting caffeic acid to ferulic acid and 5-hydroxyferulic acid to sinapic acid [74]. Both the genes were observed to be upregulated in Y3, while the former was only observed in Y2. In addition, upregulation of laccases (Os11g42200.1) was observed in both Y2 and Y3. Higher laccase activity indicates induced catalysis of oxidative coupling between phenylpropane units to form lignin [75]. Heat stress in plants could also lead to osmotic stress due to disruption of osmotic hemostasis. In order to prevent osmotic stress plants produce carotenoids [74]. Here in Y2, a gene (Os01g14630.1) geranyl diphosphate synthase (GPPs) related to osmotic stress and carotenoid synthesis was strongly downregulated compared to Y3. Upregulation of GGPS during stress conditions leading to enhanced osmotic stress tolerance was found in *Arabidopsis thaliana* [76,77]. The current result is quite opposite to previous reports. It could be due to abundance of GPPs in leaves compared to seed and differential expression of specific genes in specific tissues.

2.4.5. Starch Metabolism

Starch metabolism is a key process in regulating rice quality and germination [23,33]. Two key enzymatic processes involved in this pathway are related to starch hydrolysis and sucrose synthesis. Here, in both Y2 and Y3, few amylase genes (Os08g36910.2 and Os09g28400.1) were found to be commonly upregulated (supplementary Table S3, Metabolism overview; and Figure 5). In addition, one more gene related to amylase (Amy1) (Os02g52700.1) was also found to upregulated in Y3 alone. It was previously found that upregulation of Amy1 related genes produced chalky grains by degrading the starch reserves in the ripening grains [24–26]. It appears to be more in Y3 as multiple copies of such genes are upregulated. On the other hand, cleavage of sucrose is very important for accumulation of starch in the seeds. This is mediated by sucrose synthase (Susy) genes which converts sucrose to UDP glucose. The effect of heat stress could suppress these genes and was observed here. A gene related to the Susy family (SUS4, Os07g42490.1) was downregulated in both Y2 and Y3. Downregulation of this gene indicates a reduction in starch content and indirectly contributing to grain chalkiness along with amylases.

2.4.6. Hormone Regulated Gene Expression

Hormones play a major role in regulating stress-responsive mechanisms in plants. Several hormones are involved in the activation or inactivation of specific genes related to stress tolerance. In this experiment few hormone-regulated genes were found to be differentially expressed in Y2 and Y3. Two different SAUR class proteins OsSAUR18 (Os04g43740.1) and OsSAUR23 (Os04g56690.1) were observed in both Y2 and Y3, respectively (Supplementary Table S4, Abiotic-Biotic stress; and Figure 6). In general SAUR class genes were induced in the presence of auxins. The reduced expression of OsSAUR genes 18 and 23 in Y2 and Y3 could indicate low levels of endogenous auxins. A decrease in endogenous auxins and SAUR expression was previously observed in barley and *Arabidopsis* [78,79]. This could be due to the disruption of the auxin metabolism caused by heat stress. Previously, such effect due to heat stress was identified in rice under heat stress which could lead to inhibition of pollen tube elongation resulting in decreased spikelet fertility [80]. Further, ABA-induced GRAM domain-containing proteins like FIP1 were also found to be upregulated, indicating possible accumulation of ABA. ABA is known to regulate starch biosynthesis genes and plays a key role in grain filling under high temperatures [81]. The induced expression of the FIP1 genes could therefore suggest a heat-responsive mechanism in rice. Two F1P1 genes (os10g34730.1 and Os04g44500.1) were found to be similarly regulated in both Y2 and Y3. Additionally, genes os02g42430.1 and Os04g44510.1 were observed to be upregulated in Y3 alone indicating a stronger response to high temperatures. In addition to the above findings, unique changes in genes related to jasmonate synthesis were detected. Expression of genes related to the JA biosynthesis pathway, like lipoxygenase 2 (Os03g52860.1), 12-Oxo-PDA-reductase family genes OPR (Os06g11200.1) and OsOPR5 (Os06g11210.1) is also observed in our study. Between Y2 and Y3 the expression of these genes is opposite. Lipoxygenase-2 and OPR were downregulated in Y2, while in Y3 the lipoxygenase-2 and OsOPR5 are upregulated and OPR is not. Previously JA levels and these genes were identified to be upregulated under drought stress and downregulated under heat stress in rice [82]. Accordingly, from the current results, it can be understood that Y2 is experiencing heat stress. Whereas, Y3 might be experiencing both drought and heat stress, as downregulation of the OPR gene was also observed here.

2.4.7. Concluding Remarks

This is the first such study targeting a field situation where an experiment was designed after observing the heat, and talking to the farmers in Tsukuba city. In summary, the rice being eaten by the people has been found to have certain characteristics of heat stress-like response, and current work clarified the genome-wide changes in genes, under the indicated experimental conditions and the environment from where they were harvested. The present study also provides a unique database for the readers and scientific community to further utilize and research upon.

3. Materials and Methods

3.1. *Plant Material*

The experiment used *japonica*-type rice cultivar (*O. sativa* L. cv. Koshihikari) seeds. Dry mature seeds of cv. Koshihikari were the commercially available rice and obtained from Japan Agriculture (JA Zen-Noh, Japan) branch in Ami-town of Ibaraki prefecture (Kanto region), Japan, in September 2010. Three sets of seeds were obtained as grade 1 (labeled Y1), grade 2 (labeled Y2), and out-of-grade (Grade 3, labeled as Y3), as defined by JA Zen-Noh.

3.2. *Seed Quality Analysis*

Dry mature seeds (grain) were analyzed for its quality using a commercial rice grain analyzer service (Satake Corporation, Tokyo). The analysis consisted of the following: taste value, amylose, protein, moisture, and fatty acid degree.

3.3. *Rice Whole Genome DNA Microarray Analysis*

Dry mature seeds (12 of each grade of Y1-Y3) were used for preparing fine powders in liquid nitrogen. Briefly, the seeds were placed in a pre-chilled mortar and pestle containing liquid nitrogen, ground completely to a very fine powder with the chilled pestle in liquid nitrogen and stored at −80 °C till used for RNA extraction. For total RNA extraction, the stored sample powder (~100 mg) was transferred to a 2 mL sterile microfuge tube, followed by addition of 0.9 mL of CTAB buffer [a 10 mL volume of buffer contains 0.5 mL (50 mM) of 1 M stock Tris–HCl solution (pH 8.0), 1.0 mL (5 mM) of 500 mM ethylenediaminetetraacetic acid (EDTA, pH 8.0), 0.2 g (2%, w/v) of CTAB, 1.68 mL (0.84 M) of 5 M NaCl, and 0.1 M β-mercaptoethanol, which is added just before use of the solution]. The contents were mixed by vortexing for 30 s and incubated for 5 min at RT. After an addition of 0.8 mL of phenol-chloroform-isoamylalcohol (PCIA; 25:24:1), the homogenate was mixed well (by gentle shaking) for 5 min at RT. After centrifugation at 15,000 × g for 5 min at 4 °C, an aliquot (0.6 to 0.7 mL) of the upper phase was transferred to a 1.5-mL sterile microfuge tube, followed by addition of 1 volume of chloroform, and the mixture was centrifuged at 15,000 × g for 5 min at 4 °C. The resulting supernatant was transferred to another 1.5 mL microfuge tube and 0.033 volume of 3 M sodium acetate, pH 5.5 and 1 volume of 2-isopropanol were added. The mixture was incubated for 15 min on ice and then centrifuged at 15,000 × g for 5 to 10 min at 4 °C to collect the RNA. The supernatant was completely removed and the pellet was dissolved in 0.1 mL of RNase-free water (SDW; double sterilized distilled water) followed by using the RNeasy mini protocol for RNA cleanup exactly as described by the manufacturer (QIAGEN, Gaithersburg, MD, USA). To verify the quality of this RNA, the yield and purity were determined spectrophotometrically (NanoDrop, Wilmington, DE, USA) and visually confirmed using formaldehyde-agarose gel electrophoresis.

After extraction of high-quality total RNA from dehusked seeds using a modified CTAB extraction protocol, a rice 4 × 44K custom (eARRAY, AMAdid-017845) oligo DNA microarray chip (G2514F: Agilent Technologies, Palo Alto, CA, USA) was used for genome-wide gene profiling as described previously [83]. The flip labeling (dye-swap or reverse labeling with Cy3 and Cy5 dyes) procedure was used to nullify the dye bias associated with unequal incorporation of the two Cy dyes into cRNA [84–86]. The dye-swap approach which is well established in our laboratories and research provides a more stringent selection condition for profiling differentially expressed genes rather than simply doing only 2 or 3 replicates, which overlook the dye bias [86–92]. The experimental design for the DNA microarray analysis of rice seed transcriptome using a two-color dye-swap approach (Figure 7A) is shown along with the extracted seed total RNA quality (Figure 7B).

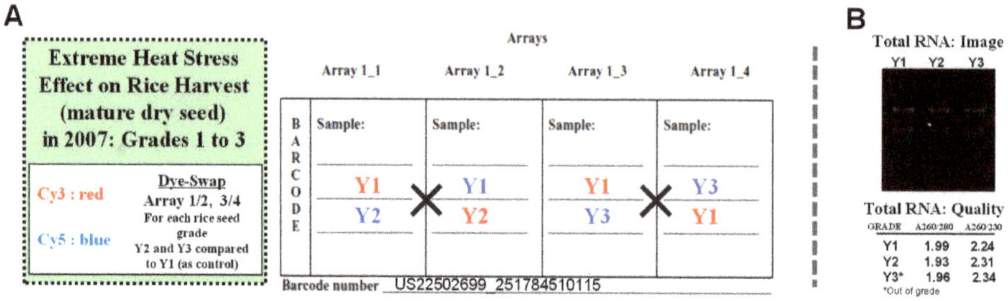

Figure 7. Experimental design for the DNA microarray analysis of high temperature affected mature dry rice (cv. Koshihikari) seed grades 2 and 3 transcriptome. A two-color dye-swap approach was used (**A**) and followed by checking the quality of extracted seed total RNA (**B**). Three sets of dry mature rice seeds (commercial) were obtained Japan Agriculture (JA Zen-Noh) branch in Ami-town of Ibaraki prefecture in September 2010, as grade 1 (labeled as Y1), grade 2 (labeled as Y2), and grade 3 (out-of-grade, labeled as Y3). Gene expressions genome-wide in grade 2 (Y2) and grade 3 (Y3) seeds over the grade 1 (Y1) was carried out using an Agilent 4 × 44K rice oligo DNA chip.

Total RNA (800 ng) for each Y1, Y2, and Y3 sample were labeled with either Cy3 or Cy5 dye using an Agilent Low RNA Input Fluorescent Linear Amplification Kit. Hybridization and wash processes were performed according to the manufacturer's instructions. Hybridized microarray chips were scanned using the Agilent Microarray Scanner G2565BA. To detect differentially expressed significant genes between control and treated samples, each slide image was processed by Agilent Feature Extraction software (version 9.5.3.1). Normalization of Cy3 and Cy5 signals was performed by LOWESS (locally weighted linear regression), which calculates the log ratio of dye-normalized Cy3- and Cy5-signals. The significance (P) value is based on the propagate error and universal error models. In this analysis, the threshold of significantly expressed differential genes was set to <0.01 (for the confidence that the feature was not differentially expressed). Lists of differentially expressed gene [up- (≥2.0 fold) and down- (≤0.5 fold) regulated genes] were generated and annotated using the Agilent GeneSpring version GX 10.

The data discussed in this publication have been deposited in NCBI's Gene Expression Omnibus (GEO) and are accessible through GEO series accession number GSE79405.

3.4. MapMan Analysis

To analyze the gene expression changes, rice genes were mapped onto metabolic pathways using MapMan tool (version 3.5.1, Max Plant Institute of Molecular Plant Physiology, Germany) [93]. To map genes onto their respective pathways MapMan uses information from its datasets (Mappings), where all annotated genes of the respective organism were classified into BINS based on their function [94]. In this work, a rice mapping file (Rice_japonica_automatic_mapping08) from MapMan database was used as a template. Out of 43,494 genes detected from our microarray data, 41,446 genes were present in the mapping file and the remaining 2048 (having ID_ TIGRv4S1) were not annotated. The annotated gene expression data was used here and pre-processed it using a PERL script to meet the locus_name identifier criteria similar to that available in the mapping file. Once this is done the expression data file with modified gene ID's were successfully classified into various BINS (Table S1-Mapman Bins) by MapMan. For the final mapping, the upregulated and downregulated genes were selected, and their fold change transformed into Log2 (fold) data was used. A total of 41,446 differentially expressed genes were sorted to select up (≥2.0 fold) and down (≤0.5 fold) regulated genes were generated and annotated using the Agilent GeneSpring version GX 10. This resulted in 161 and 490 HRDE-genes for each Y1-Y2 and Y1-Y3, respectively. This sorted gene expression data was used to map onto Mapman pathways. The genes which are differentially expressed are indicated either as highly up- or downregulated are represented in blue and red-colored data points (grid boxes), respectively. A gradient in blue or red indicate genes with medium up or downregulation.

Supplementary Materials: The following are available online at http://www.mdpi.com/2073-4433/11/5/528/s1, Table S1: Rice Japonica genes associated with Mapman BINS; Table S2: Highly expressed differentially regulated (HRDE) Genes mapped on Cell Function map; Table S3: Highly expressed differentially regulated (HRDE) Genes mapped on Metabolism Overview map; Table S4: Highly expressed differentially regulated (HRDE) Genes mapped on Abiotic and Bioic stress map.

Author Contributions: R.K.B. wrote the paper, and R.R. edited the paper along with A.S. and G.K.A., and R.R., A.S. and G.K.A. discussed the initial idea for the experiment towards an experimental approach; R.R. designed the experiment and performed the experiment with J.S.; K.C. supported with the MapMan analyses along with R.K.B., and S.K. provided the DNA microarray chip probe design; M.Y. provided the seeds and supported the rice seed quality analysis; S.S. provided the facilities and support for the DNA microarray analysis. All authors approved the paper. All authors have read and agreed to the published version of the manuscript.

Funding: This research received no external funding.

Conflicts of Interest: The authors declare no conflict of interest.

References

1. Available online: https://earthobservatory.nasa.gov/world-of-change/global-temperatures (accessed on 1 April 2020).
2. Lobell, D.B.; Gourdji, S.M. The influence of climate change on global crop productivity. *Plant Physiol.* **2012**, *160*, 1686–1697. [CrossRef] [PubMed]
3. Zhao, C.; Liu, B.; Piao, S.; Wang, X.; Lobell, D.B.; Huang, Y.; Huang, M.; Yao, Y.; Bassu, S.; Ciais, P.; et al. Temperature increase reduces global yields of major crops in four independent estimates. *Proc. Natl. Acad. Sci. USA* **2017**, *114*, 9326–9331. [CrossRef] [PubMed]
4. Bita, C.E.; Tom Gerats, T. Plant tolerance to high temperature in a changing environment: Scientific fundamentals and production of heat stress-tolerant crops. *Front. Plant Sci.* **2013**, *4*, 273. [CrossRef]
5. Qu, A.L.; Ding, Y.F.; Jiang, Q.; Zhu, C. Molecular mechanisms of the plant heat stress response. *Biochem. Biophys. Res. Commun.* **2013**, *432*, 203–207. [CrossRef] [PubMed]
6. Hatfield, J.L.; Prueger, J.H. Agroecology: Implications for Plant Response to Climate Change. In *Crop Adaptation to Climate Change*; Yadav, S.S., Redden, R.J., Hatfield, J.L., Lotze-Campen, H., Hall, A.E., Eds.; Wiley-Blackwell: West Sussex, UK, 2011; pp. 27–43.
7. Ohama, N.; Sato, H.; Shinozaki, K.; Yamaguchi-Shinozaki, K. Transcriptional regulatory network of plant heat stress response. *Trends Plant Sci.* **2017**, *22*, 53–65. [CrossRef] [PubMed]
8. Wahid, A.; Gelani, S.; Ashraf, M.; Foolad, M. Heat tolerance in plants: An overview. *Environ. Exp. Bot.* **2007**, *61*, 199–223. [CrossRef]
9. Hua, J. From freezing to scorching, transcriptional responses to temperature variations in plants. *Curr. Opin. Plant Biol.* **2009**, *12*, 568–573. [CrossRef]
10. Kim, E.H.; Kim, Y.S.; Park, S.H.; Koo, Y.J.; Do Choi, Y.; Chung, Y.Y.; Lee, I.J.; Kim, J.K. Methyl jasmonate reduces grain yield by mediating stress signals to alter spikelet development in rice. *Plant Physiol.* **2009**, *149*, 1751–1760. [CrossRef]
11. Wang, D.; Heckathorn, S.A.; Kumar Mainali, K.; Tripathee, R. Timing effects of heat-stress on plant ecophysiological characteristics and growth. *Front. Plant Sci.* **2016**, *7*, 1629. [CrossRef]
12. Apel, K.; Hirt, H. Reactive oxygen species: Metabolism, oxidative stress and signal transduction. *Annu. Rev. Plant Biol.* **2004**, *55*, 373–399. [CrossRef]
13. Andreeva, V.A. Involvement of Peroxidase in the Protective Plant Mechanism. In *Biochemical, Molecular and Physiological Aspects of Plant Peroxidases*; Lobarzewski, J., Greppin, H., Penel, C., Gaspar, T., Eds.; University of Geneva: Geneva, Switzerland, 1991; pp. 433–442.

14. Hiramoto, K.; Ojima, N.; Sako, K.; Kikugawa, K. Effect of plant phenolics on the formation of the spin-adduct of hydroxyl radical and the DNA strand breaking by hydroxyl radical. *Biol. Pharm. Bull.* **1996**, *19*, 558–563. [CrossRef] [PubMed]
15. Moller, I.M.; Jensen, P.E.; Hansson, A. Oxidative modifications to cellular components in plants. *Ann. Rev. Plant Biol.* **2007**, *58*, 459–481. [CrossRef] [PubMed]
16. Camejo, D.; Jiménez, A.; Alarcón, J.J.; Torres, W.; Gómez, J.M.; Sevilla, F. Changes in photosynthetic parameters and antioxidant activities following heat–shock treatment in tomato plants. *Func. Plant Biol.* **2006**, *33*, 177–187. [CrossRef]
17. Yoshida, S. Effects of temperature on growth of the rice plant (*Oryza sativa* L.) in a controlled environment. *Soil Sci Plant Nutr.* **1973**, *19*, 299–310. [CrossRef]
18. Jagadish, S.V.; Muthurajan, R.; Oane, R.; Wheeler, T.R.; Heuer, S.; Bennett, J.; Craufurd, P.Q. Physiological and proteomic approaches to address heat tolerance during anthesis in rice (*Oryza sativa* L.). *J. Exp. Bot.* **2010**, *61*, 143–156. [CrossRef] [PubMed]
19. Hasegawa, T.; Ishimaru, T.; Kondo, M.; Kuwagata, T.; Yoshimoto, M.; Fukuoka, M. Spikelet sterility of rice observed in the record hot summer of 2007 and the factors associated with its variation. *J. Agric. Meteorol.* **2011**, *67*, 225–232. [CrossRef]
20. Shrivastava, P.; Saxena, R.R.; Xalxo, M.S.; Verulkar, S.B. Effect of high temperature at different growth stages on rice yield and grain quality traits. *J. Rice Res.* **2012**, *5*, 29–42.
21. Fu, G.; Feng, B.; Zhang, C.; Yang, Y.; Yang, X.; Chen, T.; Zhao, X.; Jin, Q.; Tao, L. Heat stress is more damaging to superior spikelets than inferiors of rice (*Oryza sativa* L.) due to their different organ temperatures. *Front. Plant Sci.* **2016**, *7*, 163. [CrossRef]
22. Satake, T.; Yoshida, S. High temperature-induced sterility in indica rices at flowering. *Jpn. J. Crop Sci.* **1978**, *47*, 6–17. [CrossRef]
23. Sharma, K.P.; Sharma, N. Influence of high temperature on sucrose metabolism in chalky and translucent rice genotypes. *Proc. Natl. Acad. Sci. India Sect. B Biol. Sci.* **2018**, *88*, 1275–1284. [CrossRef]
24. Tashiro, T.; Wardlaw, I. The effect of high-temperature on kernel dimensions and the type and occurrence of kernel damage in rice. *Aust. J. Agric. Res.* **1991**, *42*, 485–496. [CrossRef]
25. Zakaria, S.; Matsuda, T.; Tajima, S.; Nitta, Y. Effect of high temperature at ripening stage on the reserve accumulation in seed in some rice cultivars. *Plant Prod Sci.* **2002**, *5*, 160–168.
26. Nakata, M.; Fukamatsu, Y.; Miyashita, T.; Hakata, M.; Kimura, R.; Nakata, Y.; Kuroda, M.; Yamaguchi, T.; Yamakawa, H. High temperature-induced expression of rice α-amylases in developing endosperm produces chalky grains. *Front. Plant Sci.* **2017**, *8*, 2089. [PubMed]
27. Morita, S.; Wada, H.; Matsue, Y. Countermeasures for heat damage in rice grain quality under climate change. *Plant Prod. Sci.* **2016**, *19*, 1–11. [CrossRef]
28. Hakata, M.; Kuroda, M.; Miyashita, T.; Yamaguchi, T.; Kojima, M.; Sakakibara, H.; Mitsui, T.; Yamakawa, H. Suppression of α-amylase genes improves quality of rice grain ripened under high temperature. *Plant Biotechnol. J.* **2012**, *10*, 1110–1117. [CrossRef] [PubMed]
29. Martínez-Eixarch, M.; Ellis, R.H. Temporal sensitivity of rice seed development from spikelet fertility to viable mature seed to extreme temperature. *Crop Sci.* **2015**, *55*, 354–364.
30. Yuliawan, T.; Handoko, I. The effect of temperature rise to rice crop yield in Indonesia uses Shierary Rice Model with Geographical Information System (GIS) feature. *Procedia Environ. Sci.* **2016**, *33*, 214–220. [CrossRef]
31. Available online: https://yab.yomiuri.co.jp/adv/wol/dy/opinion/society_131209.html (accessed on 1 April 2020).
32. Ohtsubo, K.; Kobayashi, A.; Shimizu, H. Quality evaluation of rice in Japan. *Jpn. Agric. Res. Q.* **1993**, *27*, 95–101.
33. Yamakawa, H.; Hirose, T.; Kuroda, M.; Yamaguchi, T. Comprehensive expression profiling of rice grain filling-related genes under high temperature using DNA microarray. *Plant Physiol.* **2007**, *144*, 258–277. [CrossRef]
34. Yamakawa, H.; Hakata, M. Atlas of rice grain filling-related metabolism under high temperature: Joint analysis of metabolome and transcriptome demonstrated inhibition of starch accumulation and induction of amino acid accumulation. *Plant Cell Physiol.* **2010**, *51*, 795–809.

35. Yasumatsu, K.; Moritaka, S. Fatty acid compositions of rice lipid and their changes during storage. *Agric. Biol. Chem.* **1964**, *28*, 257–264. [CrossRef]
36. Yamamatsu, K.; Moritaka, S.; Wada, S. Stale flavor of stored rice. *Agric. Biol. Chem.* **1966**, *30*, 483–486.
37. Wu, H.C.; Bulgakov, V.P.; Jinn, T.L. Pectin Methyl esterase's: Cell wall remodeling proteins are required for plant response to heat stress. *Front. Plant Sci.* **2018**, *9*, 1612. [CrossRef] [PubMed]
38. Rienth, M.; Torregrosa, L.; Luchaire, N.; Chatbanyong, R.; Lecourieux, D.; Kelly, M.T.; Romieu, C. Day and night heat stress trigger different transcriptomic responses in green and ripening grapevine (*vitis vinifera*) fruit. *BMC Plant Biol.* **2014**, *14*, 108. [CrossRef] [PubMed]
39. Tenhaken, R. Cell wall remodeling under abiotic stress. *Front. Plant Sci.* **2015**, *5*, 771. [CrossRef] [PubMed]
40. Gall, L.H.; Philippe, F.; Domon, J.M.; Gillet, F.; Pelloux, J.; Rayon, C. Cell wall metabolism in response to abiotic Stress. *Plants* **2015**, *4*, 112–166. [CrossRef]
41. Huang, Y.C.; Wu, H.C.; Wang, Y.D.; Liu, C.H.; Lin, C.C.; Luo, D.L.; Jinn, T.L. Pectin methyl esterase 34 contributes to Hhat tolerance through its role in promoting stomatal movement. *Plant Physiol.* **2017**, *174*, 748–763. [CrossRef]
42. Nishizawa, A.; Yabuta, Y.; Shigeoka, S. Galactinol and raffinose constitute a novel function to protect plants from oxidative damage. *Plant Physiol.* **2008**, *147*, 1251–1263. [CrossRef]
43. Egert, A.; Keller, F.; Peters, S. Abiotic stress-induced accumulation of raffinose in *Arabidopsis* leaves is mediated by a single raffinose synthase (RS5, At5g40390). *BMC Plant Biol.* **2013**, *13*, 218. [CrossRef]
44. Sengupta, S.; Mukherjee, S.; Basak, P.; Majumder, A.L. Significance of galactinol and raffinose family oligosaccharide synthesis in plants. *Front. Plant Sci.* **2015**, *6*, 656. [CrossRef]
45. Padham, A.K.; Hopkins, M.T.; Wang, T.W.; McNamara, L.M.; Lo, M.; Richardson, L.G.; Smith, M.D.; Taylor, C.A.; Thompson, J.E. Characterization of a plastid triacylglycerol lipase from *Arabidopsis. Plant Physiol.* **2007**, *143*, 1372–1384. [CrossRef] [PubMed]
46. Higashi, Y.; Okazaki, Y.; Takano, K.; Myouga, F.; Shinozaki, K.; Knoch, E.; Fukushima, A.; Saito, K. Heat inducible lipase remodels chloroplastic monogalactosyl diacylglycerol by Liberating α-Linolenic Acid in *Arabidopsis* Leaves under Heat Stress. *Plant Cell* **2018**, *30*, 1887–1905. [CrossRef]
47. Hou, Q.; Ufer, G.; Bartels, D. Lipid signaling in plant responses to abiotic stress. *Plant Cell Environ.* **2016**, *39*, 1029–1048. [CrossRef] [PubMed]
48. Mishkind, M.; Vermeer, J.E.; Darwish, E.; Munnik, T. Heat stress activates phospholipase D and triggers PIP accumulation at the plasma membrane and nucleus. *Plant J.* **2009**, *60*, 10–21. [CrossRef] [PubMed]
49. Chen, J.; Xu, W.; Burke, J.J.; Xin, Z. Role of phosphatidic acid in high temperature tolerance in Maize. *Crop Sci.* **2010**, *50*, 2506–2515. [CrossRef]
50. Escobar-Sepúlveda, H.F.; Trejo-Téllez, L.I.; Pérez-Rodríguez, P.; Hidalgo-Contreras, J.V.; Gómez-Merino, F.C. Diacylglycerol kinases are widespread in higher plants and display inducible gene expression in response to beneficial elements, metal, and metalloid ions. *Front. Plant Sci.* **2017**, *8*, 129. [CrossRef]
51. Li, Q.; Zheng, Q.; Shen, W.; Cram, D.; Fowler, D.B.; Wei, Y.; Zou, J. Understanding the biochemical basis of temperature-induced lipid pathway adjustments in plants. *Plant Cell* **2015**, *227*, 86–103. [CrossRef]
52. Dar, A.A.; Choudhury, A.R.; Kancharla, P.K.; Arumugam, N. The FAD2 Gene in Plants: Occurrence, regulation, and role. *Front. Plant Sci.* **2017**, *8*, 1789. [CrossRef]
53. Byfield, G.E.; Upchurch, R.G. Effect of temperature on delta-9 stearoyl-ACP and microsomal omega-6 desaturase gene expression and fatty acid content in developing soybean seeds. *Crop Sci.* **2007**, *47*, 1698. [CrossRef]
54. Altunoglu, Y.C.; Unel, N.M.; Baloglu, M.C.; Ulu, F.; Can, T.H.; Cetinkaya, R. Comparative identification and evolutionary relationship of fatty acid desaturase (FAD) genes in some oil crops: The sunflower model for evaluation of gene expression pattern under drought stress. *Biotechnol. Biotechnol. Equip.* **2018**, *32*, 846–857. [CrossRef]
55. Jenkins, G.M.; Richards, A.; Wahl, T.; Mao, C.; Obeid, L.; Hannun, Y. Involvement of yeast sphingolipids in the heat stress response of Saccharomyces cerevisiae. *J. Biol. Chem.* **1997**, *272*, 32566–32572. [CrossRef] [PubMed]
56. Teng, C.; Dong, H.; Shi, L.; Deng, Y.; Mu, J.; Zhang, J.; Yang, X.; Zuo, J. Serine palmitoyltransferase, a key enzyme for de novo synthesis of sphingolipids, is essential for male gametophyte development in *Arabidopsis*. *Plant Physiol.* **2008**, *146*, 1322–1332. [CrossRef] [PubMed]

57. Jin, H.; Cominelli, E.; Bailey, P.; Parr, A.; Mehrtens, F.; Jones, J.; Tonelli, C.; Weisshaar, B.; Martin, C. Transcriptional repression by AtMYB4 controls production of UV-protecting sunscreens in *Arabidopsis*. *EMBO J.* **2000**, *19*, 6150–6161. [CrossRef]
58. Roy, S. Function of MYB domain transcription factors in abiotic stress and epigenetic control of stress response in plant genome. *Plant Signal. Behav.* **2015**, *11*, e1117723. [CrossRef] [PubMed]
59. Wu, H.; Lv, H.; Li, L.; Liu, J.; Mu, S.; Li, X.; Gao, J. Genome-wide analysis of the AP2/ERF transcription factors family and the expression patterns of *DREB* genes in Moso Bamboo (*Phyllostachys edulis*). *PLoS ONE* **2015**, *10*, e0126657. [CrossRef] [PubMed]
60. Lata, C.; Prasad, M. Role of DREBs in regulation of abiotic stress responses in plants. *J. Exp. Bot.* **2011**, *63*, 4731–4748. [CrossRef] [PubMed]
61. Sharoni, A.M.; Nuruzzaman, M.; Rahman, M.A.; Karim, R.; Islam, A.K.M.R.; Hossain, M.M.; Rahman, M.M.; Parvez, M.S.; Haydar, F.M.A.; Nasiruddin, M.; et al. AP2/EREBP transcription factor family genes are differentially expressed in rice seedlings during infections with different viruses. *Int. J. Biosci.* **2017**, *10*, 1–14.
62. Wang, P.; Xu, X.; Tang, Z.; Zhang, W.; Huang, X.Y.; Zhao, F.J. OsWRKY28 regulates phosphate and arsenate accumulation, root system architecture and fertility in rice. *Front. Plant Sci.* **2018**, *9*, 1330. [CrossRef]
63. Ross, C.A.; Liu, Y.; Shen, Q.J. The *WRKY* Gene Family in Rice (*Oryza sativa*). *J. Integr. Plant Biol.* **2007**, *49*, 827–842. [CrossRef]
64. Rushton, P.J.; Somssich, I.E.; Ringler, P.; Shen, Q.J. WRKY transcription factors. *Trends Plant Sci.* **2010**, *15*, 247–258. [CrossRef]
65. Chakrabarty, D.; Trivedi, P.K.; Misra, P.; Tiwari, M.; Shri, M.; Shukla, D.; Kumar, S.; Rai, A.; Pandey, A.; Nigam, D.; et al. Comparative transcriptome analysis of arsenate and arsenite stresses in rice seedlings. *Chemosphere* **2009**, *74*, 688–702. [CrossRef] [PubMed]
66. Cai, Q.; Yuan, Z.; Chen, M.; Yin, C.; Luo, Z.; Zhao, X.; Liang, W.; Hu, J.; Zhang, D. Jasmonic acid regulates spikelet development in rice. *Nat. Commun.* **2014**, *5*, 3476. [CrossRef] [PubMed]
67. Huangfu, J.; Li, J.; Li, R.; Ye, M.; Kuai, P.; Zhang, T.; Lou, Y. The Transcription Factor OsWRKY45 Negatively Modulates the Resistance of Rice to the Brown Planthopper Nilaparvata lugens. *Int. J. Mol. Sci.* **2016**, *17*, 697. [CrossRef] [PubMed]
68. Tao, Z.; Kou, Y.; Liu, H.; Li, X.; Xiao, J.; Wang, S. OsWRKY45 alleles play different roles in abscisic acid signaling and salt stress tolerance but similar roles in drought and cold tolerance in rice. *J. Exp. Bot.* **2011**, *62*, 4863–4874. [CrossRef] [PubMed]
69. Ramakrishna, A.; Ravishankar, G.A. Influence of abiotic stress signals on secondary metabolites in plants. *Plant Signal. Behav.* **2011**, *6*, 1720–1731. [PubMed]
70. Fu, J.; Ren, F.; Lu, X.; Mao, H.; Xu, M.; Degenhardt, J.; Peters, R.J.; Wang, Q. A tandem array of ent-kaurene synthases in maize with roles in gibberellin and more specialized metabolism. *Plant Physiol.* **2015**, *170*, 742–751. [CrossRef] [PubMed]
71. Butelli, E.; Titta, L.; Giorgio, M.; Mock, H.P.; Matros, A.; Peterek, S.; Schiklen, E.G.; Hall, R.D.; Bovy Ag Luo, J.; Martin, C. Enrichment of tomato fruit with health-promoting anthocyanins by expression of select transcription factors. *Nat. Biotechnol.* **2008**, *26*, 1301–1308. [CrossRef]
72. Shirley, B.W.; Kubasek, W.L.; Storz, G.; Bruggemann, E.; Koornneef, M.; Asubel, F.M.; Goodman, H.M. Analysis of *Arabidopsis* mutants deficient in flavonoid biosynthesis. *Plant J.* **1995**, *8*, 659–671. [CrossRef]
73. Wang, H.; Fan, W.; Li, H.; Yang, J.; Huang, J.; Zhang, P. Functional characterization of dihydroflavonol-4-reductase in Anthocyanin Biosynthesis of Purple Sweet Potato Underlies the Direct Evidence of Anthocyanins Function against Abiotic Stresses. *PLoS ONE* **2013**, *8*, e78484. [CrossRef]
74. Davin, L.B.; Lewis, N.G. Phenylpropanoid metabolism: Biosynthesis of monolignols, lignans and neolignans, lignins and suberins. *Rec. Adv. Phytochem.* **1992**, *26*, 325–375.
75. Boudet, A.M.; Lapierre, C.; Grima-Pettenati, J. Biochemistry and molecular biology of lignification. *New Phytol.* **1995**, *129*, 203–236. [CrossRef]
76. Xie, Y.; Xu, D.; Cui, W.; Shen, W. Mutation of *Arabidopsis* Hy1 causes Uv-C hypersensitivity by impairing carotenoid and flavonoid biosynthesis and the down-regulation of antioxidant defense. *J. Exp. Bot.* **2012**, *63*, 3869–3883. [CrossRef]
77. Chen, W.; He, S.; Liu, D.; Patil, G.B.; Zhai, H.; Wang, F.; Stephenson, T.J.; Wang, Y.; Wang, B.; Valliyodan, B.; et al. A sweetpotato geranylgeranyl pyrophosphate synthase gene, *IbGGPS*, increases carotenoid content and enhances osmotic stress tolerance in *Arabidopsis thaliana*. *PLoS ONE* **2015**, *10*, e0137623. [CrossRef] [PubMed]

78. Sakata, T.; Oshino, T.; Miura, S.; Tomabechi, M.; Tsunaga, Y.; Higashitani, N.; Miyazawa, Y.; Takahashi, H.; Watanabe, M.; Higashitani, A. Auxins reverse plant male sterility caused by high temperatures. *Proc. Natl. Acad. Sci. USA* **2010**, *107*, 8569–8574. [CrossRef] [PubMed]
79. Higashitani, A. High temperature injury and auxin biosynthesis in microsporogenesis. *Front. Plant Sci.* **2013**, *4*, 47. [CrossRef] [PubMed]
80. Zhang, C.; Li, G.; Chen, T.; Feng, B.; Fu, W.; Yan, J.; Islam, M.R.; Jin, Q.; Tao, L.; Fu, G. Heat stress induces spikelet sterility in rice at anthesis through inhibition of pollen tube elongation interfering with auxin homeostasis in pollinated pistils. *Rice (N. Y.)* **2018**, *11*, 14. [CrossRef]
81. Mauri, N.; Fernández-Marcos, M.; Costas, C.; Desvoyes, B.; Pichel, A.; Caro, E.; Gutierrez, C. GEM, a member of the GRAM domain family of proteins, is part of the ABA signaling pathway. *Sci. Rep.* **2016**, *6*, 22660. [CrossRef]
82. Umesh, D.K.; Pal, M. Differential role of jasmonic acid under drought and heat stress in rice (Oryza sativa). *J. Pharmacogn. Phytochem.* **2018**, *7*, 2626–2631.
83. Cho, K.; Shibato, J.; Agrawal, G.K.; Jung, Y.H.; Kubo, A.; Jwa, N.S.; Tamogami, S.; Satoh, K.; Kikuchi, S.; Higashi, T.; et al. Integrated transcriptomics, proteomics, and metabolomics analyses to survey ozone responses in the leaves of rice seedling. *J Proteome Res.* **2008**, *7*, 2980–2998. [CrossRef]
84. Cho, K.; Shibato, J.; Kubo, A.; Kohno, Y.; Satoh, K.; Kikuchi, S.; Sarkar, A.; Agrawal, G.K.; Rakwal, R. Comparative analysis of seed transcriptomes of ambient ozone-fumigated 2 different rice cultivars. *Plant Signal. Behav.* **2013**, *8*, e26300. [CrossRef]
85. Altman, N. Replication, variation and normalization in microarray experiments. *Appl. Bioinform.* **2005**, *4*, 33–44. [CrossRef] [PubMed]
86. Martin-Magniette, M.L.; Aubert, J.; Cabannes, E.; Daudin, J.J. Evaluation of the gene-specific dye bias in cDNA microarray experiments. *Bioinformatics* **2005**, *21*, 1995–2000. [CrossRef] [PubMed]
87. Rosenzweig, B.A.; Pine, P.S.; Domon, O.E.; Morris, S.M.; Chen, J.J.; Sistare, F.D. Dye bias correction in dual-labeled cDNA microarray gene expression measurements. *Environ. Health Perspect.* **2004**, *112*, 480–487. [CrossRef] [PubMed]
88. Cho, K.; Kubo, A.; Shibato, J.; Agrawal, G.K.; Saji, H.; Rakwal, R. Global identification of potential gene biomarkers associated with ozone-induced foliar injury in rice seedling leaves by correlating their symptom severity with transcriptome profiling. *Int. J. Life Sci.* **2012**, *6*, 1–13. [CrossRef]
89. Hirano, M.; Rakwal, R.; Shibato, J.; Sawa, H.; Nagashima, K.; Ogawa, Y.; Yoshida, Y.; Iwahashi, H.; Niki, E.; Masuo, Y. Proteomics and transcriptomics-based screening of differentially expressed proteins and genes in brain of Wig rat: A model for attention deficit hyperactivity disorder (ADHD) research. *J. Proteome Res.* **2008**, *7*, 2471–2489. [CrossRef] [PubMed]
90. Tano, K.; Mizuno, R.; Okada, T.; Rakwal, R.; Shibato, J.; Masuo, Y.; Ijiri, K.; Akimitsu, N. MALAT-1 enhances cell motility of lung adenocarcinoma cells by influencing the expression of motility-related genes. *FEBS Lett.* **2010**, *584*, 4575–4580. [CrossRef]
91. Ogawa, T.; Rakwal, R.; Shibato, J.; Sawa, C.; Saito, T.; Murayama, A.; Kuwagata, M.; Kageyama, H.; Yagi, M.; Satoh, K.; et al. Seeking gene candidates responsible for developmental origins of health and disease (DOHaD). *Congenit. Anom.* **2011**, *51*, 110–125. [CrossRef]
92. Hori, M.; Nakamachi, T.; Rakwal, R.; Shibato, J.; Nakamura, K.; Wada, Y.; Tsuchikawa, D.; Yoshikawa, A.; Tamaki, K.; Shioda, S. Unraveling the ischemic brain transcriptome in a permanent middle cerebral artery occlusion model by DNA microarray analysis. *Dis. Models Mech.* **2011**, *5*, 270–283. [CrossRef]
93. Usadel, B.; Poree, F.; Nagel, A.; Lohse, M.; Czedik-Eysenberg, A.; Stitt, M. A guide to using MapMan to visualize and compare Omics data in plants: A case study in the crop species, Maize. *Plant Cell Environ.* **2009**, *32*, 1211–1229. [CrossRef]
94. Thimm, O.; Blaesing, O.; Gibon, Y.; Nagel, A.; Meyer, S.; Krüger, P.; Selbig, J.; Müller, L.A.; Rhee, S.Y.; Stitt, M. MAPMAN: A user-driven tool to display genomics data sets onto diagrams of metabolic pathways and other biological processes. *Plant J.* **2004**, *37*, 914–939. [CrossRef]

atmosphere
MDPI

Article

Historical Radial Growth of Chinese Torreya Trees and Adaptation to Climate Change

Xiongwen Chen

Department of Biological & Environmental Sciences, Alabama A & M University, Normal, AL 35762, USA; xiongwen.chen@aamu.edu

Received: 30 May 2020; Accepted: 29 June 2020; Published: 30 June 2020

check for updates

Abstract: Chinese Torreya is a vital crop tree with an average life span of a thousand years in subtropical China. Plantations of this tree are broadly under construction, to benefit the local economy. Information on the growth and adaptation to climate change for this species is limited, but tree rings might show responses to historical climate dynamics. In this study, six stem sections from Chinese Torreya trees between 60 and 90 years old were acquired and analyzed with local climate data. The results indicated that the accumulated radial growth increased linearly with time, even at the age of 90 years, and the average radial increment of each tree ranged from 1.9 to 5.1 mm/year. The variances of basal area increment (BAI) increased with time, and correlated with the variances of precipitation in the growing seasons. Taylor's power law was present in the radial growth, with the scaling exponents concentrated within 1.9–2.1. A "Triangle"-shaped relationship was found between the precipitation in the growing seasons and annual radial increments. Similar patterns also appeared for the standard precipitation index, maximum monthly air temperature and minimum monthly air temperature. The annual increases were highly correlated with the local climate. Slow growth, resilience to drought and multiple stems in one tree might help the tree species adapt to different climate conditions, with the implications for plantation management discussed in this paper.

Keywords: basal area increment; air temperature; precipitation; Taylor's power law; tree ring analysis

1. Introduction

Progressing global climate change, with increased frequency of extreme climatic events, increased atmospheric CO_2 concentration and related disturbance regimes, affects plant growth, survival, and range shift [1,2], which causes environmental and economic consequences [3,4]. Increasing air temperature will accelerate plant growth and increase the rate of water use; increasing atmospheric CO_2 concentration will decrease leaf stomatal conductance and lead to increased water-use efficiency, but this effect will vary with species, and will depend on soil water and nutrient status [5,6]. In order to mitigate the negative impacts of climate change on plant production and maintain sustainable food supplies, it is necessary to develop adaptation strategies and enhance plant resilience, especially for those crucial crops [7]. Since climate change's impacts on plant production may have spatial and temporal variations, due to complicated interactions among species, weather and landscape [8], environment-specific or local adaptation strategies need to be developed [9,10]. Studying some existing plants (e.g., trees) with long life spans and already-experienced climate dynamics may help provide clues in understanding plant adaptation strategies [11].

Chinese Torreya (*Torreya grandis* cv Merrillii) is an evergreen coniferous tree with light green leaves in subtropical China [12,13]. Currently, there exist only six species, with a restricted distribution globally. *T. california* Torrey and *T. taxifolia* Arn. are distributed in North America [14], *T. nucifera* (L.) Sieb. et Zucc. is in Japan and South Korean [15], and three species (*T. fargesii* Franch, *T. grandis* Fort. ex Lind. and *T. jackii* Chun) and two varieties (*T. fargesii* var. yunnanensis and *T. grandis* var.

jiulongshanensis) are found in China [13,16]. Only Chinese Torreya trees produce edible seeds, and can live for a thousand years [13,17], with some dating to the late Tang Dynasty in China. This tree was first scientifically described by R. C. Ching in 1927 [18]. Chinese Torreya has become an important economic tree cultivar for its nuts, which have been used as food in China [19]. One mature tree can produce thousands of dollars in nut crops [20]. The economic benefit has helped thousands of farmers to overcome poverty in Shaoxing [21,22]. Currently, establishing Chinese Torreya plantations and providing nuts is a strategy for poverty alleviation in some poor mountainous areas. The tree also has important medicinal value [23]. Additionally, its timber is an excellent material for high-quality ornaments, such as sculptures and furniture. An increasing number of farmers are setting up Chinese Torreya plantations in order to gain a higher income from this high-value tree species. However, this tree grows slowly, usually producing seeds after 5–10 years [12,20]. This tree species experienced an increase in annual air temperature from 15 °C to 17.5 °C during the last century in this region [11]. Under the global climate change scenario, the annual air temperature and precipitation in this region are modeled to increase by about 2–2.5 °C, and 2–12%, respectively [24]. It is unknown whether climate change will affect this tree, and whether this species can adapt to climate change. Currently, there are some studies on the short-term physiological responses of Chinese Torreya seedlings under a controlled environment [25,26]. For one-year-old seedlings, the growth in height was found mainly in May–June and September [27]. However, there is limited information concerning long-term radial growth for mature trees of this critical species.

One approach to studying the tree growth process is through tree ring analysis, which provides an understanding of tree growth dynamics and the capacity to adapt to climate change [28]. Tree rings can be used to estimate tree age and assess long-term growth patterns in tree species, which provide information on species life history [29,30]. Tree ring widths can also give information about the growth, biomass accumulation and productivity of the species [31,32]. In addition, tree rings reflect the effect of climatic and environmental conditions on tree growth, and the vulnerability of tree species [33–36]. The patterns of tree rings show the combined impact of growth responses and environmental changes on tree species. Increasing atmospheric CO_2 concentration can increase the photosynthetic rate, but this also depends on air temperature, water and nutrient condition. Climate change may lead to a reduction in rainfall but increased temperatures, which can affect some trees' growth and carbon sequestration potential [37,38]. The long life span of Chinese Torreya trees provides an opportunity to study its adaptation to climate dynamics. However, so far, there is no chronological study for Chinese Torreya.

Sustainable management of Chinese Torreya plantations needs to enhance seed production, carbon storage and livelihoods [12,20,39,40]. These objectives require a deeper understanding of the tree's growth behavior and responses to past environmental changes. However, it is challenging to acquire stem samples, because these trees have been the source of income for farmers, and it is not possible to collect samples without damaging the tree. Further, applying an increment borer can introduce diseases. Most old trees have a rotten heartwood, although they still grow well and produce seeds.

This study aims to find some tree samples and study tree rings of Chinese Torreya trees, as well as find their growth patterns. Since Chinese Torreya trees have existed in the region for more than a thousand years, it is reasonable to assume that this tree could endure climate change. Furthermore, it was found that there was an increase of inter-annual variation in seed production with time in some trees due, to environmental change [41]. Thus, it might be assumed that there was an increased inter-annual variation of tree ring growth for the Chinese Torreya. The specific objectives of this study include: (i) studying the growth rate and patterns of tree rings and testing the above assumptions; (ii) indicating relationships between tree growth and climate, and understanding the tree's adaptation; and (iii) providing suggestions for the adaptive management of Chinese Torreya forests under climate change.

2. Materials and methods

2.1. Materials

Six stem sections from six Chinese Torreya trees were used for this study, and one tree (tree #6) had two stems according to the two sets of tree rings (Table 1). All stem sections are preserved in the Torreya Museum at Zhuji County, in Zhejiang Province of China. Each stem section was acquired from the tree bottom (above the grafting point) in the nearby Chinese Torreya forest (27.7191 °N, 120.5127 °E) at Zhaojiazhen around 2015. Trees of 100 years of age can usually reach 10–15 m. These trees were cut down due to house building or wind damage. Tree #5 might be a different ecotype or mutation. This area is the central production area of Chinese Torreya trees, and the oldest Chinese Torreya tree in the area is about 1300 years old, with a height of approximately 23 m [12,20]. The local soil is yellow earth, with a pH value of 5.2–7.5. The elevation varies from about 300 m to 600 m. The region has a monsoon climate with a hot and humid summer and relatively cold and wet winter. The annual mean temperature is approximately 16 °C, and the average annual precipitation is about 1400 mm, which is mostly concentrated between May and August [11,39,40]. Intense rainfall usually occurs during the monsoon and typhoon periods, within the growing seasons. Currently, routine management practices include weeds control, tilling or fertilizer applying. However, there were limited management practices about 20 years ago, and it can thus be assumed that the growth of these sampled trees was mainly controlled by the natural condition. Since the ages of these trees were different, it is not valid to compare their tree rings directly because they might be at different life stages. However, comparing their growth patterns and general relationships with climate would be helpful to understand this tree species' adaptation to historical climate dynamics

Table 1. Tree ring information of six Chinese Torreya trees.

Item	Tree #1	Tree #2	Tree #3	Tree #4	Tree #5	Tree #6	
						Stem 1	Stem 2
Age (year)	90	89	81	86	63	75	80
Diameter (mm)	426	408	382	340	638	368	308
Average Ring width (mm)	2.4	2.3	2.4	2.0	5.1	2.5	1.9
Maximum Ring width (mm)	5	5	9	7	8	8	7
Minimum ring width (mm)	0.5	1	0.5	0.3	1	1	1

2.2. Methods

Tree ring measurement: For each stem section, the radiating rings were measured in three different directions because the tree rings were not perfect circles. Then, the average of the ring radiates, widths between rings or annual radial increments, and basal areas of increases (BAI) were estimated.

Taylor's power law: Taylor's power law is one of the most widely verified empirical relationships in ecology [42]. In this study, Taylor's power law is expressed as:

$$Variance = a \times Mean^{r} \quad (1)$$

where *Variance* is the variance of tree ring radiates, and the *Mean* is the average of tree ring radiators. After taking the logarithm, $log\ (Variance) = log\ (a) + r \times log\ (Mean)$. With the time scale increased from 1, 2 or 3 to the age of a tree, the scaling exponent (r) between the variance and average of tree ring radiators for each Chinese Torreya tree was estimated.

Climate data: Since there was limited climate information from the ground observations in the area with Chinese Torreya plantation at Zhaojiazhen, the climate data for this area were collected from the Climate Research Unit. The high-resolution gridded (0.5° × 0.5°) data of the monthly air temperature and precipitation during 1923–2015 were used in this study. The data were drawn directly from the CRU TS 4.03 dataset and related nearby observations [43]. The CRU dataset has been broadly cross-checked with ground-monitored climate data [44]. Thus, here the data were not cross-checked due to the limited ground observations.

Each year, the average monthly air temperature, maximum and minimum monthly air temperature, and the average monthly air temperature during the growing season (from April to September) were estimated. Although Chinese Torreya is an evergreen tree species, it usually stops radial growth in wintertime.

The drought was estimated by the standardized precipitation index (SPI), which is based on [45], and only the precipitation was involved.

Monthly SPI = (Monthly precipitation − the average precipitation of this month from 1923 to 2015)/the standard deviation of this month's precipitation from 1923 to 2015. If SPI > 0, it is wet; if SPI < 0, it is dry.

SPI in the growing seasons = the sum of the monthly SPI from April to September each year.

Precipitation in the growing seasons = the sum of the monthly precipitation from April to September each year.

Also, with the consideration of drought induced by heat in the growing seasons, the hydrothermal coefficient (HTC), including both air temperature and precipitation, was used here by applying Selyaninov's formula as the following [46]:

$$HTC = \Sigma p/(\Sigma t \times 10) \tag{2}$$

where $\Sigma\, p$ and $\Sigma\, t$ are the sum of precipitations and air temperatures (≥ 10 °C) in the growing seasons, respectively; when HTC > 1.0, it is considered humid; while HTC is within 0.7–1.0, it is dry; and if HTC is within 0.4–0.7, it is very dry [46].

Statistics: Pearson's correlation was used between the accumulated radial growth and time, BAI and time, log(average of radial growth) and log(variance of radial growth), the slopes of the accumulated radial growth and the scaling exponents of Taylor's power law, SPI and HTC, and log(variance of precipitation) and log(variance of BAI) in the six trees. The correlation coefficients were recorded, and the statistical test was considered as significant at $p < 0.05$.

3. Results

3.1. Tree Age and Growth

The tree ages based on the tree rings varied from approximately 63 to 90 years old, among the six trees considered under this study (Table 1). The average increment of tree rings for each tree ranged from 1.9 mm to 5.1 mm. The maximum width of the tree rings was 9 mm, and the minimum width was 0.3 mm. The increments of radial growth were mainly distributed at 1 and 2 mm. The variance of radial increments did not increase with time for all trees. However, the accumulated radial growth increased linearly with time for each tree (Figure 1), and most of them had slight changes in growth rate. For tree #3, the growth rate changed around 41 years old. For tree #6, the two stems had different rates of accumulated radial growth. The accumulated radial growth did not become stable for each tree, which means these trees were not senescent in their ages. BAI generally increased with tree, age but varied dramatically (Figure 2). The variance of BAI increased with time for all trees (Figure 3).

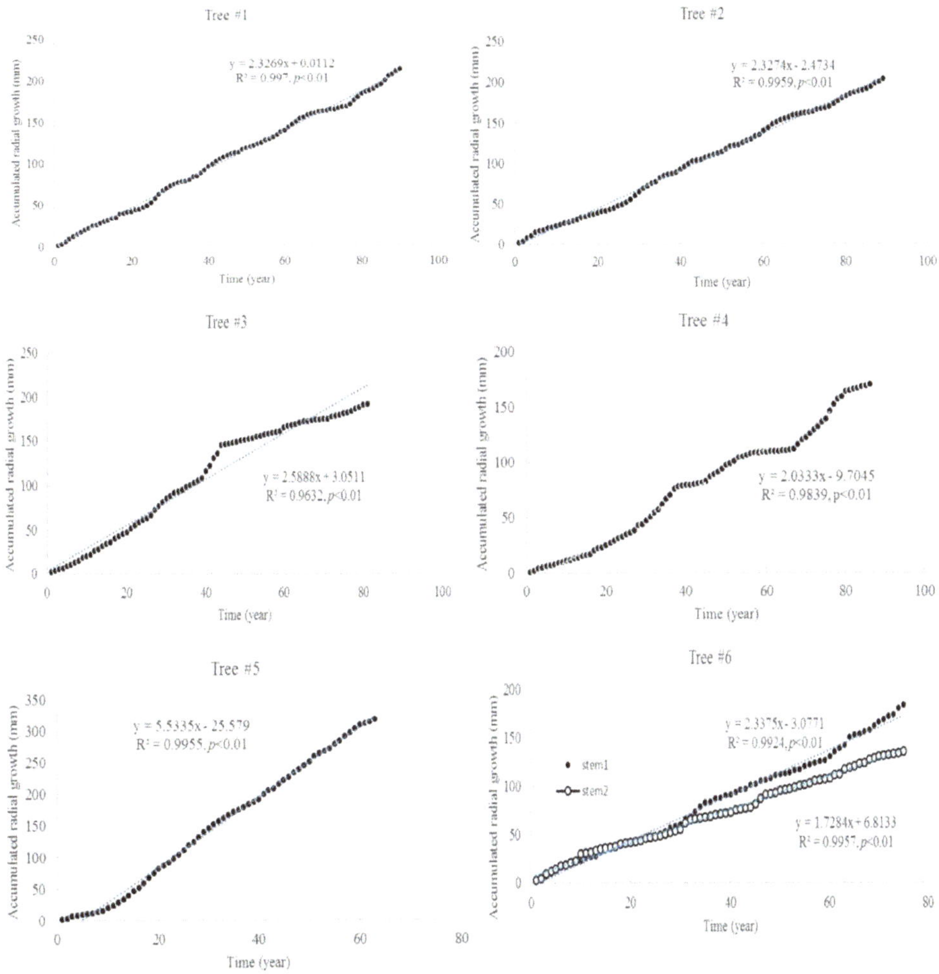

Figure 1. Accumulated radial growth with time for six Chinese Torreya trees.

3.2. Taylor's Power Law

Taylor's power law was present between the average accumulated radial growth and variance (Figure 4). The scaling exponents were concentrated within 1.9–2.1, which might indicate a similar growth regime. For tree #6, the two stems had similar scaling exponents. These scaling exponents were significantly correlated with the slopes of the fitting lines between the accumulated radial growth and time (Figure 5).

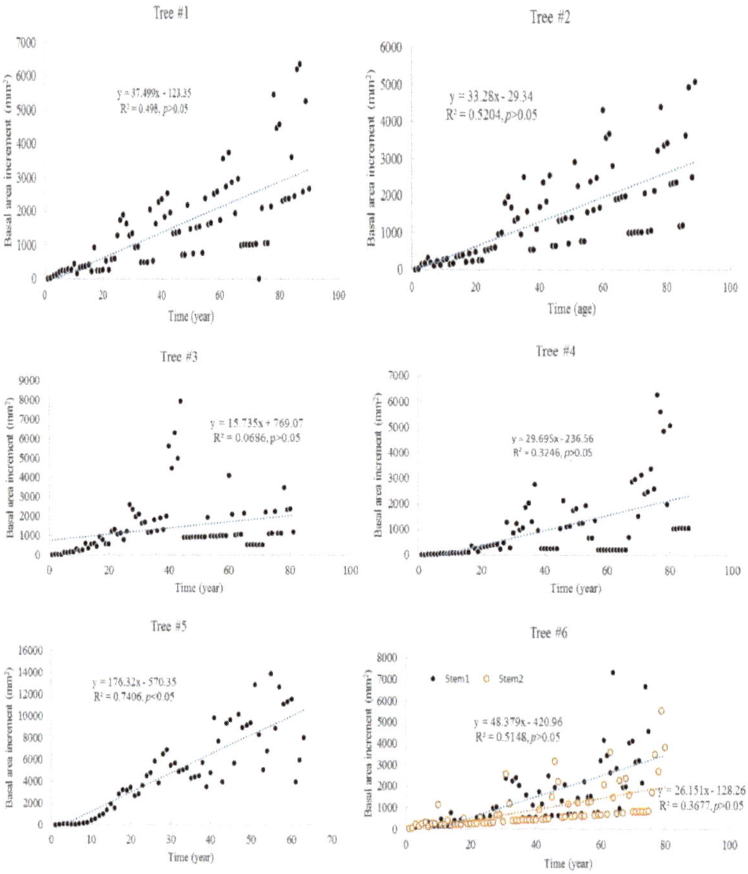

Figure 2. Basal area increments (BAI) change with time for six Chinese Torreya trees.

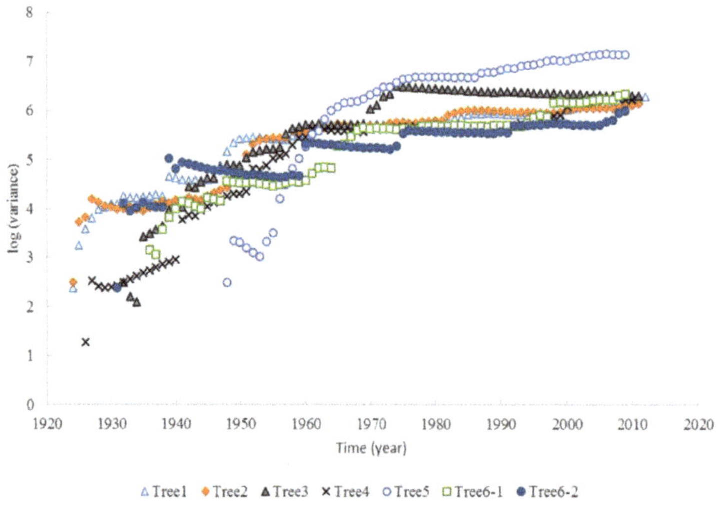

Figure 3. The variance of BAI increased with time for six Chinese Torreya trees.

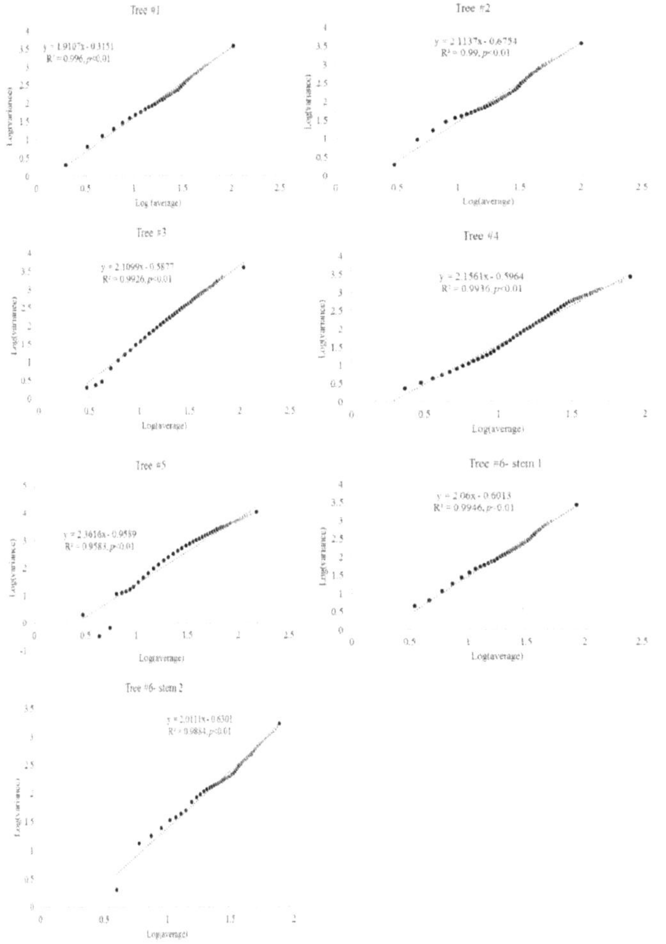

Figure 4. Taylor's power law in the radial growth for six Chinese Torreya trees.

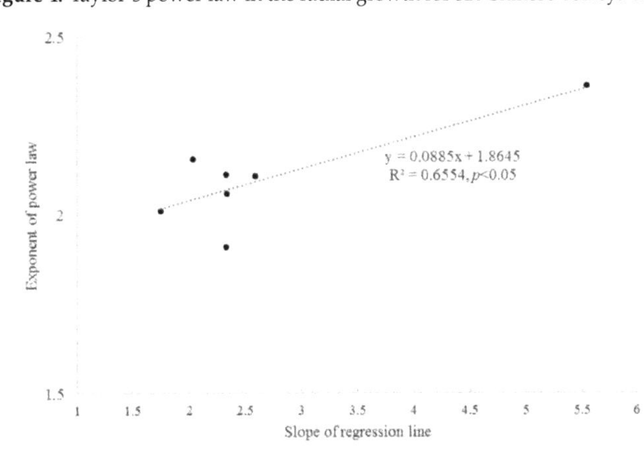

Figure 5. Correlation between the scaling exponents of Taylor's power law and the slopes of fitting lines between accumulated radial growth and time.

3.3. Climate and Radial Growth

The relationship between the average monthly air temperature during the growing season and annual radial increment was not apparent. A "Triangle"-shaped relationship existed between the precipitation in the growing seasons and annual increments. When the accumulated precipitation during the growing seasons was close to an average of 975 mm (±200 mm), there were high annual increments in radial growth (Figure 6). This pattern was similar to the SPI, which was between −2 and 2 (Figure 7). For the two stems of tree #6, stem 1 had more growth when SPI < 0, while stem 2 had more growth when SPI > 0. The sensitivity and lasting time of each tree that responded to SPI change were different; for example, tree #4 could grow at 0.3 mm each year for 10 years (Figure 8). Similar "triangle"-shaped patterns existed for the maximum monthly air temperature in the growing seasons or the minimum monthly air temperature in winters (Figure 9). When the maximum monthly air temperature was around 27.5–28.5 °C, or the winter monthly air temperature about 3–4 °C, there were high annual increments in radial growth.

There was a significant correlation between SPI and HTC ($p < 0.05$, Figure 10), which indicates the consistency between the two metrics. The trees had a high BAI when HTC was around 0.7–0.9, which was classified as a dry condition, based on HTC.

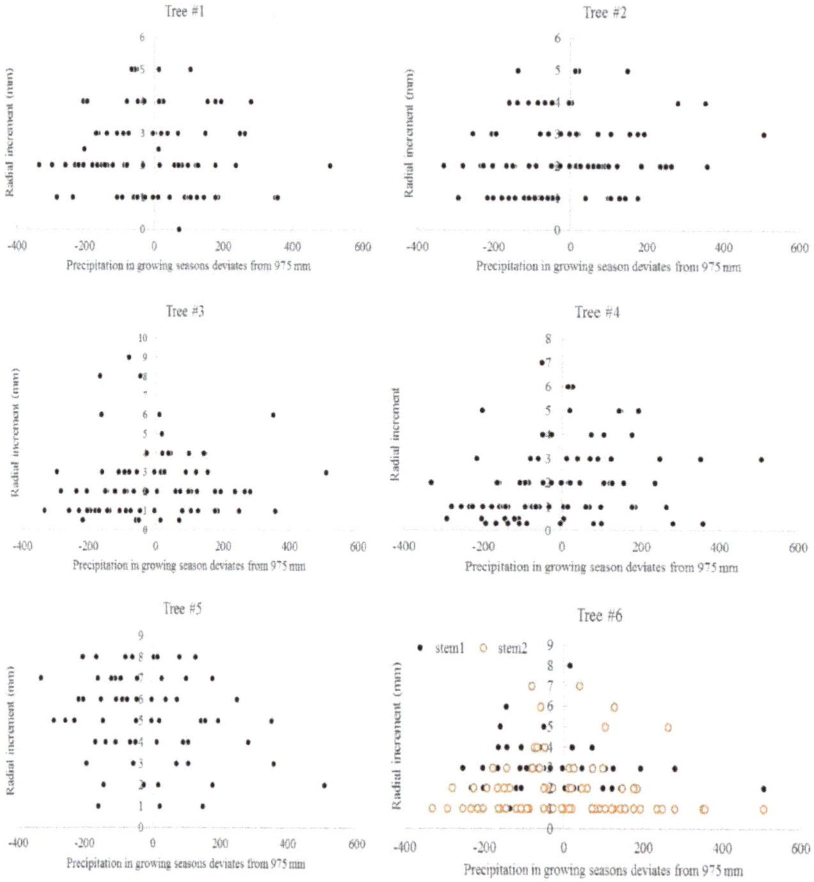

Figure 6. Precipitation in growing seasons and annual radial increments for six Chinese Torreya trees.

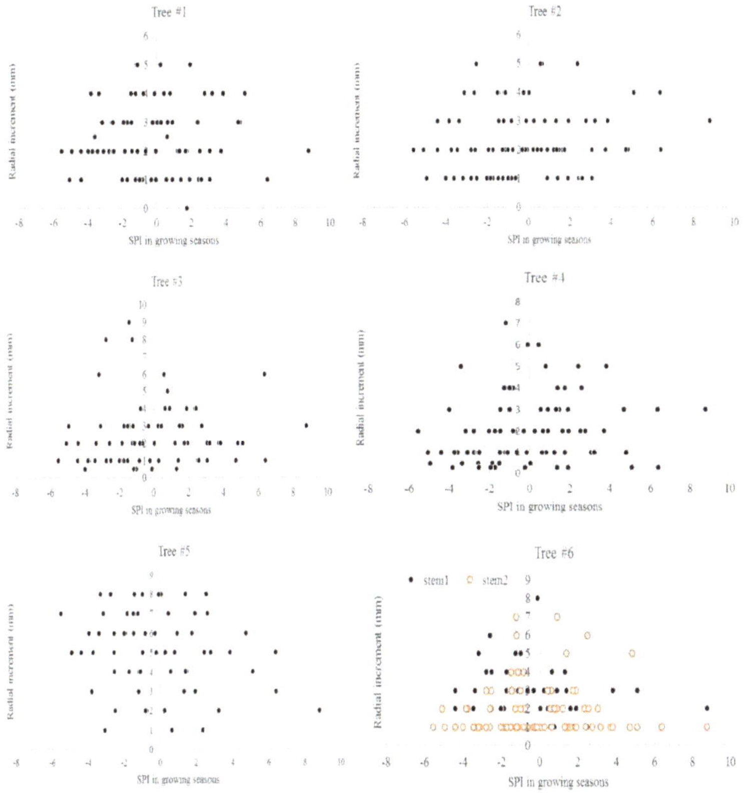

Figure 7. Standardized precipitation indices and annual radial increments for six Chinese Torreya trees.

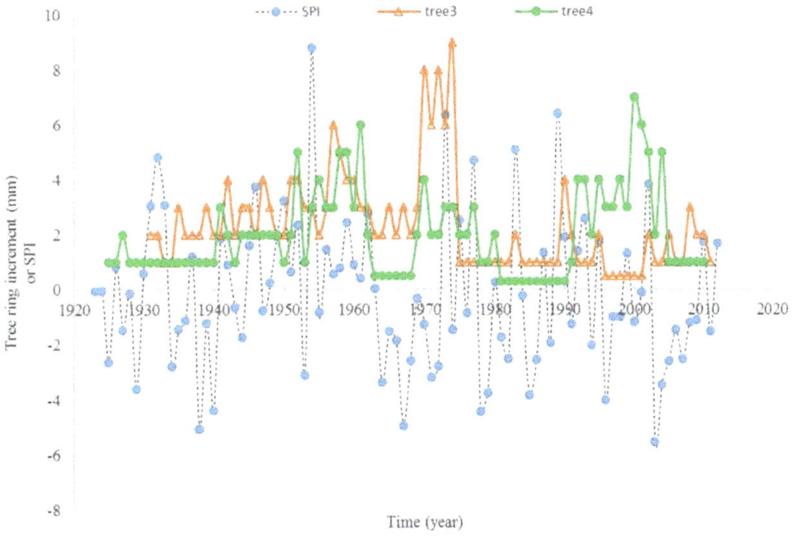

Figure 8. Response of tree ring increment to SPI change with time, for tree #3 and #4 (no negative values for tree ring increment).

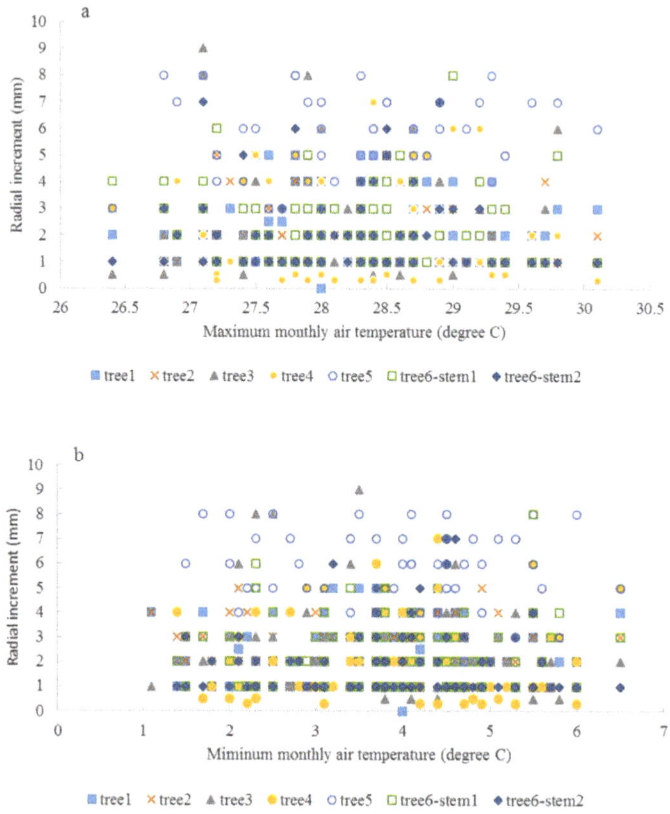

Figure 9. Maximum monthly air temperature and annual radial increments (**a**), and minimum monthly air temperature and annual radial increments (**b**).

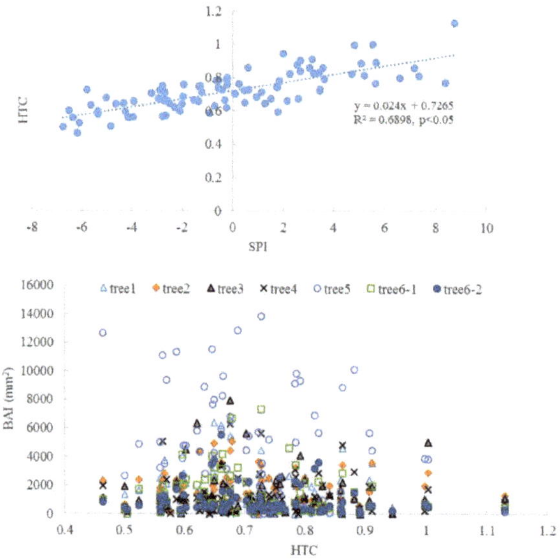

Figure 10. The correlation between SPI and HTC and the change of BAI along with HTC.

4. Discussion

The dendrochronological data of six trees indicated that the accumulated radial growth increased with time in these Chinese Torreya trees when they were cut. This result means that these trees were not at the senescent age [47]. Usually, the life span of this tree can reach hundreds of years, or even more. Currently, the oldest in the region is approximately 1300 years old [13]. This result provided more evidence that slow-growth trees were associated with long life spans [48]. The long life span is advantageous for this tree species, as it helps create a long-term income from the seeds. Further, the slow growth and long life span of the species make it suitable for carbon sequestration. Other tree species grow fast but die young, so the periods over which they store carbon are short [49].

Taylor's power law can be observed in the accumulated radial growth, and the scaling exponents concentrated at 1.9–2.1, although the rates of accumulated radial growth varied among these trees. This means that the radial growth of these trees was under a similar regime (e.g., the same area with similar biological and environmental interactions) [50]. The historical climate dynamics did not constitute a significant regime shift in the radial growth for these trees. This study also confirmed the high tolerance of this tree species [11], that is, Tree #3 and #4 could endure limited radial growth for about 10 years. The slopes of the accumulated radial growth with time are significantly correlated with the scaling exponents of Taylor's power law. Both these slopes may reflect the biological and environmental interactions (e.g., self-organization) of these trees. Most annual radial increments were approximately 1 or 2 mm/year in these trees. The average annual radial increment of these trees was above 2 mm/year, and tree #5 even reached 5 mm/year. This rate is considerably larger than the original reported value, of 1.1 mm/year, from a 1500-year-old tree by the Beijing Natural History Museum in 2012 [17]. These different rates and lasting times might be linked to small-scale local environmental conditions that support individual trees. However, the growth rates found here are still very low compared to those of other trees, such as loblolly pine in southeastern USA, which can reach 12 mm/year [51]. Thus, the first assumption of the slow growth rate being related to adaptation to the environment is valid.

Despite the approximate increase of 70 ppm in global average CO_2 concentration (increased from 315 ppm to 390 ppm from the 1920s to 2015), and the increase of approximately 2.5 °C in the local annual air temperature [11], there was no clear trend of increasing radial increment in these trees. Furthermore, the variance in the radial increment did not increase for all individual trees. However, the variance of BAI generally increased for all trees. The second assumption related to the increase of variance is valid for BAI; it is more accurate for describing tree growth than radial increment, because the same radial increment can indicate a different BAI if the radiating rings are different. For some trees, there was a significant correlation between the variance of growing season precipitation and the variance of BAI (Figure 11). The variance of growing season precipitation explains about 40–66% of the variance of BAI. The CO_2 fertilization effect might be counterbalanced by other environmental stressors, such as light condition, insufficient precipitation, and extremely high or low air temperature [52]. Different from [44], the tree ring widths here were not correlated with the annual rainfall.

There were "triangle"-shaped relationships between the annual radial increments and climate factors, such as the precipitation in growing seasons, the maximum monthly air temperature, and the minimum monthly air temperature in winter. This result means that tree growth was limited by many factors. The outlier points can be fitted by polynomials with a power of 2. The optimum values of these factors were found in this study: the optimum precipitation during the growing seasons was approximately 975 mm (±200 mm), SPI was between −2 and 2, the maximum monthly air temperature was about 27.5–28.5 °C, the minimum monthly air temperature in winter was around 3–4 °C, and HTC was approximately 0.7–0.9. A climate within these limits may be better for growing Chinese Torreya trees. Climate conditions that are too dry, too wet, too warm or too cold may affect the radial increments of Chinese Torreya, which are believed to be linked with seed production. The trees could also tolerate a drought condition, with SPI around −6, or HTC of 0.46 in the growing seasons. This climate envelope may help in finding suitable areas to introduce Chinese Torreya into for industrial plantations. The result also provides a base for the agricultural insurance policy in this region. If trees cannot produce a normal amount of seeds under unfavorable weather conditions, then an agricultural insurance company will make a specific payment to the farmers to cover the loss. This result is different from [44], wherein rainfall augmented tree ring width, and hot temperature reduced tree ring growth in *Scutia buxifolia*. A positive correlation with precipitation, and a negative correlation with extreme summer temperatures, were usually observed for trees in Africa and Australia [53,54]. However, growth ceased at a certain threshold, rather than showing a continuous linear decline, was also observed in ponderosa pine [55].

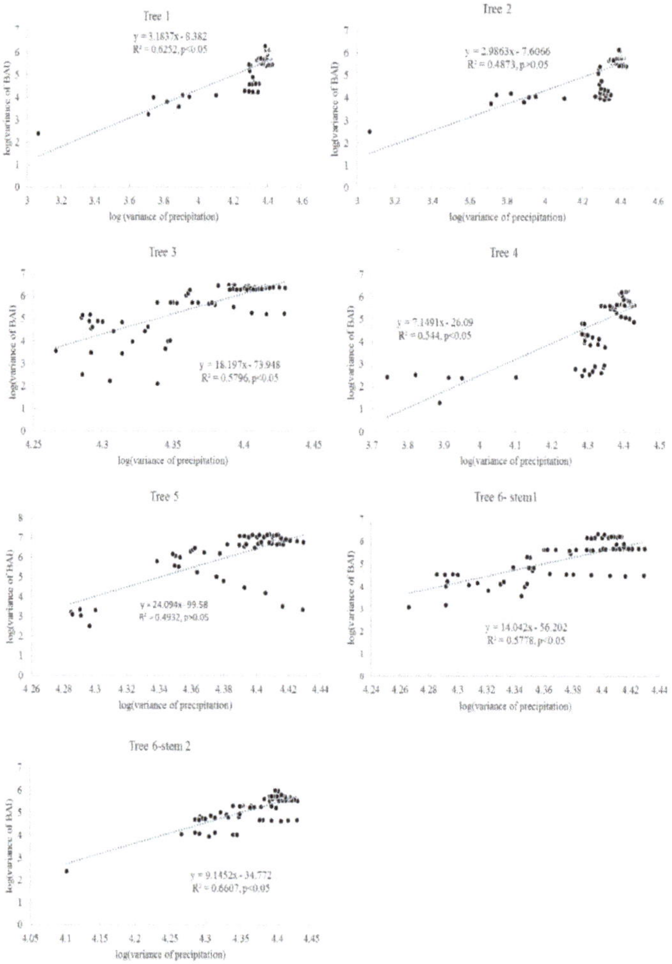

Figure 11. Correlation between BAI variances and the variances of precipitation in the growing seasons for six Chinese Torreya trees.

The phenomenon of multiple stems living on one tree is very popular in Chinese Torreya. In this study, it was found that (i) the two stems had quite different radial increment rates; (ii) the two stems shared a similar growth regime (e.g., similar scaling exponents in Tayler's power law); and (iii) the two stems had different climate adaptations (e.g., dry or wet climate). The heterogeneous growth rates might be related to different positions, or the adaptation to different microclimate conditions. The phenomena may show the ability of an individual tree to achieve maximum growth under different environmental conditions. Morphological changes in trees are an adaptation strategy to climate change [56]. A tree with multiple stems may have the advantage of using solar radiation at different positions, and maintaining productivity under environmental variation, over a tree with a single stem. It is not known whether more stems is always better. However, Chinese Torreya trees with a couple of stems are popular in the central production area. Usually, a tree with multiple stems may have a big canopy, which may be easily damaged by typhoons.

This study provides several implications for the adaptive management of Chinese Torreya plantations. First, these trees still had radial growth at the age of 90 years. Since the life span of this tree species could reach a thousand years, it indicated that these trees were still at the early growth stage, which means Chinese Torreya trees can lead to long-term economic and biological income. However, the farmers still need to wait for 5–10 years for seedlings to produce the first seeds. Second, the timber growth rate of most trees was slow. However, some trees could have a relatively high growth rate, such as tree #5. This result might be related to a genetic mutation or suitable local environmental setting (e.g., soil, water). This deviation makes it possible to identify the fast-growing tree varieties or select appropriate sites for a plantation. Regular citizen science or community activities may help select the variety with a high timber or seed production. Local citizen scientists should be trained to preserve genetic variations and other natural resources that are particularly adapted to climate change [12]. Some trees grow slowly, but may have a high endurance to climate change. Third, selecting trees with multiple stems in plantations may offer the advantage of maintaining productivity under environmental change. The stems can take advantage of the light at different positions, and adapt to different conditions. Fourth, the effect of climate change on tree growth could be estimated via precipitation in growing seasons, and the highest or the lowest monthly air temperature. Further, the climate conditions identified by this study may be useful for introducing Chinese Torreya trees to new locations. Besides, human management practices, such as cutting grasses and shrubs and digging surface soil, could alter tree growth through the changing soil water and air temperature. However, these management practices may not be a good way to maintain proper environmental conditions at certain stages, and may lead to environmental problems [12,39,40]. Finally, based on the radial growth patterns, it is possible to estimate the tree canopy size and possible seed production through allometric scaling relationships [49,57,58]. This algorithm may help forest farmers assess potential income and benefits, and also helps them manage Chinese Torreya plantations more effectively.

5. Conclusions

Chinese Torreya is an important crop tree with a long life span. The dendrochronological data of Chinese Torreya trees are very scarce. This study is the first one that provides relevant information for understanding growth patterns and the adaptation of this tree to climate dynamics. In this central production area, the long-term persistence of Chinese Torreya could indicate its adaptation to the historical climate fluctuation and atmospheric CO_2 enrichment. The slow radial growth rate of Chinese Torreya (on average, form 1.9 to 5.1 mm/year) might help this species to adapt to unfavorable conditions. The change of precipitation in growing seasons could impact the variation of basal area increment. The radial growth regime did not change significantly during their life spans, because similar Taylor's power law exponents existed in the accumulated radial growths. "Triangle"-shaped relationships, which indicate optimum values, occurred between the climate and annual radial increments, especially the extreme air temperature (maximum and minimum monthly air temperatures), precipitation in growing seasons, and drought, which could affect its growth rate as well as the variance. The trees could tolerate the drought conditions with SPI of approximately −6, or HTC of 0.46 in growing seasons, and being either too dry or too wet both decreased tree growth. The presence of multiple stems in one tree could help it adapt more effectively to the local environment. It is necessary to conduct long-term intensive monitoring projects across different environmental settings, which can help in collecting data to find the relationships between tree growth and environmental change. It will be helpful to derive the core increment data from some old trees, although this may affect these trees.

However, these long dendrochronological data could provide beneficial information about adaptive response and endurance. Local agencies need to organize citizen scientists to collect sample trees and preserve genetic diversity, since some trees (e.g., tree #5) could grow much faster, which may provide the opportunity for developing fast-growing trees. In contrast, other trees (e.g., tree #4) might have high endurance. The current introduction of this tree to new areas for plantation should be based on the environmental requirements. Regional planners and decision-makers need to know the tree growth ecology, and consider the possible unfavorable environmental conditions, before introducing or developing large-scale plantations of Chinese Torreya.

Funding: This study was partially supported by the USDA Mc-Stennis project (1008643).

Acknowledgments: The author thanks anonymous reviewers and Troy Bowman for their helpful suggestions. Hangbiao Jin and Jinchang Li from Zhuji Forestry Academy provided relevant information, and Xi Chen assisted in measurements.

Conflicts of Interest: The author declares no conflict of interest.

References

1. Pretzsch, H.; Biber, P.; Schütze, G.; Uhl, E.; Rötzer, T. Forest stand growth dynamics in Central Europe have accelerated since 1870. *Nat. Commun.* **2014**, *5*, 4967. [CrossRef]
2. Brandl, S.; Paul, C.; Knoke, T.; Falk, W. The influence of climate and management on survival probability for Germany's most important tree species. *For. Ecol. Manag.* **2020**, *458*, 117652. [CrossRef]
3. Vossen, P. Olive oil: History, production, and characteristics of the world's classic oils. *HortScience* **2007**, *42*, 1093–1100. [CrossRef]
4. Seidl, R.; Schelhaas, M.-J.; Rammer, W.; Verkerk, P.J. Increasing forest disturbances in Europe and their impact on carbon storage. *Nat. Clim. Chang.* **2014**, *4*, 806–810. [CrossRef] [PubMed]
5. Chapman, S.C.; Chakraborty, S.; Dreccer, M.F.; Howden, S.M. Plant adaptation to climate change—opportunities and priorities in breeding. *Crop. Pasture Sci.* **2012**, *63*, 251–268. [CrossRef]
6. van der Sleen, P.; Groenendijk, P.; Vlam, M.; Anten, N.P.R.; Boom, A.; Bongers, F.; Pons, T.L.; Terburg, G.; Zuidema, P.A. No growth stimulation of tropical trees by 150 years of CO_2 fertilization but water-use efficiency increased. *Nat. Geosci.* **2014**, *8*, 4. [CrossRef]
7. Maryinez-Feria, R.A.; Basso, B. Unstable crop yields reveal opportunities for site-specific adaptations to climate variability. *Sci. Rep.* **2020**, *10*, 2885. [CrossRef]
8. Chen, X.; Brockway, D.G.; Guo, Q. Characterizing the dynamics of cone production for longleaf pine forests in the southeastern United States. *For. Ecol. Manag.* **2018**, *429*, 1–6. [CrossRef]
9. Lobell, D.B.; Burke, M.B.; Telbaldi, C.; Mastrandrea, M.D.; Falcon, W.P.; Naylor, R.L. Prioritizing climate change adaptation needs for food security in 2030. *Science* **2008**, *319*, 607–610. [CrossRef]
10. Hammer, G.L.; McLean, G.; Chapman, S.; Zheng, B.; Doherty, A.; Harrison, M.T.; van Oosterom, E.; Jordan, D. Crop design for specific adaptation in variable dryland production environments. *Crop. Pasture Sci.* **2014**, *65*, 614–626. [CrossRef]
11. Chen, X.; Niu, J. Evaluating the adaptation of Chinese Torreya plantations to climate change. *Atmosphere* **2020**, *11*, 176. [CrossRef]
12. Chen, X.; Jin, H. Review of cultivation and development of Chinese torreya in China. *For. Trees Liveli.* **2019**, *28*, 68–78. [CrossRef]
13. Li, Z.; Dai, W. *Chinese Torreya*; Science Press: Beijing, China, 2007. (in Chinese)
14. Roy, D.F. Torreya Arn. Torreya. In *Seeds of Woody Plants in the United States*; Schopmeyer, C.S., Ed.; Agriculture Handbook No. 450 USDA Forest Service: Washington, DC, USA, 1974.
15. Shin, S.; Lee, S.G.; Kang, H. Spatial distribution patterns of old-growth forest of dioecious tree Torreya nucifera in rocky Gotjawal terrain of Jeju Island, South Korea. *J. Ecol. Environ.* **2017**, *41*, 31. [CrossRef]
16. Farjon, A. *A Handbook of the World's Conifers*; Brill Academic Publishing: Leiden, Netherlands, 2010.
17. Wang, B.; Ming, Q.-W. *Zhejiang Shaoxing Kuaijishan Guxiangfeiqun*; China Agricultural Press: Beijing, China, 2015. (in Chinese)
18. Hu, H.-H. Synoptical study of Chinese Torreyas: With supplemental notes on the distribution and habitat by R. C. Ching. *Contrib. Biol. Lab. Sci. Soc. China* **1927**, *3*, 1–37.

19. Cheng, X.; Li, Z.; Yu, W.; Dai, W.; Fu, Q. Distribution and ecological characteristics of *Torreya grandis* in China. *J. Zhejiang For. Coll.* **2007**, *24*, 383–388, (in Chinese with English abstract).
20. Chen, X.; Jin, H. A case study of enhancing sustainable intensification of Chinese torreya forest in Zhuji of China. *Environ. Nat. Res. Res.* **2019**, *9*, 53–60. [CrossRef]
21. People's Government of Shaoxing City. *Kuanjishan Ancient Chinese Torreya Community*; Proposal for Global Important Agricultural Heritage System Initiative: Shaoxing, China, 2013.
22. Chen, X.; Chen, H. Dynamics in production of four heritage foods at the mountainous region of Shaoxing City, China. *Emir. J. Food Agric.* **2019**, *31*, 645–653. [CrossRef]
23. Huang, Y.; Wang, J.; Li, G.; Zheng, Z.; Su, W. Antitumor and antifungal activities in endophytic fungi isolated from pharmaceutical plants *Taxus mairei*, *Cephalataxus fortunei* and *Torreya grandis*. *FEMS Immunol. Med. Microbiol.* **2001**, *31*, 163–167. [CrossRef]
24. Chen, X.; Zhang, X.; Li, B.-L. The possible response of life zones in China under global climate change. *Glob. Planet. Chang.* **2003**, *38*, 323–337. [CrossRef]
25. Li, T.; Hu, Y.; Du, X.; Tang, H.; Shen, C.; Wu, J. Salicylic acid alleviates the adverse effects of salt stress in Torreya grandis cv. Merrillii seedlings by activating photosynthesis and enhancing antioxidant systems. *PLoS ONE* **2014**, *9*, e109492. [CrossRef]
26. Lin, J.; Zhang, R.; Hu, Y.; Song, Y.; Hanninen, H.; Wu, J. Interactive effects of drought and shading on *Torreya grandis* seedlings: Physiological and growth responses. *Trees* **2019**, *33*, 951–961. [CrossRef]
27. Wang, Z.; Xu, W.-Z. Seedling cultivation of Chinese Torreya and observation on height growth in one-year-old seelings. *Appl. For. Tech.* **2006**, *4*, 18–19. (in Chinese).
28. Boden, S.; Kahle, H.P.; Wilpert, K.V.; Spiecker, H. Resilience of Norway spruce (*Picea abies* (L.) Karst) growth to changing climatic conditions in Southwest Germany. *For. Ecol. Manag.* **2014**, *315*, 12–21. [CrossRef]
29. Costa, M.S.; Ferreira, K.E.B.; Botosso, P.C.; Callado, C.H. Growth analysis of five Leguminosae native tree species from a seasonal semidecidual lowland forest in Brazil. *Dendrochronologia* **2015**, *36*, 23–32. [CrossRef]
30. Worbes, M.; Staschel, R.; Roloff, A.; Junk, W.J. Tree ring analysis reveals age structure, dynamics and wood production of a natural forest stand in Cameroon. *For. Ecol. Manag.* **2003**, *173*, 105–123. [CrossRef]
31. Mbow, C.; Chhin, S.; Sambou, B.; Skole, D. Potential of dendrochronology to assess annual rates of biomass productivity in savanna trees of West Africa. *Dendrochronologia* **2012**, *31*, 41–51. [CrossRef]
32. Shimamoto, C.Y.; Botosso, P.C.; Marques, M.C.M. How much carbon is sequestered during the restoration of tropical forests? Estimates from tree species in the Brazilian Atlantic Forest. *For. Ecol. Manag.* **2014**, *329*, 1–9. [CrossRef]
33. Granato-Souza, D.; Adenesky-Filho, E.; Esemann-Quadros, K. Dendrochronology and climatic signals in the wood of Nectandra oppositifolia from a dense rain forest in southern Brazil. *J. For. Res.* **2018**, *30*, 545–553. [CrossRef]
34. Natalini, F.; Correia, A.C.; Vazquez-Pique, J.; Alejano, R. Tree rings reflect growth adjustments and enhanced synchrony among sites in Iberian stone pine (*Pinus pinea* L.) under climate change. *Ann. For. Sci.* **2015**, *72*, 1023–1033. [CrossRef]
35. Prestes, A.; Klausner, V.; Silva, I.R.; Ojeda-Gonzalez, A.; Lorensi, C. *Araucaria* growth response to solar and climate variability in South Brazil. *Ann. Geophys. Dis.* **2018**, *36*, 717–719. [CrossRef]
36. Rahman, M.; Islam, R.; Islam, M. Long-term growth decline in *Toona ciliata* in a moist tropical forest in Bangladesh: Impact of global warming. *Acta Oecol.* **2017**, *80*, 8–17. [CrossRef]
37. Allen, C.D.; Breshears, D.D.; McDowell, N.G. On underestimation of global vulnerability to tree mortality and forest die-off from hotter drought in the Anthropocene. *Ecosphere* **2015**, *6*, 1–55. [CrossRef]
38. Corlett, R.T. The impacts of droughts in tropical forests. *Trends Plant. Sci.* **2016**, *21*, 584–593. [CrossRef] [PubMed]
39. Chen, X.; Chen, H. Comparing environmental impacts of Chinese Torreya plantations and regular forests using remote sensing. *Environ. Dev. Sustain.* **2020**, in press. [CrossRef]
40. Chen, X.; Xiao, P.; Niu, J.; Chen, X. Evaluating soil and nutrients (C, N, and P) loss in Chinese Torreya plantations. *Environ. Pollut.* **2020**, in press. [CrossRef]
41. Pearse, I.S.; LaMontagne, J.M.; Koenig, W.D. Inter-annual variation in seed production has increased over time (1900–2014). *Proc. Royal Soc. B* **2017**, *284*, 20171666. [CrossRef]
42. Taylor, L.R. Aggregation, variance and the mean. *Nature* **1961**, *189*, 732–735. [CrossRef]

43. Harris, I.; Jones, P.D.; Osborn, T.J.; Lister, D.H. Updated high-resolution grids of monthly climatic observations—The CRU TS3.10 Dataset. *Int. J. Climatol.* **2014**, *34*, 623–642. [CrossRef]
44. Lucas, C.; Puchi, P.; Profumo, L.; Ferreira, A.; Muñoz, A. Effect of climate on tree growth in the Pampa biome of Southeastern South America: First tree-ring chronologies from Uruguay. *Dendrochronologia* **2018**, *52*, 113–122. [CrossRef]
45. McKee, T.B.; Doesken, N.J.; Kliest, J. The relationship of drought frequency and duration to time scales. In Proceedings of the 8th Conference of Applied Climatology, Anaheim, CA, USA, 17–22 January 1993; American Meteorological Society: Boston, MA, USA, 1993; pp. 179–184.
46. Evarte-Bundere, G.; Evarts-Bunders, P. Using of the hydrothermal coefficient (HTC) for interpretation of distribution of non-native tree species in Latvia on example of cultivated species of genus *Tilia*. *Acta Biol. Univ. Daugavp.* **2012**, *12*, 135–148.
47. Fritts, H. *Tree Rings and Climate*; Academic: San Diego, CA, USA, 1976.
48. Black, B.A.; Colbert, J.J.; Pederson, N. Relationships between radial growth rates and lifespan within North American tree species. *Ecoscience* **2008**, *15*, 349–357. [CrossRef]
49. Chen, X. Carbon storage traits of main tree species in natural forests in Northeast China. *J. Sustain. For.* **2006**, *23*, 67–84. [CrossRef]
50. Chen, X.; Guo, Q.; Brockway, D.G. Power laws in cone production of longleaf pine across its native range in the United States. *Sustain. Agric. Res.* **2017**, *4*, 64–73. [CrossRef]
51. Chen, X. Will more tree diversity bring back more income from timber? A case study from Alabama of USA. *For. Lett.* **2017**, *110*, 20–25.
52. Wagner, F.H.; Hérault, B.; Bonal, D.; Stahl, C.; Anderson, L.O.; Baker, T.R.; Becker, G.S.; Beeckman, H.; Boanerges, S.D.; Botosso, P.C.; et al. Climate seasonality limits leaf carbon assimilation and wood productivity in tropical forests. *Biogeosciences* **2016**, *13*, 2537–2562. [CrossRef]
53. Baker, P.J.; Palmer, J.G.; D'Arrigo, R. The dendrochronology of Callitris intratropica in northern Australia: Annual ring structure, chronology development and climate correlations. *Aust. J. Bot.* **2008**, *56*, 311–320. [CrossRef]
54. Trouet, V.; Coppin, P.; Beeckman, H. Annual growth ring patterns in Brachystegia spiciformis reveal influence of precipitation on tree growth. *Biotropica* **2006**, *38*, 375–382. [CrossRef]
55. McCullough, I.M.; Davis, F.W.; Williams, A.P. A range of possibilities: Assessing geographic variation in climate sensitivity of ponderosa pine using tree rings. *For. Ecol. Manag.* **2017**, *402*, 223–233. [CrossRef]
56. McDowell, N.G.; Allen, C.D.; Anderson-Teixeira, K.; Bond-Lamberty, B.; Chini, L.; Clark, J.S.; Dietze, M.; Grossiord, C.; Hanbury-Brown, A.; Hurt, G.C.; et al. Pervasive shifts in forest dynamics in a changing world. *Science* **2020**, *368*, eaaz9463. [CrossRef]
57. Chen, X. Diverse scaling relationships of tree height and diameter in five tree species. *Plant. Ecol. Divers.* **2018**, *11*, 147–155. [CrossRef]
58. Chen, X.; Chen, H. Analyzing patterns of seed production for Chinese Torreya. *HortScience* **2020**, *55*, 778–786. [CrossRef]

atmosphere

MDPI

Article

Ground Level Isoprenoid Exchanges Associated with *Pinus pinea* Trees in A Mediterranean Turf

Zhaobin Mu [1,2,*], Joan Llusià [1,2] and Josep Peñuelas [1,2]

1 Consejo Superior de Investigaciones Científicas, Global Ecology Unit CREAF-CSIC-UAB, E08193 Bellaterra, Catalonia, Barcelona, Spain; j.llusia@creaf.uab.cat (J.L.); josep.penuelas@uab.cat (J.P.)

2 Centre de Recerca Ecològica i Aplicacions Forestals, E08193 Cerdanyola, Catalonia, Spain

* Correspondence: zhaobin@creaf.uab.cat

check for updates

Received: 10 July 2020; Accepted: 28 July 2020; Published: 31 July 2020

Abstract: The emissions of isoprenoids, a kind of biogenic volatile organic compounds (BVOCs), from soils is not well characterized. We quantified the exchange of isoprenoids between soil with litter and atmosphere along a horizontal gradient from the trunks of the trees, in a Mediterranean *Pinus pinea* plantation with dry and green needle litter to open herbaceous turf during mornings at mid-summer. Further, potential associated drivers were identified. Isoprenoid emissions were greatest and most diverse, and also can be roughly estimated by litter dry weight near the trunk, where the needle litter was denser. The composition of emitted isoprenoid by needle litter was different than the composition previously described for green needles. Low exchange rates of isoprenoids were recorded in open turf. Isoprenoid exchange rates were correlated positively with soil temperature and negatively with soil moisture. Given the variations in ground emissions with soil, vegetation, microorganisms, and associated interactions, we recommend widespread extensive spatio-temporal analysis of ground level BVOC exchanges in the different ecosystem types.

Keywords: isoprenoid exchanges; ground; litter emissions; soil; *Pinus pinea*; distance gradient; Mediterranean turf

---

1. Introduction

Biogenic volatile organic compounds (BVOCs) represent 90% of total volatile organic compound (VOC) emitted into the atmosphere [1], impacting the atmospheric chemistry and climate processes [2–4]. BVOC emission profiles from terrestrial ecosystems tend to be driven by plant species composition [5] which is linked to phenology and climate [2,6] and are usually dominated by isoprenoids [5] with blends of other carbon-based compounds, such as alkanes, alkenes, carbonyls, alcohols, esters, ethers, and acids [5,6]. This key role of terrestrial plants in BVOCs has received much research attention [7–10]; however, there is emerging evidence that a wide range of BVOCs are also released from terrestrial ecosystem ground [10] regardless of level of vegetation [8,10,11].

Ground level emission of BVOCs from natural and semi-natural ecosystems may derive from organic litter and soil where plant root systems and microorganisms are major sources [11,12] and sometimes also by understory vegetation [13,14]. Most of the ground measurements do not distinguish the emissions from plant roots, decomposing litter, or the microbes themselves [15,16]. Litter has often been suggested as the main BVOC source in the forests besides vegetation [14,17–19], in fact, the decomposing litter has been assumed to be the main BVOCs source in the forest ground [8,14,20,21]. It is evident that both decomposers and the decomposing material affect the quantities and types of VOC productions [21], and also that VOCs released through the decomposition processes are strongly dependent on litter type, climate and soil microbial composition [21,22]. Differentiating each of the soil component responsible for these emissions is very complex [10,18,23]. For example, the assessment of the contribution of root emissions to the overall soil VOC fluxes is difficult because of their linkage with soil microbes owing to root exudates can boost microbial activity, which can either increase the production or consumption of VOCs [10]. Some soil microbes, particularly fungi, are capable of producing terpenoid compounds [24], but plant roots are likely to be the dominant source of these compounds [8,23,25]. Isoprenoids are commonly emitted from litters and soils [8,12,22], and are likely adsorbed on the living leaf surfaces which are covered by a lipophilic cuticle layer [14]. Soil and litter microbes can also modify VOC emissions by metabolizing plant-emitted VOCs [26], which may cause low isoprenoid fluxes measured from soil with dense understorey vegetation cover [15]. Some understorey vegetation (grasses, shrubs, mosses, lichens, and other vegetation) [13,14] can also contribute to the exchange of BVOCs by emitting them [7].

Ground level emissions to the atmosphere are often 1–2 orders of magnitude lower than those from aboveground vegetation [10]. Moreover, they may represent up to 50% of net canopy BVOC flux, depending on the type of ecosystem, litter and soil [19], environmental conditions [10,27] and season of the year [19], particularly in coniferous forests that produce large amounts of litter [4,10]. Nevertheless, some studies suggest that these emissions play an insignificant role because they constitute a very low fraction with respect to the total ecosystem emissions [12,19,20,28]. In addition, soil VOCs also have important ecological roles [8,10,29], affecting microbial process such as methane oxidation, nitrification, nitrogen mineralization, and aerobic respiration [8,23,25] and biological interactions as key compounds in communication among soil microorganism and plant roots [4] that release carbon-rich root exudates and thus feed associated populations of bacteria, fungi, arthropod and nematode within the rhizosphere [8].

Soils are considered to be sources and sinks of BVOCs [8,10,12,22] with very low exchanges in Mediterranean-type ecosystems [17,29,30]. Maybe microbial processes play most important roles in atmosphere-soil exchanges of BVOCs [7,12,19]. In this line, there are studies showing the lower VOC emission rates in the litter plus soil treatments indicating many litter VOCs appear to be metabolized in soil [8,12,22], meanwhile, litter VOCs represent an important carbon source to soil and elevate soil microbial activity [22]. In addition, the consumption of some specific VOCs in soils result from microbial activities [7,22] depending on the type of compound and soil [7]. Besides, abiotic processes like adsorption to soil particles [19,31], dissolution in soil water [19], and reactions with soil chemicals [32] are also the mechanism behind the soil uptake [7,19].

Deposition and emission of ground level BVOCs is strongly influenced by environmental conditions [19,29]. Soil temperature and moisture seems to be the most important factors since they control physiological processes both in plants and microorganisms [30]. Temperature affects VOC production [10,12] through the temperature dependence of enzyme production and activity in VOC synthesis [2,21], while soil water content can determine which microbial groups are most active [21,33], which means both the physiological activity and community composition of decomposer microorganisms can be affected by environmental conditions [4,34]. Soil temperature and moisture affect, moreover, soil BVOC physical processes, including dissolution in soil water [7], and physico-chemical processes [2,21], such as diffusion and volatility. The over-arching effects of climate warming on increasing soil temperature and decreasing soil moisture will contribute higher BVOC volatilization from soil into the atmosphere [31], and may influence composition of vegetation and distribution of the associated soil microorganisms, and cause further variations in BVOC exchange profiles [4,8,34].

Isoprenoids are produced by all conifers and are stored in the needles [35] where they readily volatize from needle storing tissues [22]. The distance to the conifer tree can be a qualitative and quantitative determinant of ground level BVOC exchange profiles [17]. Here, we aimed to quantify emissions and exchange of isoprenoids and its potential drivers along a horizontal gradient from dense *Pinus pinea* litter to open herbaceous turf to improve understanding of spatial-temporal differences in ground level BVOC exchange to the atmosphere.

2. Material and Methods

2.1. Study Site and Experimental Design

We selected four isolated, similar sized (mean trunk circumference at breast height: 1.20 ± 0.06 m) *Pinus pinea* L. trees in a managed herbaceous turf on a silty-clay Typic Calcixerept soil, with a high proportion of carbonates (pH: 8) [17], near the campus of the Autonomous University of Barcelona (41°30′ N, 2°6′ E). Ground vegetation was dominated by legumes, such as *Trifolium repens* L., *Psoralea bituminosa* L., *Medicago minima* (L.) Bartal, with other herbs, such as *Plantago lanceolata* L. and grasses (*Lolium perenne* L., *Brachypodium phoenicoides* R. and S., and *Bromus intermedius* Guss). Sampling points (N = 11) were arranged every meter along a single 10 m transect from the trunk of each tree, and avoided canopy effects of other trees, where point 1 was as close to the trunk as possible. Leaf litter was present within 4 m, and most dense within 2 m from the trunk where litter covered ground totally.

2.2. Isoprenoid Sampling

Sampling was carried out during the summer when climate warming effects are most pronounced and BVOCs emissions are greatest at this region [9,29], from 18 July to 8 August of 2018, on sunny or slightly cloudy days, between 09:00 and 13:00 hrs. Emitted isoprenoids were collected with a Teflon® soil VOC chamber and retained in stainless steel tubes (89 mm in length with 6.4 mm external diameter, Markes International Inc. Wilmington, NC, USA) manually filled with adsorbents (115 mg of Tenax TA and 230 mg of SulfiCarb, Markes International Inc. Wilmington, NC, USA) separated by sorbent-retaining springs that were fixed using gauze-retaining springs and closed with air-tight caps. Flow was generated using a Q-MAX air-sampling pump (Supelco, Bellefonte, PA, USA) and measured using a Bios Defender 510 flow meter (Bios International Corporation, Butler, PA, USA) and sampling time was 20 min. This dynamic system was also connected to ambient air with a Teflon® tube of 3 mm of inner diameter and air inside chamber was homogenized using a small fan. The flow rate across the sampling cartridges was adjusted at around 200 mL min^{-1} [17]. Although the studied emission could be slightly influenced by an addition from ambient air under this situation, it can be counteracted by the subtraction of blank measurements whose emissions were collected prior to the measurement of each sample using Tedlar® PVF film between ground and the chamber (Figure S1). Each point cost around 1 h including twice measurements (blank and sample), the time for operation and movement to next point, and the order of sampling at points along the transect was randomized for every tree and varied with sampling period (Table S1). Soil temperature and moisture content around the soil chamber were measured using a Pt100 4.5 × 150 mm probe (Jules Richard Instruments-ICT, SL, Fesches-le-Chatel, France) and a ML3 ThetaProbe sensor connected to a ML3 ThetaKit (Delta-T Devices, Cambridge, UK), respectively. The litter below the soil chamber was collected after sampling at 0, 1, 2 and 3 m from tree trunks and oven-dried at 60 °C to a constant weight. The sampled cartridges were stored at 4 °C until analysis.

2.3. GC-MS Analyses of BVOCs

BVOCs were analyzed using a GC-MS system (7890A GC-system interfaced with a 5975C VL MSD and a Triple-Axis detector; Agilent Technologies, Palo Alto, CA, USA). An automated thermal desorption unit (Ultra 2 and Unity 2; Markes International Ltd, Llantrisant, UK) was used for desorption of sampled cartridges. Desorbed BVOCs were cryofocused at −25 °C for 2 min, then, the cryotrap was rapidly heated to 320 °C and conducted into a 30 m × 0.25 mm × 0.25 μm film capillary column (HP-5, Crosslinked 5% pH Me Silicone; Supelco, Bellefonte, PA, USA). Carrier gas was helium and column flow was 1 mL min^{-1}. Total run time was 30 min, where initial oven temperature was held at 35 °C for 5 min, then programmed to increase by 15 °C min^{-1} to 150 °C for 5 min, then by 15 °C min^{-1} to 250 °C for another 3 min, and finally by 30 °C min^{-1} to 280 °C for 2 min [17]. Terpenes were identified by comparing retention times with those of standards from Fluka (Buchs, Switzerland) and published spectra from the Wiley275 and NIST05a mass-spectral library using GCD ChemStation G1074A HP.

Isoprenoid concentrations were determined by reference to trapped standards of α-pinene, 3-carene, β-pinene, limonene and sabinene every five analyses, and their calibration curves were made using three terpene concentrations (relationship between signal and terpene concentrations: $r^2 > 0.99$) [9,12]. The most abundant isoprenoids, such as α-pinene, 3-carene, β-pinene, limonene and sabinene, had similar sensitivities, with <5% differences among calibration factors.

The exchange rates were expressed as differences between the emission rates of the sample and the corresponding blank in μg m^{-2} h^{-1}. When the values are positive, they indicate BVOC emission from ground to atmosphere, and the exchange rates are then referred to be "emission rates". When the values are negative, they indicate BVOC adsorption to ground, and the exchange rates are then referred to be "adsorption rates".

2.4. Statistical Analyses

Differences in terpene exchange along the horizontal gradient were tested using one-way analysis of variance at $P < 0.05$ in Statistica v.8.0 (StatSoft, Inc., Tulsa, OK, USA) and covariance in terpene exchanges with soil environmental conditions was analyzed using partial least squares (PLS) regression using the plsdepot package in R v. 3.3.3. A comparison of emission profile was made between the ground covering dense litter and green needle of *P. pinea* according to the data from this study and Staudt et al (2000) [36].

3. Results

3.1. Soil Environmental Conditions

Mean soil temperatures along the transects of the four trees were 28.5 ± 0.80°C, 28.8 ± 1.16 °C, 33.1 ± 1.01 °C and 34.3 ± 1.46 °C, and soil moisture (*v*/*v*) was 12.2 ± 1.06%, 16.7 ± 0.91%, 4.7 ± 0.70% and 2.2 ± 0.39%. There was some variation in soil environmental conditions among the trees due to precipitation, but within-transect variation was lower owing to the randomization of sampling. Mean soil temperature ranged between 28.4 ± 1.00 °C (at 1 m) and 33.2 ± 1.88 °C (at 8 m), and mean soil moisture (*v*/*v*) ranged between 7.1 ± 2.23% (at 8 m) and 10.5 ± 3.09% (at 6 m) (Figure S2). The BVOC exchanges at same distance can be considered paralleled in terms of similar environmental conditions, which makes the average value can represent exchange at the distance; the conditional environment for all samplings was concentrated in certain scope which is also optimum to analyse its relationship with exchange.

3.2. Terpene Exchange

There were no detectable isoprene emissions, while the terpene emissions varied greatly in terms of amount and composition along the transects. Terpene emissions varied most significantly along the horizontal gradient under litter, where they were greater at 1 m (371.4 ± 71.1 μg m^{-2} h^{-1}) than at 0 m (135.2 ± 22.9 μg m^{-2} h^{-1}; $P < 0.05$) and 2 m (13.1 ± 1.8 μg m^{-2} h^{-1}; $P < 0.01$), and greater at 0 m than at 2 m ($P < 0.01$) (Figure 1); emissions were 7.3 ± 2.0 μg m^{-2} h^{-1} at 3 m. Litter was absent from 4 m along the transect and terpene exchange became irregular, where terpenes were emitted or adsorbed at low rates (average < 2 μg m^{-2}·h^{-1}); greatest adsorption rates were recorded at 4m (1.6 μg m^{-2} h^{-1}) (Figure 1).

Nine monoterpenes and two sesquiterpenes were detected from the transects, all of which were detected for emission at only 0 and 1 m and, with the exception of camphene, tended to be emitted in greater quantities at 1 m than at 0 m (Figure 2). Limonene, β- and γ-terpinene with sabinene, and α-pinene together account for around 90% of total emissions at dense litter zone, while the spectrum of emission of green needles from *P. pinea* is dominated by trans-β-ocimene, followed by limonene, linalool, and 1,8-cineole, that together, accounted for around 90% (Table S2; Figure 3) of all emissions in summer mornings [36]; α-pinene and limonene account for higher proportion at dense litter zone than green needles, while myrcene showed similar proportion (Table S2; Figure 3).

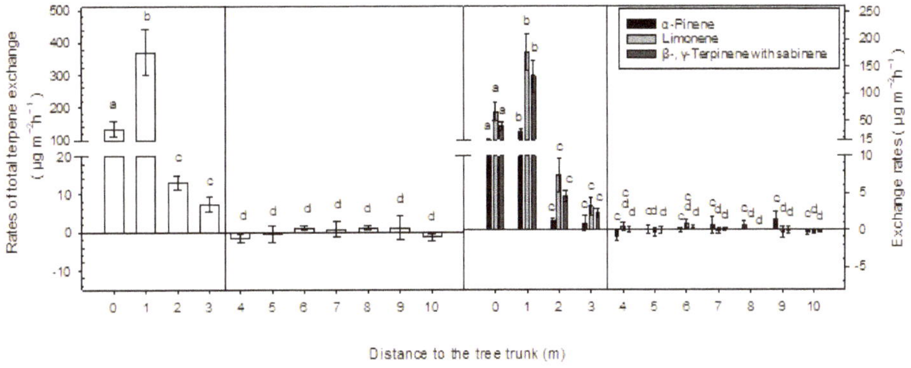

Figure 1. Rates of total terpene, α-pinene, limonene, β- and γ-terpinene with sabinene exchange along the transects. Data are means ± SE; n = 4. Different letters indicate differences among distances ($P < 0.05$).

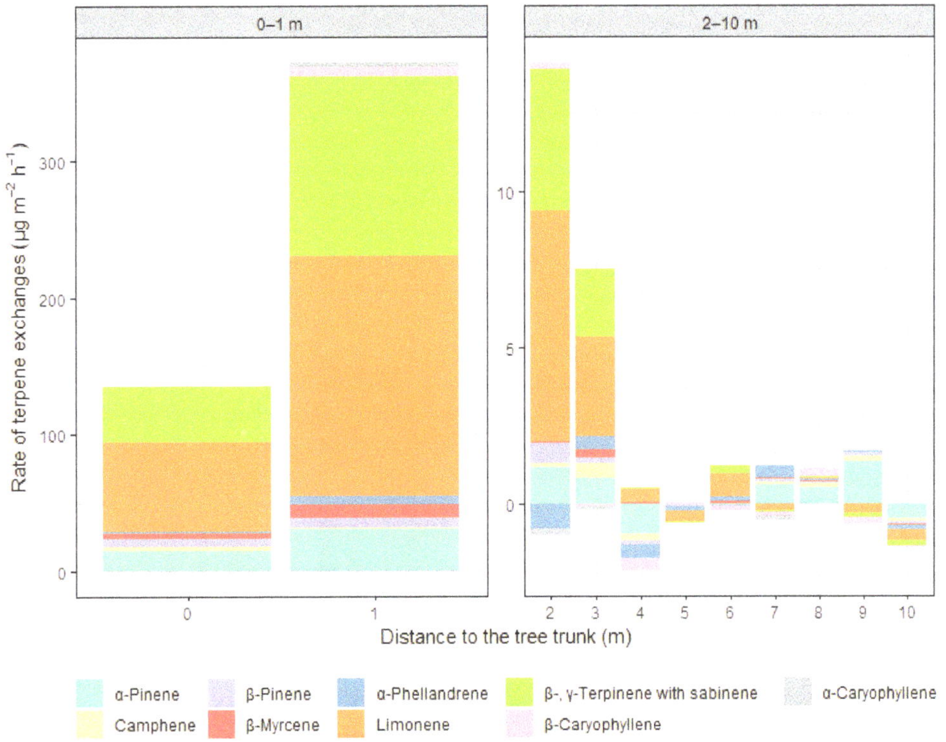

Figure 2. Distribution of terpene exchange along the transects.

Limonene was the most dominant compound, followed by β- and γ-terpinene with sabinene, and α-pinene (Figure 1), and their emissions along the transect were similar to those of total terpenes in the litter zone (greater concentrations at 1 m than 0 and 2 m, and greater at 0 m than at 2 m; $P < 0.05$) (Figure 1), and there were relatively high emissions of other terpenes at 0 and 1 m (<10 μg m^{-2} h^{-1}), but lower emissions (<1 μg m^{-2} h^{-1}) at the other distances (Figure 2). While two sesquiterpenes β-caryophyllene and α-caryophyllene were emitted at around 7.5 and 2.5 μg m^{-2} h^{-1} at 1 m, they were emitted at <0.5 μg m^{-2} h^{-1} at other distances, and were barely detected after 4 m (Figure 2). Almost all terpenes were adsorbed at 4 and 10 m, particularly α-pinene (Figure 2).

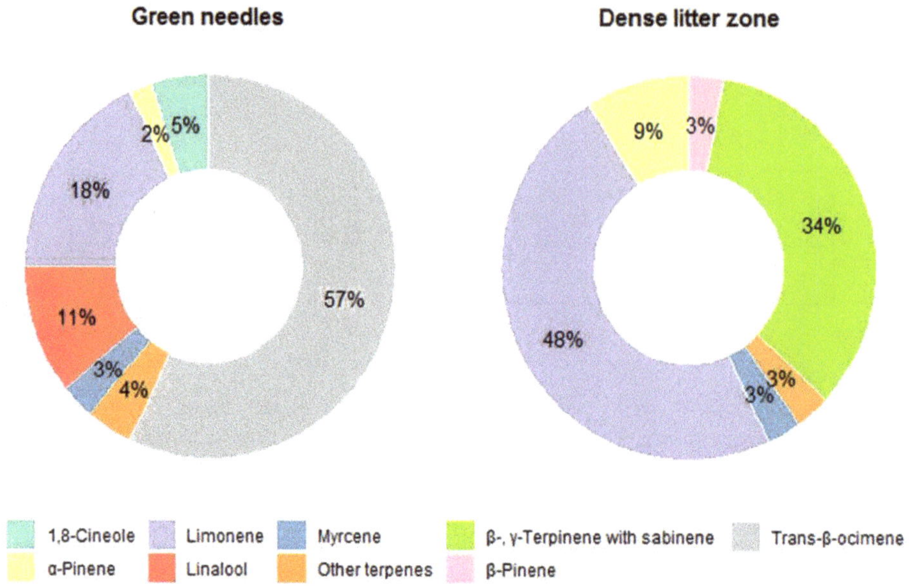

Figure 3. The composition of terpene emissions for green needles [36] and dense litter zone of our study (≤1 m) in summer mornings.

The litter dry weight was largest at 1 m (4.768 g), followed by 0 m (3.282 g), 2 m (2.128 g) and 3 m (0.492 g), and it showed strong exponential relationship with terpene emission (Figure 4a). The detected compounds were correlated positively with soil temperature and negatively with soil moisture content and distance to the tree (Figure 4b).

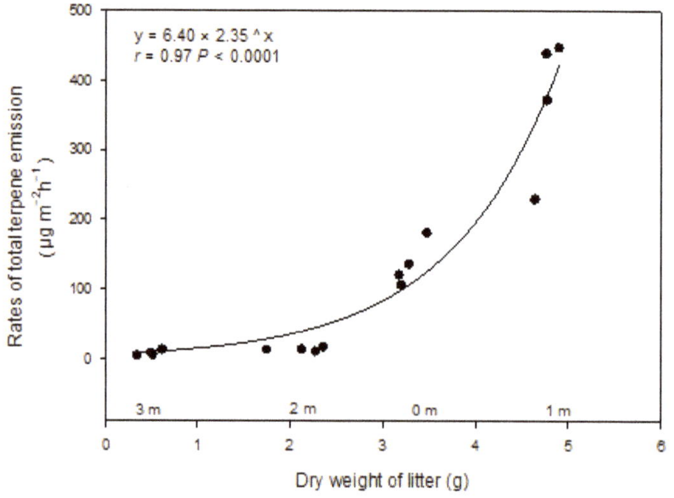

(a)

Figure 4. *Cont.*

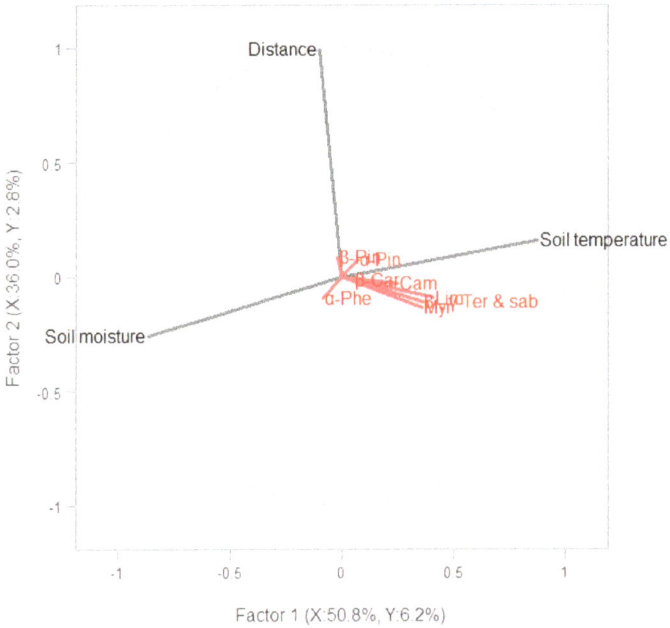

(**b**)

Figure 4. Relationship between total terpene emissions and litter dry weight within 4 m of the study trees (**a**). Partial least squares (PLS) regression between soil temperature or soil moisture content or distance to the trees and terpene exchange along the transects in turf zone (≥4 m) (**b**). Black represents soil temperature, soil moisture content and distance to the trees as independent variables (X), red represents exchange rates of individual terpenes as dependent variables (Y). α-Pin, α-pinene; Cam, camphene; β-Pin, β-pinene; Myr, myrcene; α-Phe, α-phellandrene; Lim, limonene; β-, γ-ter and sab, β-, γ-terpinene with sabinene; β-Car, β-caryophyllene.

4. Discussion

Terpene exchanges varied with the distance to the trunk of *Pinus pinea*, a storage species for these compounds [35]. The ground showed totally different emission pattern due to litter coverage or not, with different magnitude, and it is possible to divide the exchange profiles into three groups: short distances (<2 m) covering dense litter, medium distances covering moderate litter (2–4 m), turf (≥4 m), where emission rates were greatest at <2 m and lowest at ≥4 m, reflecting the degree of canopy coverage of the ground. Previous studies also found evidence of a gradient from high levels of monoterpenes in the vicinity of the tree trunk to lower levels at the farthest distance [16,17,37]. These studies suggested that the large source of volatiles result predominantly from a large amount of litter or roots/rhizosphere activity in the soil near trunks [16,19].

The BVOC emissions at short and medium-distances were dominated by litter (Figure 4a), and also probably released by microbial metabolism and sparse ground vegetation, especially by roots owing to emissions decreasing with increasing distance from the tree trunk. However, the points at 0m with less litter, maybe owing to uneven ground near trunk, also showed obviously less emissions compare to that of the points at 1 m, where is supposed to be farther from and have less quantities of underground roots for the species of taproot system. The roots were reported to increase [38] or decrease [29] soil emissions, both with low fluxes indicating root-rhizosphere activity [10,30,39] would be a much smaller source compared to litter in this study. However, roots may represent a strong terpene source for *Pinus* spp. as well [17,37] and are a non-negligible source of VOCs for some species, like Arabidopsis [30,39]. Soil microbial activity has been shown to correlate with VOC emissions over a range of different forest soils [8,12,16], however, the soil moisture recorded were very low which has probably strongly hampered microbial metabolism [33]. Although activity of soil microorganisms and roots were both decreased in summer [29], the changes towards a decrease in the ratio of microorganisms/roots activities in the rhizosphere was found [30], which may suggest the emission from microbe activity may be a smaller source compared to roots.

The strongly positive relationship between emission rate and litter dry weight (Figure 4a) also indicated that the aboveground plant litter was likely the dominant terpene source [8,16,18] as also reported for aboveground litter of other species storing terpenes [28,34,40]. Turf is a "simpler" ecosystem compared to other ecosystems because it has less vegetation mass and associated interactions, than forests, grasslands or croplands [28,41]. The strong relationship (Figure 4a) may actually indicate that the aboveground plant litter was the most dominant terpene source, while other biotic sources like roots and microorganism, and abiotic factors like soil properties and environmental conditions may play less important roles in ground level isoprenoid emissions. This makes that the emissions can be estimated by litter dry weight in this type of ecosystem while it may be instead unrealistic for other ecosystems.

The quantity of terpenes is thus dominantly linked to the amount of needle litter at short and medium-distances which might mask the variation caused by environmental conditions, although effects of temperature on emissions elicit changes in transport resistance along the diffusion path from the litter [5] and temperature and humidity are always supposed to be main factors acting on terpene emission in Mediterranean summer daytime [29,42]. However, litter emissions associated with microbial decomposition of organic matter have been reported to be quantitatively more dominant than emissions caused only by abiotic factors as temperature and humidity [8,34,40] as suggested by the strong correlation between VOC production and microbial CO_2 production [8].

The emitted compounds found in this study follow a pattern similar to other studies of ground VOC emissions which consist of very few abundant compounds associated with several less representative ones (Figure 3) [11,22,29]. The spectrums of emission vary significantly between dense litter zone and green needles from *P. pinea* (Figure 3), but both contain high amounts of limonene [36]. Emission rates of trans-β-ocimene, linalool and 1,8-cineole are light-dependent, and carbon dioxide exerts a particularly positive influence on the emission rates of trans-β-ocimene [35]. Trans-β-ocimene is directly synthesized in chloroplasts and follow a different metabolic path to other monoterpenes which are stored in resin ducts [35], this can be proved by its presence only in the sampling of green needle. However, this variation was also found in another typical Mediterranean pine species *Pinus halepensis* whose litter showed remarkably high sesquiterpenes (β-caryophyllene, followed by α-caryophyllene) emissions [17] which represent less than 5% of the total emissions of green needles [43,44]. Although the relative composition of terpenes in needle litter is related to that of green needles, terpene concentrations may change with time during decomposition processes [17]. The increased proportion of limonene and α-pinene and similar proportion of myrcene emission in dense litter zone compared with green needles may showed soil microbes readily consume a diverse array of BVOCs with different ability of utilization which also varied from distinct microbial communities [8,24,34], representing an important sink of BVOCs in terrestrial ecosystems dominated by plants that store terpenes [8]. On the other hand, the high proportion of β- and γ-terpinene with sabinene emissions in dense litter zone could indicate soil microbes producing terpenes that are not emitted by plants [8,17].

We found that terpene exchanges were very low (Figure 1) which was in agreement with the previous studies [17,20,29,30] and not correlated with the distance from the trunk overall (Figure 4b) at the greatest distance from the trees in the herbaceous turf, where there was a lack of needle litter and too far to be influenced by roots as well. Despite potential terpene content in grasses [17,45], there was a negligible impact on exchange rates owing to the small biomass compared with the pine, supporting research that shows terpene emissions from ground in close proximity to trees derives from litter and plant roots [17]. Much less research has been directed towards the more intensively anthropogenic managed turf soils than forest, grassland or cropland systems that have been studied [28,40]. Our findings showed that turf soils produced negligible BVOC emissions, which were much lower than forest soil in Mediterranean summer [30]. Further, BVOC exchange profiles depend on soil type [10,28], and influenced by environmental conditions [10,27]. However, most of the measured fluxes from forest soil probably originated from understory vegetation [21,34]. Previous research suggests that biotic factors affecting the emissions of VOCs from soil are 5–10 times stronger than the abiotic ones [8,27], and soil environmental conditions affect both sources by altering volatility of VOCs and the activity and community composition of microorganisms [34]. In this study, α-caryophyllene was lacking at the greatest distances from the trunks where turf dominated, and emissions of β-pinene were not related to soil temperature or moisture content but positively related to the distance (Figure 4b). However, the emission of the rest of emitted compounds including the most abundant compound, limonene, along with β- and γ-terpinene, sabinene, and α-pinene were correlated positively with soil temperature and negatively with soil moisture content except for α-phellandrene which showed an opposite trend (Figure 4b). Soils in this study emitted a variety of terpenes that varied as a function of soil temperature and moisture [4,32], and slight trend can be found for total terpene exchange (Figure S3). The positive correlation with temperature and negative correlation with moisture of BVOC emissions are also in agreement with previous study in Mediterranean holm oak forest soil [29] and high arctic soil [6]. The diversity of compounds found in this study, although not very high, gives an idea, of the various factors for VOC emissions that can be taken into account, such as temperature, moisture and their interaction. In addition, the type of soil and low vegetation also influences. Mediterranean soil behaves more as a sink than as a source of BVOCs since total soil BVOC adsorption overcame emission over the year [29,30]. However, our results show that soil VOC exchange with the atmosphere might greatly change in response to climate change, with likely increased emissions under the warmer and drier summers expected for the coming decades in the Mediterranean region [41].

5. Conclusions

To conclude, the presence of aboveground litter was the dominant source of ground level terpene emissions in the proximity of *Pinus pinea* trees and the emission amounts can be estimated by litter dry weight while the components vary significantly from green needle. In addition, soils act as a source or sink of terpenes in managed Mediterranean turf environments with negligible terpene exchanges and the exchange rates of total terpenes were correlated positively with soil temperature and negatively with soil moisture. The soil terpene emissions are expected to increase by climate change in the Mediterranean region. Given the variations in ground emissions with soil, vegetation, microorganisms, and associated interactions, we recommend further spatio-temporal analysis of ground level BVOC exchanges in a wider range of ecosystem types.

Supplementary Materials: The following are available online at http://www.mdpi.com/2073-4433/11/8/809/s1: Figure S1. Schematic of the isoprenoid sampling; Figure S2. Mean soil temperature and moisture content along the transects. Data are means ±SE; n = 4; Figure S3. Relationships for the rates of total terpene exchange with soil temperature and soil moisture along the transects in turf zone (≥4 m); Table S1. Randomized sampling plan for the four studied trees during the morning. T1-T4 are first to fourth study trees, respectively; Table S2. The component of terpene emissions of *P. pinea* for green needle (a) from the study of Staudt et al. (2000) and dense litter zone (≤1 m) (b) from this study in summer mornings.

Author Contributions: Conceptualization: J.L. and J.P.; Data curation: Z.M.; Formal analysis: Z.M.; Investigation: Z.M.; Methodology: J.L. and J.P.; Visualization: Z.M.; Writing–original draft: Z.M.; Writing–review & editing: J.L. and J.P. All authors have read and agreed to the published version of the manuscript.

Funding: This research was financially supported by the Spanish Government (project PID2019-110521GB-I00), the Catalan government (project SGR2017-1005), and the ERC Synergy project, SyG-2013-610028 IMBALANCE-P. We are also grateful for the financial support from the China Scholarship Council.

Conflicts of Interest: The authors declare no conflict of interest. The funders had no role in the design of the study; in the collection, analyses, or interpretation of data; in the writing of the manuscript, or in the decision to publish the results.

References

1. Atkinson, R. Atmospheric chemistry of VOCs and NO(x). *Atmos. Environ.* **2000**, *34*, 2063–2101. [CrossRef]
2. Peñuelas, J.; Staudt, M. BVOCs and global change. *Trends Plant Sci.* **2010**, *15*, 133–144. [CrossRef]
3. Shindell, D.T.; Faluvegi, G.; Koch, D.M.; Schmidt, G.A.; Unger, N.; Bauer, S.E. Improved attribution of climate forcing to emissions. *Science* **2009**, *326*, 716–718. [CrossRef]
4. Svendsen, S.H.; Priemé, A.; Voriskova, J.; Kramshøj, M.; Schostag, M.; Jacobsen, C.S.; Rinnan, R. Emissions of biogenic volatile organic compounds from arctic shrub litter are coupled with changes in the bacterial community composition. *Soil Biol. Biochem.* **2018**, *120*, 80–90. [CrossRef]
5. Kesselmeier, J.; Staudt, M. Biogenic Volatile Organic Compunds (VOC): An Overview on Emission, Physiology and Ecology. *J. Atmos. Chem.* **1999**, *33*, 23–88. [CrossRef]
6. Svendsen, S.H.; Lindwall, F.; Michelsen, A.; Rinnan, R. Biogenic volatile organic compound emissions along a high arctic soil moisture gradient. *Sci. Total. Environ.* **2016**, *573*, 131–138. [CrossRef]
7. Albers, C.N.; Kramshøj, M.; Rinnan, R. Rapid mineralization of biogenic volatile organic compounds in temperate and Arctic soils. *Biogeosciences* **2018**, *15*, 3591–3601. [CrossRef]
8. Leff, J.W.; Fierer, N. Volatile organic compound (VOC) emissions from soil and litter samples. *Soil Biol. Biochem.* **2008**, *40*, 1629–1636. [CrossRef]
9. Mu, Z.; Llusià, J.; Liu, D.; Ogaya, R.; Asensio, D.; Zhang, C.; Peñuelas, J. Seasonal and diurnal variations of plant isoprenoid emissions from two dominant species in Mediterranean shrubland and forest submitted to experimental drought. *Atmos. Environ.* **2018**, *191*, 105–115. [CrossRef]
10. Peñuelas, J.; Asensio, D.; Tholl, D.; Wenke, K.; Rosenkranz, M.; Piechulla, B.; Schnitzler, J.P. Biogenic volatile emissions from the soil. *Plant Cell Environ.* **2014**, *37*, 1866–1891. [CrossRef]
11. Potard, K.; Monard, C.; Le Garrec, J.L.; Caudal, J.P.; Le Bris, N.; Binet, F. Organic amendment practices as possible drivers of biogenic Volatile Organic Compounds emitted by soils in agrosystems. *Agric. Ecosyst. Environ.* **2017**, *250*, 25–36. [CrossRef]
12. Asensio, D.; Peñuelas, J.; Ogaya, R.; Llusià, J. Seasonal soil VOC exchange rates in a Mediterranean holm oak forest and their responses to drought conditions. *Atmos. Environ.* **2007**, *41*, 2456–2466. [CrossRef]
13. He, N.P.; Han, X.G.; Pan, Q.M. Variations in the volatile organic compound emission potential of plant functional groups in the temperate grassland vegetation of inner Mongolia, China. *J. Integr. Plant Biol.* **2005**, *47*, 13–19. [CrossRef]
14. Mäki, M.; Heinonsalo, J.; Hellén, H.; Bäck, J. Contribution of understorey vegetation and soil processes to boreal forest isoprenoid exchange. *Biogeosciences* **2017**, *14*, 1055–1073. [CrossRef]
15. Aaltonen, H.; Aalto, J.; Kolari, P.; Pihlatie, M.; Pumpanen, J.; Kulmala, M.; Nikinmaa, E.; Vesala, T.; Bäck, J. Continuous VOC flux measurements on boreal forest floor. *Plant Soil* **2013**, *369*, 241–256. [CrossRef]
16. Bäck, J.; Aaltonen, H.; Hellén, H.; Kajos, M.K.; Patokoski, J.; Taipale, R.; Pumpanen, J.; Heinonsalo, J. Variable emissions of microbial volatile organic compounds (MVOCs) from root-associated fungi isolated from Scots pine. *Atmos. Environ.* **2010**, *44*, 3651–3659. [CrossRef]
17. Asensio, D.; Owen, S.M.; Llusià, J.; Peñuelas, J. The distribution of volatile isoprenoids in the soil horizons around Pinus halepensis trees. *Soil Biol. Biochem.* **2008**, *40*, 2937–2947. [CrossRef]
18. Hellén, H.; Hakola, H.; Pystynen, K.H.; Rinne, J.; Haapanala, S. C2-C10 hydrocarbon emissions from a boreal wetland and forest floor. *Biogeosciences* **2006**, *3*, 167–174. [CrossRef]
19. Tang, J.; Schurgers, G.; Rinnan, R. Process Understanding of Soil BVOC Fluxes in Natural Ecosystems: A Review. *Rev. Geophys.* **2019**, *57*, 966–986. [CrossRef]
20. Hayward, S.; Muncey, R.J.; James, A.E.; Halsall, C.J.; Hewitt, C.N. Monoterpene emissions from soil in a Sitka spruce forest. *Atmos. Environ.* **2001**, *35*, 4081–4087. [CrossRef]
21. Mäki, M.; Aaltonen, H.; Heinonsalo, J.; Hellén, H.; Pumpanen, J.; Bäck, J. Boreal forest soil is a significant and diverse source of volatile organic compounds. *Plant Soil* **2019**, *441*, 89–110. [CrossRef]
22. Ramirez, K.S.; Lauber, C.L.; Fierer, N. Microbial consumption and production of volatile organic compounds at the soil-litter interface. *Biogeochemistry* **2010**, *99*, 97–107. [CrossRef]
23. Smolander, A.; Ketola, R.A.; Kotiaho, T.; Kanerva, S.; Suominen, K.; Kitunen, V. Volatile monoterpenes in soil atmosphere under birch and conifers: Effects on soil N transformations. *Soil Biol. Biochem.* **2006**, *38*, 3436–3442. [CrossRef]

24. Stahl, P.D.; Parkin, T.B. Microbial production of volatile organic compounds in soil microcosms. *Soil Sci. Soc. Am. J.* **1996**, *60*, 821–828. [CrossRef]
25. Paavolainen, L.; Kitunen, V.; Smolander, A. Inhibition of nitrification in forest soil by monoterpenes. *Plant Soil* **1998**, *205*, 147–154. [CrossRef]
26. Farré-Armengol, G.; Filella, I.; Llusia, J.; Peñuelas, J. Bidirectional interaction between phyllospheric microbiotas and plant volatile emissions. *Trends Plant Sci.* **2016**, *21*, 854–860. [CrossRef]
27. Rossabi, S.; Choudoir, M.; Helmig, D.; Hueber, J.; Fierer, N. Volatile Organic Compound Emissions From Soil Following Wetting Events. *J. Geophys. Res. Biogeosciences* **2018**, *123*, 1988–2001. [CrossRef]
28. Greenberg, J.P.; Asensio, D.; Turnipseed, A.; Guenther, A.B.; Karl, T.; Gochis, D. Contribution of leaf and needle litter to whole ecosystem BVOC fluxes. *Atmos. Environ.* **2012**, *59*, 302–311. [CrossRef]
29. Asensio, D.; Peñuelas, J.; Llusià, J.; Ogaya, R.; Filella, I. Interannual and interseasonal soil CO2 efflux and VOC exchange rates in a Mediterranean holm oak forest in response to experimental drought. *Soil Biol. Biochem.* **2007**, *39*, 2471–2484. [CrossRef]
30. Asensio, D.; Peñuelas, J.; Filella, I.; Llusià, J. On-line screening of soil VOCs exchange responses to moisture, temperature and root presence. *Plant Soil* **2007**, *291*, 249–261. [CrossRef]
31. Van Roon, A.; Parsons, J.R.; Te Kloeze, A.M.; Govers, H.A.J. Fate and transport of monoterpenes through soils. Part I. Prediction of temperature dependent soil fate model input-parameters. *Chemosphere* **2005**, *61*, 599–609.
32. Insam, H.; Seewald, M.S.A. Volatile organic compounds (VOCs) in soils. *Biol. Fertil. Soils* **2010**, *46*, 199–213. [CrossRef]
33. Veres, P.R.; Behrendt, T.; Klapthor, A.; Meixner, F.X.; Williams, J. Volatile Organic Compound emissions from soil: Using Proton-Transfer-Reaction Time-of-Flight Mass Spectrometry (PTR-TOF-MS) for the real time observation of microbial processes. *Biogeosci. Discuss.* **2014**, *11*, 12009–12038. [CrossRef]
34. Gray, C.M.; Monson, R.K.; Fierer, N. Emissions of volatile organic compounds during the decomposition of plant litter. *J. Geophys. Res. Biogeosci.* **2010**, *115*, 115. [CrossRef]
35. Loreto, F.; Nascetti, P.; Graverini, A.; Mannozzi, M. Emission and content of monoterpenes in intact and wounded needles of the Mediterranean Pine, Pinus pinea. *Funct. Ecol.* **2000**, *14*, 589–595. [CrossRef]
36. Staudt, M.; Bertin, N.; Frenzel, B.; Seufert, G. Seasonal variation in amount and composition of monoterpenes emitted by young Pinus pinea trees—Implications for emission modeling. *J. Atmos. Chem.* **2000**, *35*, 77–99. [CrossRef]
37. Lin, C.; Owen, S.M.; Peñuelas, J. Volatile organic compounds in the roots and rhizosphere of *Pinus* spp. *Soil Biol. Biochem.* **2007**, *39*, 951–960. [CrossRef]
38. Rinnan, R.; Gierth, D.; Bilde, M.; Rosenørn, T.; Michelsen, A. Off-season biogenic volatile organic compound emissions from heath mesocosms: Responses to vegetation cutting. *Front. Microbiol.* **2013**, *4*, 1–10. [CrossRef]
39. Steeghs, M.; Bais, H.P.; De Gouw, J.; Goldan, P.; Kuster, W.; Northway, M.; Fall, R.; Vivanco, J.M. Proton-transfer-reaction mass spectrometry as a new tool for real time analysis of root-secreted volatile organic compounds in Arabidopsis. *Plant Physiol.* **2004**, *135*, 47–58. [CrossRef]
40. Gray, C.M.; Fierer, N. Impacts of nitrogen fertilization on volatile organic compound emissions from decomposing plant litter. *Glob. Chang. Biol.* **2012**, *18*, 739–748. [CrossRef]
41. Karl, T.G.; Spirig, C.; Rinne, J.; Stroud, C.; Prevost, P.; Greenberg, J.; Fall, R.; Guenther, A. Virtual disjunct eddy covariance measurements of organic compound fluxes from a subalpine forest using proton transfer reaction mass spectrometry. *Atmos. Chem. Phys.* **2002**, *2*, 279–291. [CrossRef]
42. Mu, Z.; Llusià, J.; Liu, D.; Ogaya, R.; Asensio, D.; Zhang, C.; Peñuelas, J. Profile of foliar isoprenoid emissions from Mediterranean dominant shrub and tree species under experimental nitrogen deposition. *Atmos. Environ.* **2019**, *216*, 116951. [CrossRef]
43. Blanch, J.S.; Peñuelas, J.; Llusià, J. Sensitivity of terpene emissions to drought and fertilization in terpene-storing *Pinus halepensis* and non-storing *Quercus ilex*. *Physiol. Plant.* **2007**, *131*, 211–225. [CrossRef] [PubMed]
44. Simon, V.; Dumergues, L.; Solignac, G.; Torres, L. Biogenic emissions from *Pinus halepensis*: A typical species of the Mediterranean area. *Atmo. Res.* **2005**, *74*, 37–48. [CrossRef]
45. Tava, A.; Berardo, N.; Odoardi, M. Composition of essential oil of tall fescue. *Phytochemistry* **1991**, *30*, 1455–1458. [CrossRef]

atmosphere

MDPI

Article

Arable Weed Patterns According to Temperature and Latitude Gradient in Central and Southern Spain

María Luisa Gandía, Carlos Casanova, Francisco Javier Sánchez, José Luís Tenorio and María Inés Santín-Montanyá *

Environment and Agronomy Department, National Institute for Agricultural and food Research and Technology (INIA), 28040 Madrid, Spain; gandia.mluisa@inia.es (M.L.G.); casanova@inia.es (C.C.); fsanchez@inia.es (F.J.S.); tenorio@inia.es (J.L.T.)

* Correspondence: isantin@inia.es

Received: 1 July 2020; Accepted: 11 August 2020; Published: 13 August 2020

Abstract: (1) Background: In agro-ecosystems, the success of the crops has a strong connection to biodiversity in the landscape. In the face of climate change, it is important to understand the response to environmental variation of weed species by means of their distribution. In the last century, biodiversity has been impacted due to a variety of stresses related to climate change. Although the composition of vegetation tends to change at a slower rate than climate change, we hypothesize species present in weed communities are distributed in diverse patterns as a response to the climate. Therefore, the general aim of this paper is to investigate the effect of temperature, using latitude as an indicator, on the composition and distribution of weed communities in agro-ecosystems. (2) Methods: Weeds were monitored in georeferenced cereal fields which spanned south and central Spanish regions. The graphic representation according to latitude allowed us to identify groups of weeds and associate them to a temperature range. We classified weeds as generalist, regional, or local according to the range of distribution. (3) Results: The monitoring of species led to the classification of weeds as generalist, regional or local species according to latitude and associated temperature ranges. Three weed species that were present in all latitude/temperature regions, were classified as generalist (*Linaria micrantha* (Cav) Hoffmanns & Link, *Sonchus oleraceous* L., and *Sysimbrium irium* L.). The species were classified as regional or local when their presence was limited to restricted latitude/temperature ranges. One weed, *Stellaria media* (L.) Vill., was considered a local species and its distribution dynamics can be considered an indicator of temperature. (4) Conclusions: The novel methodology used in this study to assign weed distribution as an indicator of climatic conditions could be applied to evaluate climate gradients around the world.

Keywords: agro-ecosystems; biodiversity; climate change; weed communities

1. Introduction

Biodiversity plays an important role in maintaining the processes and functions of ecosystems, including farming [1,2]. Plants respond to environmental conditions based on their needs in the habitat and physiological tolerances, which influences the composition of weed communities, their structure and resilience. In the last century, biodiversity has been impacted due to a variety of stresses related to climate change which have affected how and where species live, reproduce, and interact with each other [3,4]. The composition of vegetation tends to change at a slower rate than climate change, but nevertheless species present in weed communities have been distributed in diverse patterns according to their response to the climate [5].

Because climate change has influenced the annual productivity of many crops [6,7], as well as the composition and distribution of weed communities in agro-ecosystems [8,9], we can say that climate change presents both a threat to biodiversity and a cost in terms of weed control, for agriculture. Weeds are one of the main limiting factors of crop production worldwide, with part of their success being due to their plasticity: weed flora can adapt to changes in the environment [10–13]. This plasticity may respond to climate change at the local and regional level [14–16], and can act as a filter to soften the negative effects of a rapid change of climatic conditions in a particular area. In the short term, plasticity in the weeds allows them to cope with changes of environmental conditions.

Most of the agronomic actions in an agro-ecosystem has been designed to reduce the overall weed density [17–19]. However, the sustainability of cropping systems should be assessed not only in terms of crop yield but also adequate levels of biodiversity within cropland [20]. Agricultural practices, which vary according to local climate, cause a disturbance to the agro-ecosystem, and change the dynamics of the weed community, creating ecological niches [2,5,21,22]. These niches formed by weed communities affect plant diversity in the field [23,24]. Extreme weather combined with agronomic practices can lead to empty niches which present an opportunity for the establishment of new weed species [25,26]. The way that these niches are filled depends on the level of biodiversity. In agro-ecosystems with a greater diversity of weeds, there are more species available to fill these niches, which prevents any one species becoming dominant [27–29]. Logically, in agro-ecosystems with less diversity, it is easier for one weed species to take over a niche created by agricultural activity and become dominant. Weed diversity provides ecosystems services for the upper trophic levels in cropland ecosystems [23,30]. These changes in distribution of the weed community (abundance and richness) are often difficult to predict so monitoring changes is recognized as crucial both in stable and unstable environments. In the face of this challenge, several authors have argued that monitoring arable weed species can be used as biodiversity indicators in agro-ecosystems [31,32].

Studies on global warming have predicted a change in general temperature and rainfall patterns [33,34], but there are high levels of uncertainty about the nature of local changes. In Spain, temperature trend analysis confirms that there has been a widespread rise in annual average temperature since the mid-1970s, with warming being more apparent in winter (1.9 °C), and this increase in temperature in winter has given rise to longer growing seasons. One of the least explored aspects of global warming is its possible impact on the geographic distribution of agricultural weed species [35,36]. Climate change, and the consequent longer growing seasons, has led to the appearance of weed species which are more common in warmer conditions [24,37,38]. In this context, we have seen that the weed distribution patterns found in agro-ecosystems vary according to latitude, as influenced by temperatures gradients.

We hypothesize that climatic conditions are linked to weed species distribution. Although, studies have been conducted on the effects of climatic conditions on weeds, to our knowledge, relatively few studies have focused on weed distribution on regional and local scales. The objectives of this study were: (1) to implement a novel methodology; (2) to analyze the distribution pattern of weed species with respect to climate gradients and identify weed species as potential indicator of climate changes; and (3) to provide data regarding expected weed distribution changes due to global warmer for agricultural managers, in order to maximize yield parameters by maintaining biodiversity.

2. Material and Methods

In this study, we identified the weed species in cereal agro-ecosystems found in UTM quadrants of 10 km^2 (grid zones 29 and 30+, from latitude bands S and T). These quadrants were established within a North-South latitude range between Madrid (central Spain) 40.51 °N and Seville (southern Spain) 37.24 °N. We categorized plant species by latitude in a total of 50 quadrats.

This geographical range covers three latitudinal communities. The first two communities, located within Madrid and Castile La Mancha regions, are under Central Iberian plateau conditions. The altitude of the plateau is responsible for the existence of a continental Mediterranean climate. The most significant characteristics in these areas are severe winters, hot summers, summer drought, irregular rainfall, strong thermal oscillations and remarkable aridity. These features have been the result of the interrelations between geographical factors such as latitude, the situation of the region within the Iberian Peninsula, the relief layout and altitude. The annual thermal amplitude (difference between the average temperature of the coldest month and the hottest month) is very high, normally between 18 and 20 °C due to continentality. In July, the average monthly temperature is above 24 °C in most of the regions. The southern-most latitudinal community in the study, Andalusia region (Seville), has also semiarid conditions, similar to the continental Mediterranean climate but with more temperate conditions due to the coastal proximity.

The weed monitoring was carried out on georeferenced cereal fields, in the established route, during April 2018. The "time window" for the monitoring was decided according to the crop maturation stage. The cereal booting stage was selected because it is one of the critical competition stages in cereals [39], and the best time to identified weed flora representative of agro-ecosystems. All of the fields monitored, from central Spain were between 600 and 800 m in altitude and southern Spain fields were at an altitude range from 6 to 400 m. All the fields monitored showed the standard conventional management for each region.

To display the weed distribution along the latitude gradient a novel ordination methodology was performed on weed species present in the time of monitoring. A binary code was assigned to the presence or absence of weed species. The data was presented according to latitude (North–South) and the Tmax and Tmin were ordered from lowest to highest. Then, we represented these weed arrangements graphically (see the figures): where the peaks on the chart indicate the presence of weed species in a certain latitude or temperature. Also, the Spearman rank order correlation (Supplementary Materials) was used to obtain the relation between the weed species ordered by latitude and temperatures at the time window for monitoring, in the month of April.

In terms of temperatures, average maximum and minimum temperatures in April from the WorldClime database were reviewed, and data were adjusted to sea level. This method allowed us to classify and catalogue weed species according to environmental conditions, specifically different temperatures. This data analysis allowed for the identification of weeds that are able to tolerate high temperatures, these species, known as thermophilic, were found in latitude ranges that correspond to higher average temperatures. Other species were found not to thrive in the same conditions and weeds that are particularly sensitive to low temperatures and therefore have a reduced range of distribution were identified.

3. Results

A total of sixty-six weed species were observed in agro-ecosystems along our established route. The results of the process of categorizing weed species provided valuable information about distribution of weeds according to latitude and temperature. For the timing of weed monitoring, April 2018, the correlation between latitude and Tmax and Tmin were 0.749 and 0.790, respectively (in Supplementary Materials).

Figure 1 represents the findings of the sampling surveys. The peaks in the graph show presence of the weed species in the corresponding area. For this graphic representation we classified the weeds according to latitude. We can see three latitude ranges of 3 degrees (from 40.5° to 37.2°) which coincided with three regions of Spain called latitudinal communities: Madrid, Castile La Mancha, and Andalusia. We classified the weeds as local, regional and generalist, according to their range of dispersion:

(1) Local weed species were defined as restricted to a single latitudinal community (either Madrid, Castile La Mancha, or Andalusia), less than 33% of the geographical range.

(2) Species were classified as regional (two latitudinal communities) if they had a higher dispersion than local, between 33 and 66% of the geographical range.

(3) Generalists weeds were found in all areas of study, over three latitudinal communities, over 66% of the geographical range.

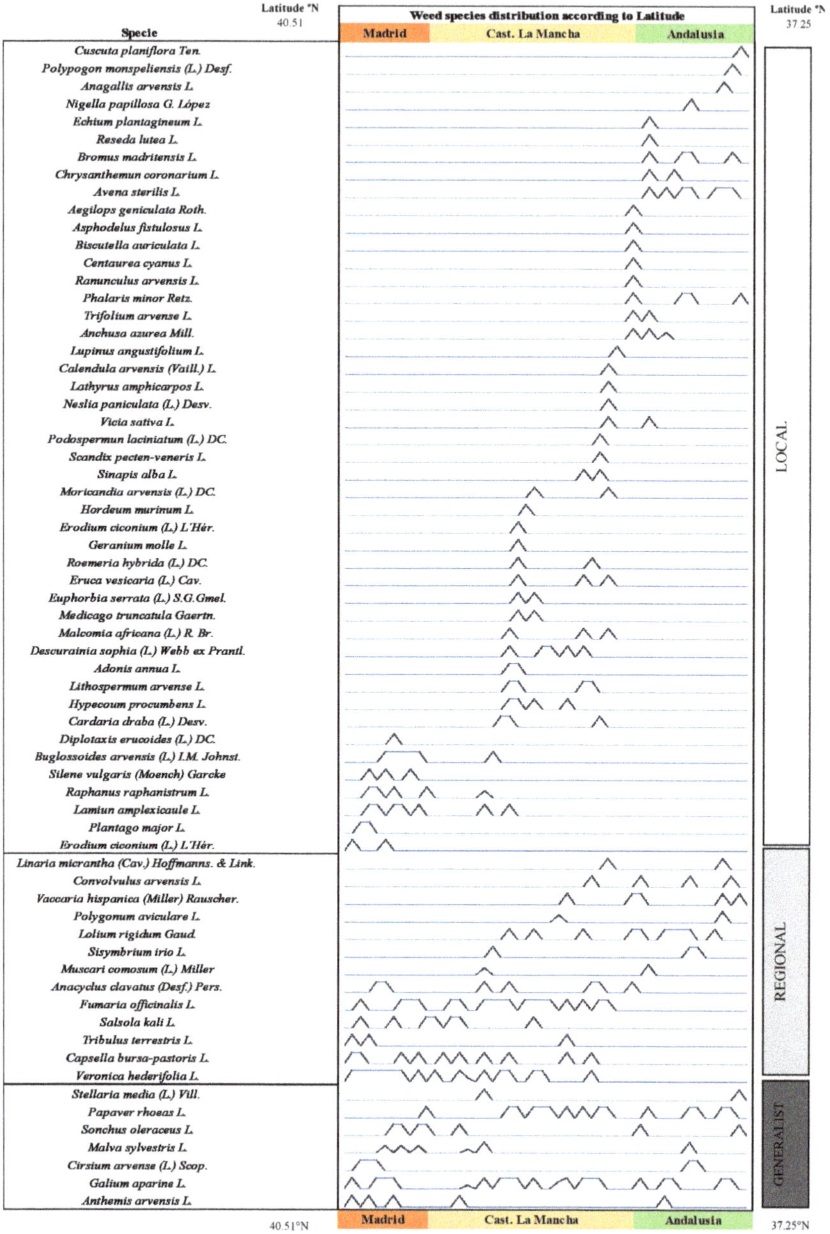

Figure 1. Graphical pattern of arable weed species distribution found in the studied areas according to latitude from Central to South of Spain (the peaks show presence of the weed species in the corresponding area).

Weed species were also divided into four groups by their thermal amplitude (species that appeared in up to 18, 36, 54 and 100% of Tmax and Tmin ranges), Figures 2 and 3 illustrate this. Figure 2 categorizes weed species according Tmax in April. The Tmax range was 4.7 °C (from 20.9 °C to 25.7 °C), and Figure 3 classifies weed species observed within a Tmin range of 4.1 °C (from 10.2 °C to 14.3 °C). The weed species were divided into four groups according to temperature: (1) weed species dispersed within a narrow temperature range, less than 18% of the total range; (2) weed species which were found within the 18–36% of temperature range; (3) weed species with more thermal amplitude, between 36 and 54%, and finally, (4) weed species present in all Tmax ranges. The classification of local, regional and generalist species had different results according to each of the three figures. We considered a species to be an indicator of temperature if its classification was the same in all three figures.

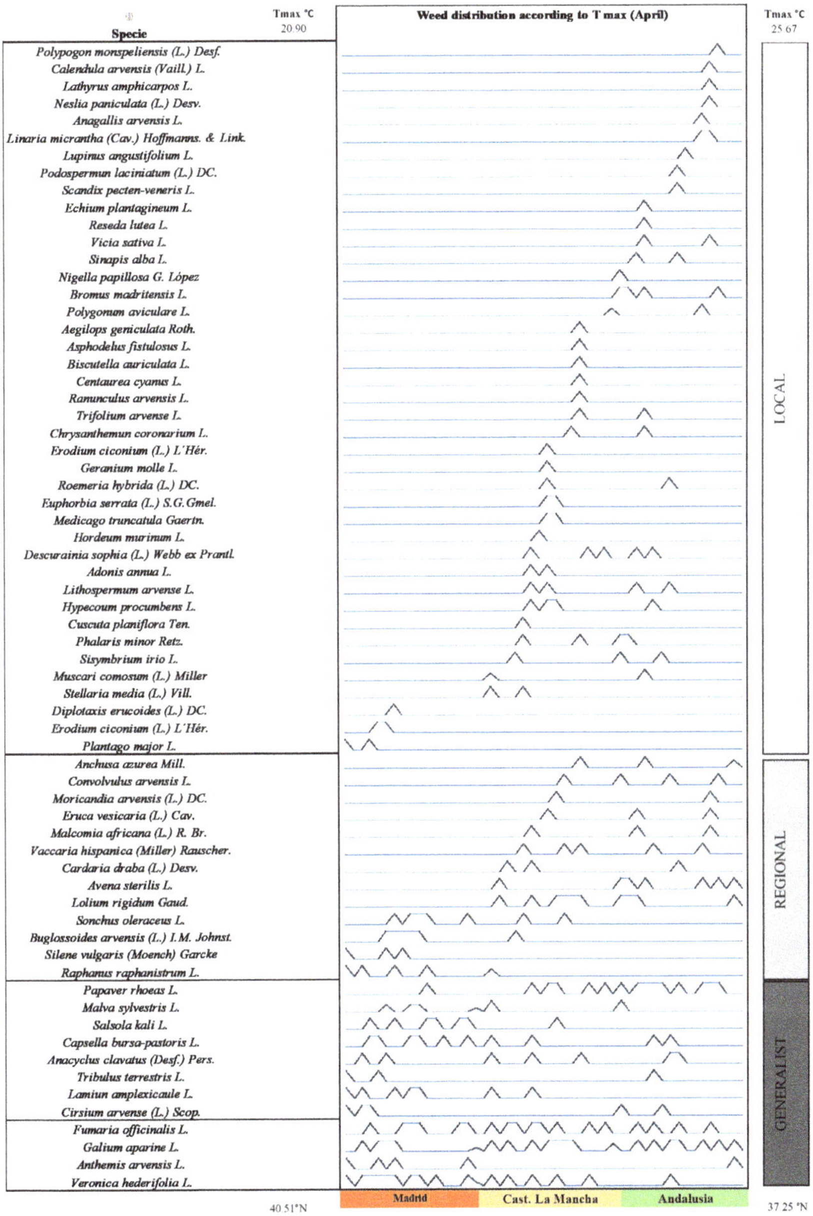

Figure 2. Graphical pattern of arable weeds distribution according to the Tmax in April.

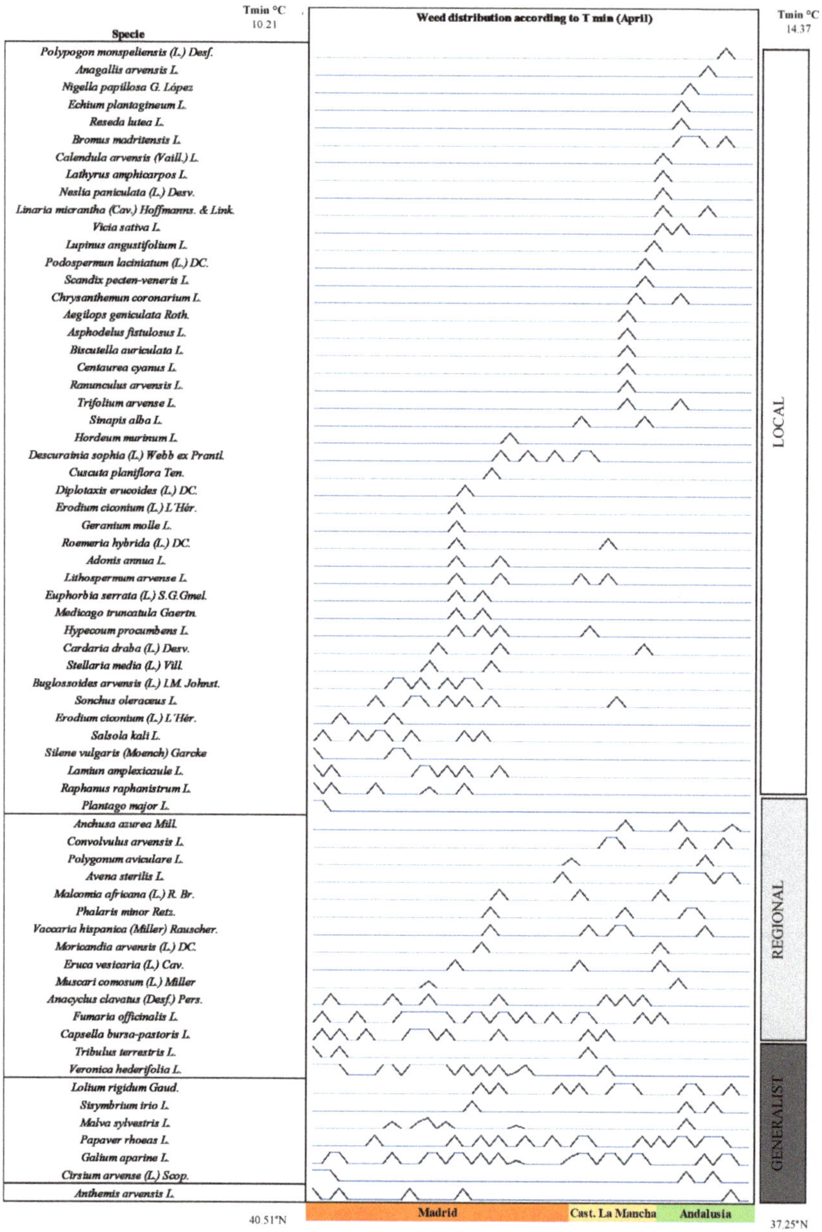

Figure 3. Graphical pattern of arable weeds distribution according to the Tmin in April.

According to latitude, *Anchusa azurea* Mill. and *Avena sterilis* L. were local species in Andalusia. *Moricandia arvensis* (L.) DC., *Eruca vesicaria* (L.) Cav. and *Malcomia africana* (L.) R.Br. were local species in Castile La Mancha. We consider these species to be potential indicators because their latitude ranges were narrow (≤33% of range) and also their dispersion of Tmax and Tmin range were between 18 and 36% (Figures 2 and 3, respectively). Other local species found were *Cardaria draba* (L.) Desv., *Blugossoides arvensis* (L.) I.M. Johnst., *Silene vulgaris* (Moench) Garcke, *Raphanus raphanistrum* L., and *Lamium amplexicaule* (L.). All of them were recorded in the latitude community of Madrid, 40° of latitude-north, with distribution within a geographical range corresponding to 33%, and 18% of dispersion in Tmin range. Local species were distributed in narrow ranges in terms of latitude and minimum temperature, however, a rise in average Tmin could favor their spread into more northern geographical ranges.

Linaria micrantha (Cav.) Hoffmanns. and Link. appeared as regional species, between 33 and 66% of dispersion within latitude range, and the variation in Tmax and Tmin ranges were narrower, ≤18%. In climate gradients where average temperatures are too high or too low, this weed species was not present. Then, global warming could completely change the geographical distribution of this weed species. *Sysimbrium irio* L. was observed as regional species, with similar variation in latitude and Tmin, but its Tmax range was very low (≤18%). When the maximum temperature is too high this weed species was not present. The opposite occurred with *Sonchus oleraceous* L., similar latitude and Tmax ranges variations, and Tmin range of ≤18%. When the average temperature was too low, this weed species was not present.

Stellaria media (L.) Vill. appeared in the highest latitude range (66–99%), but the narrowest Tmax and Tmin ranges (≤18%). We can state that this generalist weed is sensitive to temperature.

4. Discussion

Some weed species can share response patterns to particular environmental circumstances and hence affect their geographical distribution. Global warming, which implies longer growing seasons, may favor the appearance of weed species in regions which used to be colder [24,38]. Weeds sensitive to temperatures such as thermophilic weeds, or species with late emergence and opportunistic species can thrive now in some farming systems due to a rise in temperatures [5,8,9]. Numbers of these weed species have been increasing in northern areas because they are able to adapt to warmer conditions, and the growth cycle of these weeds has changed, accelerating the flowering stage [36–38,40–42].

On the other hand, climate change indirectly influences weeds that adapt to different agronomic practices. Frequently, we find that weeds are closely associated with the cropping system, and climate conditions have, therefore, an influence on the occurrence of weeds through crop management and land use [18,19]. From the above, we can deduce that the weeds that are present today in the agro-ecosystems are not necessarily the weeds that will be of concern in the future. However, further compilation of data regarding climatic conditions and the identification of weed flora in a determined area is necessary [26].

In this paper, we propose the implementation of a monitoring program for weeds in cereal systems, at a regional or local scale. It is important to facilitate the detection of changes within agro-ecosystems and allow farm managers the chance to predict the effects of climate change and reduce the impact on crop parameters. However, there is little information of the distribution of weed species [24,43], within the cereal agro-ecosystems, according to variables as latitude and temperature changes. Our study about weed distribution offers a resource to observe and compare weed species distribution in semiarid agro-ecosystems at the local level.

We found *Anchusa azurea* Mill., *Avena sterilis* L., *Moricandia arvensis* (L.) DC., *Eruca vesicaria* (L.) Cav., and *Malcomia africana* (L.) R.Br. as local species. We could consider these species as marker of warm conditions. If average minimum and maximum temperature increase, can be facilitated a thermophilic movement of these species and we would find these weeds relocated in other latitudes. *Avena* spp. has been categorized as aggressive grass in northern countries [12]. *Avena sterilis* L. is a widely spread weed by the Iberian Peninsula and has been principally controlled by herbicides. Its presence in the monitoring reflects its ability to adapt to warmer conditions. Therefore, the dispersal mechanisms of these species will have a great influence on their distribution and the plant traits of these local species should be object of future consideration.

Cardaria draba (L.) Desv., *Blugossoides arvensis* (L.) I.M. Johnst., *Silene vulgaris* (Moench) Garcke, *Raphanus raphanistrum* L., and *Lamium amplexicaule* L., were local species disperse in Tmin range narrowest. So, we think that Tmin can sorted the weed species within the observed latitude. Some authors [44,45] have found as local species *Silene noctiflora* due to propagule transport mainly relies on biological dispersal mechanisms and the habitat fragmentation prevent the dispersal of species.

Linaria micrantha (Cav.) Hoffmanns. and Link., *Sysimbrium irio* L. and *Sonchus oleraceous* L. were species more spread than others mentioned in this study. However, too low or too high temperatures may challenge their survival within their latitude, and so we could consider these three weeds (*Linaria micrantha, Sysimbrium irio,* and *Sonchus oleraceous*) as opportunistic because their presence is a function of optimal temperatures. Several *Sysimbrium* spp. species has been found in oilseed rape fields in Germany [38]. *Stellaria media* (L.) Vill. would also be considered a generalist species, widespread in all latitudes, but sensitive to changes in temperatures. *Stellaria media* (L.) Vill., as a nitrophilous species with shading tolerance has become relevant species in northern areas [12,46].

Temperature change reveals both the threat to biodiversity and the cost of weeds for agriculture. Because temperature has influenced the annual productivity of many crops, as well as the composition and distribution of weed communities in agro-ecosystems. These results demonstrate some degree of plasticity in response to environmental variation of weed species by means of their spatial distribution. Of course, the plants are integrated in agro-ecosystems, and these responses are influenced by other factors such as agronomic practices.

In view of the results, climate change poses an uncertainty about the best way to design weed management strategies. A static style of management cannot be assumed any longer, and an adequate management of weeds, in the future, must take into account temperature change, land use, and human activity. Changes in the composition of the weed community can reduce the effectiveness of existing control strategies, as well as yield and economic cost to producers due to uncontrolled weeds. Also, important ecosystem services provided by weeds can be compromised if the composition of the community evolves with climate change. The different weed distribution patterns found could, in the long term, lead to variations of ecosystem functions. Therefore, the estimation of the damage of the weeds in the agro-ecosystems will be very important to reduce its impact and develop management strategies, current and future, effective against climate change.

5. Conclusions

Our research supports the common view that the monitoring of biodiversity is a means to obtain information on the state and dynamics of the agro-ecosystem. The application of the novel methodology proposed has made it possible to visualize clear links between latitude and temperature ranges related to weed distribution. The monitoring method presented here can be a promising tool to supply information in bioclimatic distribution models of species that needs to be validated with empirical data on weeds under changing climatic conditions. Also, this methodology proposed may be applied to the study of climate gradients around the world.

We think that there are temperature sensitive weeds which can be used in further studies as indicators of climate change by comparing distribution to local and regional data. Considering that weed community changes are not always noticeable in the short term, we recommend establishing long-term monitoring to detect changes in the biodiversity of agro-ecosystems. Furthermore, any changes in weed distribution in the agro-ecosystem due to temperature changes would affect the crops. Therefore, any information about shifts of weed dynamics related to temperature changes is going to be especially important in the future for crop management.

Supplementary Materials: The following are available online at http://www.mdpi.com/2073-4433/11/8/853/s1, Figure S1: Distribution of weed species according Latitude (from South to North in *X*-axis) and temperatures. Spearman rank order correlation values with temperatures in April month = 0.749 (Tmax) and 0.7906 (Tmin).

Author Contributions: All authors contributed to the monitoring of species. M.I.S.-M. and M.L.G. performed the original draft preparation of the manuscript; M.I.S.-M. and C.C. conceived and designed the experiments and revised the whole manuscript; M.L.G., J.L.T., and F.J.S. contributed to the analysis de data and the design of the experiments. All authors have read and agreed to the published version of the manuscript.

Funding: This work has been funded by projects AT2017-003 and RTA2017-00006-C03-01 (Spanish Ministry of Science and Innovation) and by Biodiversity Foundation of Spain (Ministry for the Ecological Transition) and INIA by means of the project PRCV00590 ("The role of resilience in the composition of weed communities of cereal agro-ecosystems. Adaptive responses of flora to climate change").

Conflicts of Interest: The authors declare no conflict of interest.

References

1. Reidsma, P.; Tekelenbur, T.; van den Berg, M.; Alkemade, R. Impacts of land-use change on biodiversity: An assessment of agricultural biodiversity in the European Union. *Agric. Ecosyst. Environ.* **2006**, *114*, 86–102. [CrossRef]
2. Sala, O.E.; Chapin, F.S.; Armesto, J.J.; Berlow, E.; Bloomfield, J.; Dirzo, R.; Huber-Sanwald, E.; Huenneke, L.F.; Jackson, R.B.; Kinzig, A.; et al. Biodiversity—Global biodiversity scenarios for the year 2100. *Science* **2000**, *287*, 1770–1774. [CrossRef]
3. Brock, W.A.; Carpenter, S.R.; Scheffer, M. Regime Shifts, Environmental Signals, Uncertainty, and Policy Choice. In *Complexity Theory for a Sustainable Future*; Norberg, J., Cumming, G.S., Eds.; Columbia University Press: New York, NY, USA, 2008; pp. 180–206.
4. Samhouri, J.F.; Levin, P.S.; Ainsworth, C.H. Identifying thresholds for ecosystem-based management. *PLoS ONE* **2010**, *5*, e8907. [CrossRef]
5. Pautasso, M.; Dehnen-Schmutz, K.; Holdenrieder, O.; Pietravalle, S.; Salama, N.; Jeger, M.J.; Lange, E.; Hehl-Lange, S. Plant health and global change—Some implications for landscape management. *Biol. Rev.* **2010**, *85*, 729–755. [CrossRef]
6. Schneider, A.; Havlík, P.; Schmid, E.; Valin, H.; Mosnier, A.; Obersteiner, M.; Böttcher, H.; Skalsky´, R.; Balkovič, J.; Sauer, T.; et al. Impacts of population growth, economic development, and technical change on global food production and consumption. *Agric. Syst.* **2011**, *104*, 204–215. [CrossRef]
7. Ray, D.K.; Mueller, N.D.; West, P.C.; Foley, J.A. Yield trends are insufficient to double global crop production by 2050. *PLoS ONE* **2013**, *8*, e66428. [CrossRef]
8. Petit, J.R.; Jouzel, J.; Raynaud, D.; Barkov, N.I.; Barnola, J.M.; Basile, I.; Bender, M.; Chappellaz, J.; Davis, M.; Delaygue, G.; et al. Climate and atmospheric history of the past 420,000 years from the Vostok ice core, Antarctica. *Nature* **1999**, *399*, 429–436. [CrossRef]
9. Loss, S.R.; Terwilliger, L.A.; Peterson, A.C. Assisted colonization: Integrating conservation strategies in the face of climate change. *Biol. Conserv.* **2011**, *144*, 92–100. [CrossRef]
10. Fried, G.; Norton, L.R.; Reboud, X. Environmental and management factors determining weed species composition and diversity in France. *Agric. Ecosyst. Environ.* **2008**, *128*, 68–76. [CrossRef]
11. Potts, G.R.; Ewald, J.A.; Aebischer, N.J. Long-term changes in the flora of the cereal ecosystem on the Sussex Downs, England, focusing on the years 1968–2005. *J. Appl. Ecol.* **2010**, *47*, 215–226. [CrossRef]
12. Andreasen, C.; Streibig, J.C. Evaluation of changes in weed flora in arable fields of Nordic countries—Based on Danish long-term surveys. *Weed Res.* **2011**, *51*, 214–226. [CrossRef]
13. Salonen, J.; Hyvönen, T.; Kaseva, J.; Jalli, H. Impact of changed cropping practices on weed occurrence in spring cereals in Finland—A comparison of surveys in 1997–1999 and 2007–2009. *Weed Res.* **2013**, *53*, 110–120. [CrossRef]
14. Post, E.; Forchhammer, M.C.; Stenseth, N.C.; Callaghan, T.V. The timing of life-history events in a changing climate. *Proc. R. Soc. Lond. Ser. B* **2001**, *268*, 15–23. [CrossRef]
15. Nogues-Bravo, D. Predicting the past distribution of species climatic niches. *Glob. Ecol. Biogeogr.* **2009**, *18*, 521–531. [CrossRef]
16. Estrella, N.; Sparks, T.H.; Menzel, A. Effects of temperature, phase type and timing, location, and human density on plant phenological responses in Europe. *Clim. Res.* **2009**, *39*, 235–248. [CrossRef]
17. Kaukoranta, T.; Hakala, K. Impact of spring warming on sowing times of cereal, potato and sugar beet in Finland. *Agric. Food Sci.* **2008**, *17*, 165–176. [CrossRef]
18. Fleming, A.; Vanclay, F. Farmer responses to climate change and sustainable agriculture. A review. *Agron. Sustain. Dev.* **2010**, *30*, 11–19. [CrossRef]
19. Daccache, A.; Keay, C.A.; Jones, R.J.A.; Weatherhead, E.K.; Stalham, M.A.; Knox, J.W. Climate change and land suitability for potato production in England and Wales: Impacts and adaptation. *J. Agric. Sci.* **2012**, *150*, 161–177. [CrossRef]
20. Gerowitt, B.; Bertke, E.; Hespelt, S.-K.; Tute, C. Towards multifunctional agriculture—Weeds as ecological goods? *Weed Res.* **2003**, *43*, 227–235. [CrossRef]

21. Grime, J.P. Evidence for the existence of three primary strategies in plants and its relevance to ecological and evolutionary theory. *Am. Nat.* **1977**, *111*, 1169–1194. [CrossRef]
22. Fuhrer, J. Agroecosystem responses to combinations of elevated CO2, ozone, and global climate change. *Agric. Ecosyst. Environ.* **2003**, *97*, 1–20. [CrossRef]
23. Marshall, E.J.P.; Brown, V.K.; Boatman, N.D.; Lutman, P.J.W.; Squire, G.R.; Ward, L.K. The role of weeds in supporting biological diversity within crop fields. *Weed Res.* **2003**, *43*, 77–89. [CrossRef]
24. Petit, S.; Boursault, A.; Le Guilloux, M.; Munier-Jolain, N.; Reboud, X. Weeds in agricultural landscapes. A review. *Agron. Sustain. Dev.* **2011**, *31*, 309–317. [CrossRef]
25. Clements, D.R.; Weise, S.F.; Swanton, C.J. Integrated weed management and weed species diversity. *Phytoprotection* **1994**, *75*, 1–18. [CrossRef]
26. Fried, G.; Petit, S.; Reboud, X. A specialist-generalist classification of the arable flora and its response to changes in agricultural practices. *BMC Ecol.* **2010**, *10*, 1–11. [CrossRef]
27. Chapin, F.S., III; Zavaleta, E.S.; Eviner, V.T.; Naylor, R.L.; Vitousek P., M.; Reynolds, H.L.; Hooper, D.U.; Lavorel, S.; Sala, O.E.; Hobbie, S.E.; et al. Consequences of changing biodiversity. *Nature* **2000**, *405*, 234–242. [CrossRef]
28. Booth, B.D.; Swanton, C.J. Assembly theory applied to weed communities. *Weed Sci.* **2002**, *50*, 2–13. [CrossRef]
29. Eriksson, O. Species pools in cultural landscapes—Niche construction, ecological opportunity and niche shifts. *Ecography* **2013**, *36*, 403–413. [CrossRef]
30. Hawes, C.; Haughton, A.J.; Bohan, D.A.; Squire, G.R. Functional approaches for assessing plant and invertebrate abundance patterns in arable systems. *Basic Appl. Ecol.* **2009**, *10*, 34–42. [CrossRef]
31. Urruty, N.; Deveaud, T.; Guyomard, H.; Boiffin, J. Impacts of agricultural land use changes on pesticide use in French agriculture. *Eur. J. Agron.* **2016**, *80*, 113–123. [CrossRef]
32. Hyvönen, T.; Huusela-Veistola, E. Arable weeds as indicators of agricultural intensity—A case study from Finland. *Biol. Conserv.* **2008**, *141*, 2857–2864. [CrossRef]
33. Dukes, J.S.; Pontius, J.; Orwig, D.; Garnas, J.R.; Rodgers, V.L.; Brazee, N.; Cooke, B.; Theoharides, K.A.; Stange, E.E.; Harrington, R.; et al. Responses of insect pests, pathogens, and invasive plant species to climate change in the forests of northeastern North America: What can we predict? *Can. J. For. Res.* **2009**, *39*, 231–248. [CrossRef]
34. Singer, A.; Travism, J.M.J.; Johst, K. Interspecific interactions affect species and community responses to climate shifts. *Oikos* **2013**, *122*, 358–366. [CrossRef]
35. Malavasi, M.; Santoro, R.; Cutini, M.; Acosta, A.T.R.; Carranza, M.L. The impact of human pressure on landscape patterns and plant species richness in Mediterranean coastal dunes. *Plant Biosyst.* **2016**, *150*, 73–82. [CrossRef]
36. Bloomfield, J.P.; Williams, R.J.; Gooddy, D.C.; Cape, J.N.; Guha, P. Impacts of climate change on the fate and behaviour of pesticides in surface and groundwater—A UK perspective. *Sci. Total Environ.* **2006**, *369*, 163–177. [CrossRef] [PubMed]
37. Walck, J.L.; Hidayati, S.N.; Dixon, K.W.; Thompson, K.; Poschlod, P. Climate change and plant regeneration from seed. *Glob. Chang. Biol.* **2011**, *17*, 2145–2161. [CrossRef]
38. Hanzlik, K.; Gerowitt, B. Occurrence and distribution of important weed species in German winter oilseed rape fields. *J. Plant. Dis. Prot.* **2012**, *119*, 107–120. [CrossRef]
39. Zimdahl, R.L. *Weed-Crop Competition: A Review*, 2nd ed.; Blackwell: Ames, IA, USA, 2004; p. 220.
40. Jump, A.S.; Peñuelas, J. Running and stand still: Adaptation and the response of plants to rapid climate change. *Ecol. Lett.* **2005**, *8*, 1010–1020. [CrossRef]
41. Cimalova, S.; Lososova, Z. Arable weed vegetation of the northeastern part of the Czech Republic: Effects of environmental factors on species composition. *Plant Ecol.* **2009**, *203*, 45–57. [CrossRef]
42. Silc, U.; Vrbnicanin, S.; Bozi, C.D.; Carni, A.; Stevanovic, Z.D. Weed vegetation in the northwestern Balkans: Diversity and species composition. *Weed Res.* **2009**, *49*, 602–612. [CrossRef]
43. Hulme, P.E.; Barrett, S.C.H. Integrating trait- and niche-based approaches to assess contemporary evolution in alien plant species. *J. Ecol.* **2013**, *101*, 68–77. [CrossRef]
44. Lososova, Z.; Chytry, M.; Kühn, I.; Hájeka, O.; Horáková, V.; Pysek, P.; Tichy, L. Patterns of plant traits in annual vegetation of man-made habitats in Central Europe. *Perspect. Plant. Ecol. Evol. Syst.* **2006**, *8*, 69–81. [CrossRef]

45. Hyvönen, T.; Luoto, M.; Uotila, P. Assessment of weed establishment risk in a changing European climate. *Agric. Food Sci.* **2012**, *21*, 348–360. [CrossRef]
46. Peters, K.; Breitsameter, L.; Gerowitt, B. Impact of climate change on weeds in agriculture: A review. *Agron. Sustain. Dev.* **2014**, *34*, 707–721. [CrossRef]

atmosphere

Article

Tree Surface Temperature in a Primary Tropical Rain Forest

Qinghai Song [1,2,3,*], Chenna Sun [1,2,3,4], Yun Deng [1,2,4], He Bai [1,2,3,4], Yiping Zhang [1,2,3,*], Hui Yu [1,2,3,4], Jing Zhang [1,2,3,4], Liqing Sha [1,2,3], Wenjun Zhou [1,2,3] and Yuntong Liu [1,2,3]

1 CAS Key Laboratory of Tropical Forest Ecology, Xishuangbanna Tropical Botanical Garden, Chinese Academy of Sciences, Menglun 666303, China; sunchenna19@mails.ucas.ac.cn (C.S.); dy@xtbg.org.cn (Y.D.); baihe19@mails.ucas.ac.cn (H.B.); yuhui@xtbg.ac.cn (H.Y.); zhangjing1@xtbg.ac.cn (J.Z.); shalq@xtbg.ac.cn (L.S.); zhouwj@xtbg.ac.cn (W.Z.); liuyuntong@xtbg.ac.cn (Y.L.)

2 Center for Plant Ecology, Core Botanical Gardens, Chinese Academy of Sciences, Xishuangbanna 666303, China

3 Global Change Research Group, Xishuangbanna Tropical Botanical Garden, Chinese Academy of Sciences, Menglun 666303, China

4 Department of Life Sciences, University of Chinese Academy of Sciences, Beijing 100049, China

* Correspondence: sqh@xtbg.ac.cn (Q.S.); yipingzh@xtbg.ac.cn (Y.Z.); Tel.: +86-15925111979 (Q.S.)

Received: 29 June 2020; Accepted: 23 July 2020; Published: 29 July 2020

Abstract: As one of the important factors affecting plant productivity and plant distribution, temperature also affects the physiological and ecological characteristics of plants to a large extent. We report canopy leaf temperature distribution over a 36 m tall primary tropical rain forest and samplings of 28 tree species in SW China by means of two high resolution thermal cameras (P25, Flir systems, Wilsonville, OR, USA). The leaf temperature of dominant tree Species *Pometia tomentosa* was the highest (31.8 °C), 10.2 °C higher than that of tree species *Mezzettipsis creaghii* (21.6 °C). The mean leaf to air temperature difference (T_c–T_a) of *Pometia tomentosa* was the highest (6.4 K), the second highest was *Barringtonia pendula* (6.1 K), and *Mezzettipsis creaghii* had the lowest (T_c–T_a) (1.9K). (T_c–T_a) of tree species with smaller leaves and larger stomatal conductance was lowly sensitive to climate factors. Leaf size and stomatal conductance together decided the effect of climate change to (T_c–T_a) of the different tree species. We have shown that the composition of tree species in tropical rain forest areas is important to the climate through our research.

Keywords: leaf temperature; infrared thermography; thermal imagery; tropical rain forest

1. Introduction

There are many factors that affect plant productivity and plant distribution, such as climate, topography, water, soil, microorganisms, and so on. However, as one of the important factors affecting plant productivity and plant distribution, temperature also affects the physiological and ecological characteristics of plants to a large extent. Temperature influences rates of plant photosynthesis and respiration, litter decomposition and microbial activity [1], and other biological processes will affect the fixation and release of carbon dioxide [2]. On the other hand, leaf temperature is also affected by stomatal control of transpiration [3] and traits affecting heat exchange [4]. For example, leaf size, leaf shape, petiole length, and other traits can affect leaf temperature [5–7].As a result, global warming is expected to affect carbon pools on land, increasing the amount of carbon dioxide in the atmosphere [8,9]. The rate of ecosystem respiration will increase with increasing temperature. At the same time, rising temperatures may lead to plant stomata closing, thus reducing the primary productivity of tropical rainforest ecosystems.

With the global biodiversity crisis becoming increasingly serious, research on plant diversity and its role in ecosystem function is becoming increasingly important in ecology. Due to the good combination of precipitation and temperature in the tropical rain forest area, the plant types in the tropical rain forest area are also very rich. Moist rainforests cover about 6~7% of the earth's surface, but they are home to more than half of all life on earth [10]. According to research, the tropical rainforest contains the largest collection of living plant species in the world [9]. They contain 40% of the world's forest biomass [10] and soil carbon [11]. Especially on a small scale, dominant tree species in a tropical rain forest community have a great influence on forest carbon sequestration [12].

Due to the incomplete development of technology, the early research mostly focused on the microhabitat temperature measurement of single tree species. Recently, new digital technologies in combination with thermal (IR) transmission lenses have been developed to accurately measure actual temperature regimes in canopies. For example, Kumar et al. conducted field experiments on farms in central and southern India (18°9′ N, 74°28′ E). They used thermal imaging to assess the canopy temperature differences among different genotypes of soybeans to further differentiate the soybean's ability to withstand water stress [13]. Padhi et al. used infrared thermal imaging technology to measure the canopy temperature of cotton fields at Kingsthorpe Research Station (27°30′44″ S,151°46′55″ E), thus providing a basis for assessing crop water deficit pressure using stomatal conductance index [14]. Daniel and Körner assessed, by using a combination of IR imagery, both surface and root zone temperatures on a landscape scale in the Swiss Alps [13]. In Switzerland, leaf surface temperatures vary widely between species in mixed deciduous forests and urban environments [14,15]. Stomatal conductance, the key factor controlling leaf temperature, varies greatly in ten tropical forests [3]. Thus, we can see that different plants have different canopy temperatures. Moreover, the temperature of the crown is closely related to the stomatal conductance and water content of the leaves. Canopy temperatures have a series of uncertainties in their variations, and controlling environmental factors [3,4]. Therefore, we hypothesized that canopy temperatures of broadleaved trees in primordial tropical forests may differ significantly among species. According to the 2007 report by the Intergovernmental Panel on Climate Change (IPCC), global temperatures are expected to rise from 2.4 to 5.5 °C as carbon dioxide levels in the atmosphere increase [16]. This global warming may be related to changes in solar radiation, precipitation, and other micrometeorological factors on a regional or ecosystem scale. Studying the responses of different tree species to these climatic stress factors is helpful to understand the dynamics of vegetation in the context of global warming [14,15].

Little is known about how species-scale variation in leaf temperature influences community-scale variation in canopy temperature [14]. Our study has two main purposes (1) to study the spatial and temporal distribution of species under specific canopy temperature, (2) to discuss the possible influence of future climate change on canopy temperature difference (CTD). Through this study, we want to explore the significance of forest tree species' composition on canopy heat accumulation.

2. Materials and Methods

2.1. Site Description and Studied Species

Our experiment was conducted in a tropical rain forest in Xishuangbanna, southwestern China (21°55′39″ N, 101°15′ 55″ E, elevation 750 m). Xishuangbanna Nature Reserve Authority is mainly responsible for protecting the reserve forest sites. Our institute workers were approved by Xishuangbanna Nature Reserve Authority to conduct experiments in the site. There were no specific permissions required for the activities. The height of the rainforest canopy is about 36 m. The stand has a stem density of 964 trees ha^{-1} (diameter ≥ 5 cm) and a total basal area of 32.28 m^2 ha^{-1}; the number of tree species in the plot was 179 in 2007 [17].

2.2. Dominant Mature Tree Species

The important value of the five tree species ranked in the top five over the whole tree species [17]. So, we selected the top five canopy mature tree species in this rainforest which is dominated by *Pometia tomentosa, Barringtonia macrostachya, Gironniera subaequalis, Ardisia tenera,* and *Mezzettiopsis creaghi,* respectively [18]. The canopy temperature characteristics of the five tree species are shown in Table 1. The sites did not involve endangered or protected species.

2.3. Saplings of 28 Tree Species

We also selected samplings of 28 tree species in the lower layer of the forest that could be adequately replicated. Five tree samples of each tree species were measured.

2.4. Thermal Imaging of the Canopy Dominant Tree Species

A 70 m tower was established at the center of the plot. In this study, two thermal imaging cameras (P25, Flir systems, Wilsonville, OR, USA) with a resolution of 320 × 420 pixels were used to determine the canopy temperature of the dominant tree species. Two thermal imaging cameras (P25, Flir systems, Wilsonville, OR, USA) were mounted 3 to 5 m above the canopy. In addition, exposed and certain sized canopies were selected to measure the average canopy surface temperature. The camera (P25, Flir systems, Wilsonville, OR, USA) software was used to analyze the canopy surface temperature of the measured images. This provided us with 76,800 temperature data with a resolution of 0.1K under sunny conditions. We only cut three leaves in a tree to analyze the leaf stomatal conductance.

2.5. Thermal Imaging and Stomatal Conductance of 28 Tree Species Samplings

By measuring the temperature of the leaves and the temperature of the air, we calculated the difference between the two average temperatures. In addition, we measured the leaf stomatal conductance of samplings of 28 tree species. All measurements were conducted on 13–15 June 2018, three meteorologically similar days (from 09:30 to 13:00). We also used two thermal cameras (P25, Flir systems, Wilsonville, OR, USA) to measure the leaf temperature. Then, the leaf stomatal conductance was measured with a portable photosynthesis system (LI-6400, Li-COR, Lincoln, NE, USA).

2.6. Environmental Data and Soil Moisture

In the corresponding period, wind speed (A100R, Vector, UK), air temperature and humidity (HMP45C, Vaisala, Vantaa, Finland), photosynthetically active radiation (LI-190SB, Li-COR, Lincoln, NE, USA), net radiation (CNR-1, Kipp & Zonen, The Netherlands), soil temperature (TCAV, Scientific Inc., Logan, UT, USA), and soil moisture (CS616, Campbell Scientific Inc., Logan, UT, USA) were measured simultaneously. All these factors were sampled at 0.5 Hz and the data were stored in the data logger. The 30 min average was also calculated and stored by the data logger (CR1000, Campbell Scientific Inc., Logan, UT, USA).

The canopy temperature difference was calculated by using the canopy surface temperature and air temperature at the same time.

2.7. Evaluate the Canopy Temperature Change

In order to explore the possible influence of future climate change on the canopy temperature difference (CTD), we used Function (1) to evaluate the canopy temperature change [19].

$$T_c - T_a = \frac{P_a(R_n - G) - g_c\lambda\,VPD + \left(g_c/g_H\right)P_a(R_n - G)]}{g_cC_p + g_HC_pVPD} \tag{1}$$

where P_a is atmospheric pressure (kPa), R_n is net radiation (W m^{-2}), G is soil heat flux (W m^{-2}), g_c is canopy stomatal conductance (mmol m^{-2} s^{-1}), g_H is boundary layer heat conductance (mmol m^{-2} s^{-1}),

C_p is the specific heat of air at constant pressure (J mol^{-1} K^{-1}), λ is latent heat of vaporization (J mol^{-1}), VPD is the vapor pressure deficit (kPa), and P_a is atmospheric pressure (kPa).

2.8. Statistical and Analytical Methods

All data processing and statistical analyses were conducted using the Statistical Analysis System (SPSS 26.0 Software, IBM, Armonk, NY, USA). Significant differences between means were tested using one-way analysis of variance (ANOVA). Significant effects of the main meteorological elements on the canopy temperature difference rates were determined by multiple linear regressions.

3. Results

3.1. Environmental Conditions

Figure 1 showed diurnal variations in the main meteorological elements during the observational period. The maximum air temperature was 27.6 °C. The maximum photosynthetic active radiation (PAR) was 1380 μmol m^{-2} s^{-1}. Wind speed was very low in this site.

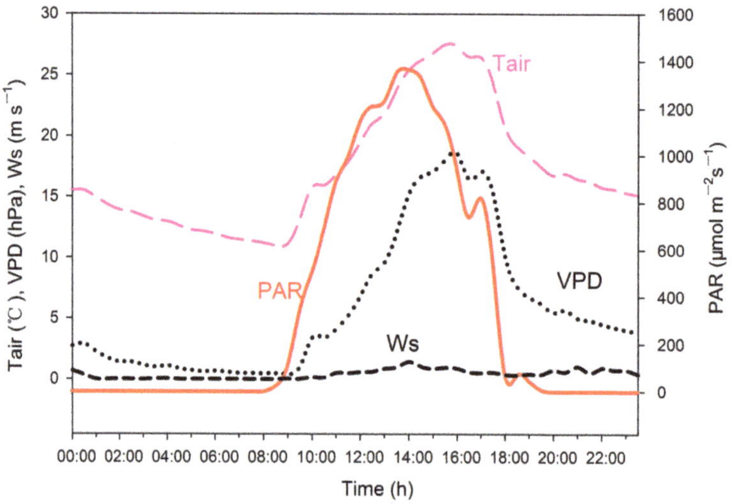

Figure 1. Air temperature (T_a), vapor pressure deficit (VPD), photosynthetic active radiation (PAR) and wind speed (Ws) during the canopy temperature measurements.

Soil moisture has a strong influence on the plant surface temperature. During the measurement, the soil moisture content at 5 cm depth was 12.5%. This is an area with very low soil moisture throughout the year, which means trees can tolerate drought.

3.2. Spatial and Temporal Temperature Distribution of the Dominant Mature Tree Species

Canopies were scanned on 4 February, 2018 from 13:55 to 14:05 true local time. Figure 2 shows the pattern of the mean temperature of the three layers at 14:00. The mean leaf temperature of *Pometia tomentosa* in upper layer was the highest (31.8 °C), 10.2 °C higher than that of the tree species *Mezzettipsis creaghii* (21.6 °C) in the lower layer.

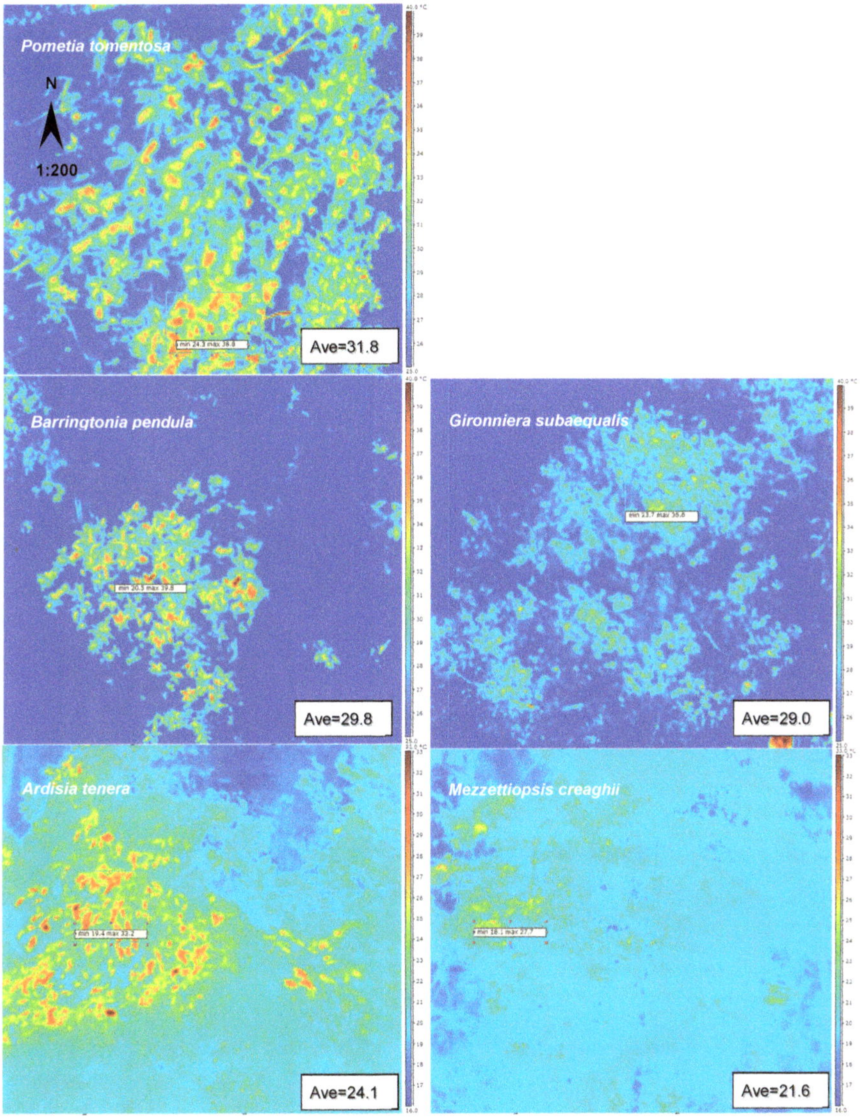

Figure 2. The false color thermal image of part of the canopy of five tree species taken at 14:00 on 4 February 2018, shows the canopy surface temperature of five different tree species. The squares show some selected frames that have been used (Table 1 and Figure 5).

Table 1. The average canopy temperature repeat number n, canopy temperature difference (CTD), temperature range, minimum and maximum values of the five rainforest tree species selected in the study (letter a~e represents the difference from high to low at the significant level of 0.05 ($p < 0.05$)).

Species	T_c–T_a (K)	T-Range (K)	T_{min} (°C)	T_{max} (°C)	n
Pometia tomentosa	6.4 ^{a}	14.5 ± 1.51	24.3 ± 0.37	38.8 ± 1.14	9
Barringtonia pendula	6.1 ^{b}	11.7 ± 2.04	23.8 ± 0.61	35.5 ± 1.43	9
Gironniera subaequalis	4.7 ^{c}	8.7 ± 1.26	25.6 ± 0.25	34.3 ± 1.01	9
Ardisia tenera	4.4 ^{d}	13.8 ± 1.05	19.4 ± 0.18	33.2 ± 0.87	9
Mezzettiopsis creaghii	1.9 ^{e}	10.9 ± 1.07	16.8 ± 0.26	27.7 ± 0.81	9

The maximum stomatal conductance of the five species ranged from 93~120 mmol m^{-2} s^{-1} (Figure 3), and there were significant differences between the five species. The leaf size of the five species ranged from 14~173 cm^2 (Figure 3). With the increase in temperature, the relative frequency of *Gironniera subaequalis* changed most obviously, and reached the maximum at 28 °C. The change of the other four trees was relatively gentle (Figure 3).

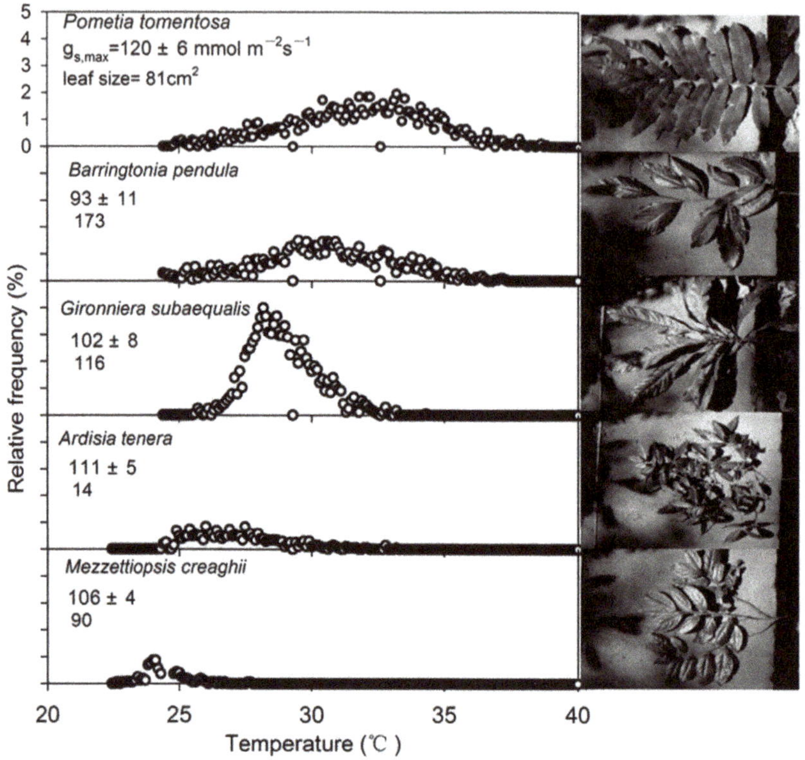

Figure 3. Leaf temperature distributions of the five tree species showed (photos). Maximum g_s and leaf size values are inserted.

We measured the mean (T_c–T_a) on 4 February, 2018. The mean (T_c–T_a) consisted with the photo flux density patterns (Figures 3 and 4). The five species of the three layers monitored on that day showed significant differences in (T_c–T_a). *Pometia tomentosa* had the highest (T_c–T_a) (6.4 K), the second highest was *Barringtonia pendula* (6.1 K), and *Mezzettipsis creaghii* had the lowest (T_c–T_a) (1.9K) (Figure 4).

We found that the temperature change within the canopies was very significant ($p < 0.01$) (Figure 4). The canopy temperature difference of Pometia tomentosa and Barringtonia macrostachya reached the highest at about 13:00. However, the canopy temperature difference of Gironniera subaequalis reached its maximum two hours later (at about 15:00) (Figure 4).

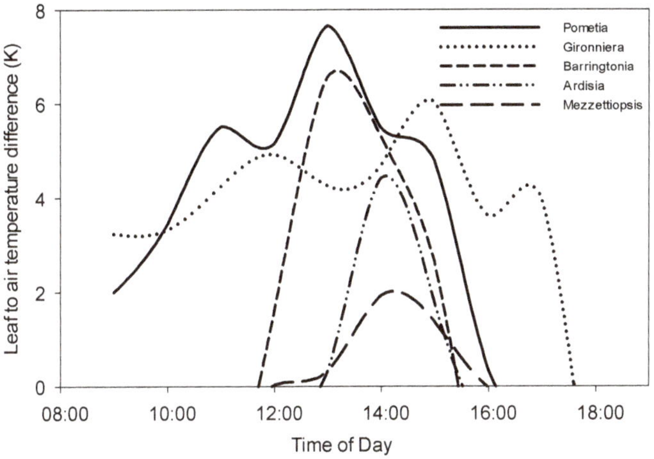

Figure 4. Leaf to air temperature difference of the five tree species.

3.3. (T_c–T_a) Variations in a Changing Climate

(T_c–T_a) ranged from nearly 0 K (*Barringtonia racemosa* (L.) *Spreng.*) to 3 K (*Swietenia mahagoni* (L.) *Jacq.*), showing that the mean leaf temperatures of 28 tree species had a highly species-specific manner (Table 2). The mean leaf temperature cannot be fully or solely explained by stomatal conductance or leaf area. This is because of the complex feedback effect between leaf temperature and climatic conditions or leaf function [20].

Table 2. Leaf-to-air temperature (T_c–T_a), leaf area, and maximum stomatal conductance (g_{smax}) of 28 tree species.

No.	Species Name	T_c–T_a °C	Leaf Area cm^2	g_{smax} mmol m^{-2} s^{-1}	n
1	*Mezzettiopsis creaghii*	0.99 ± 0.86	32.83 ± 6.47	173.29 ± 49.90	5
2	*Swietenia mahagoni* (L.) Jacq.	2.93 ± 2.00	44.32 ± 5.10	162.93 ± 27.76	5
3	*Dipterocarpus turbinatus* Gaertn. f.	1.80 ± 0.51	106.38 ± 7.44	222.04 ± 17.58	4
4	*Cleistanthus sumatranus* (Miq.) Muell. Arg.	0.81 ± 0.11	15.98 ± 5.93	401.26 ± 29.47	5
5	*Dalbergia odorifera* T. Chen	0.57 ± 0.48	14.51 ± 3.83	221.41 ± 57.08	5
6	*Pterocarpus indicus Willd.*	0.79 ± 0.15	70.93 ± 30.95	188.18 ± 65.16	5
7	*Artocarpus heterophyllus* Lam.	1.43 ± 1.92	97.80 ± 11.60	448.82 ± 44.31	5
8	*Hopea hainanensis Merr. et Chun*	0.83 ± 0.28	74.03 ± 12.96	480 ± 31.26	5
9	*Saraca dives Pierre*	1.27 ± 0.11	100.88 ± 30.05	166.59 ± 33.11	5
10	*Woodfordia fruticose* (Linn.) *Kurz*	1.15 ± 1.73	20.10 ± 3.70	548.88 ± 48.58	5
11	*Aquilaria agallocha Roxb*	0.31 ± 0.08	29.64 ± 4.58	217.26 ± 21.22	5
12	*Magnolia rostrata W. W. Smith*	0.59 ± 0.51	253.10 ± 116.03	170.23 ± 49.83	5
13	*Mesua ferrea L.*	0.63 ± 0.11	22.28 ± 7.27	127.65 ± 26.22	5
14	*Rauwolfia yunnanensis Tsiang*	0.23 ± 0.22	23.86 ± 4.01	88.26 ± 43.74	5
15	*Oroxylum indicum*	0.30 ± 0.72	59.16 ± 23.86	355.58 ± 139.85	5
16	*Millettia rubiginosa Wight et Arn.*	1.68 ± 1.06	46.97 ± 16.90	235.17 ± 102.95	5
17	*Ficus curtipes*	0.26 ± 0.38	70.38 ± 17.56	209.97 ± 78.20	5
18	*Bauhinia Linn.*	0.16 ± 0.17	65.65 ± 9.33	183.91 ± 53.83	5
19	*Mayodendron igneum* (Kurz) *Kurz*	0.32 ± 0.22	41.82 ± 8.00	274.25 ± 77.56	6
20	*Dracaena cambodiana Pierre ex Gagnep.*	0.93 ± 0.16	95.00 ± 21.43	80.78 ± 33.25	5
21	*Baccaurea ramiflora Lour*	0.67 ± 0.19	76.53 ± 8.38	121.56 ± 21.29	5
22	*Barringtonia racemosa* (L.) *Spreng.*	0.08 ± 0.24	178.54 ± 51.97	219.04 ± 79.81	5
23	*Moghania macrophylla* (Willd.) *O Ktze.*	1.57 ± 0.46	115.02 ± 25.24	212.76 ± 70.90	5
24	*Plukenetia volubilisLinneo*	2.61 ± 1.69	94.54 ± 19.97	550.80 ± 93.84	5
25	*Terminalia bellirica* (Gaertn.) *Roxb.*	0.38 ± 0.66	57.87 ± 11.24	269.73 ± 46.20	5
26	*Camptotheca acuminata.*	0.76 ± 0.23	144.90 ± 34.76	398.22 ± 93.55	5
27	*Cinnamomum japonicum Sieb.*	0.07 ± 0.42	23.21 ± 4.00	187.48 ± 28.19	5
28	*Clerodendrum bungei Steud.*	1.20 ± 0.12	171.80 ± 47.19	638.84 ± 43.46	5

It is estimated that global warming will strengthen the water cycle and increase the demand for evaporation in ecosystems. Under drought stress, stomatal conductance of plants decreases, thus reducing transpiration and increasing canopy temperature. Therefore, in order to explore the differences in canopy temperature among tree species in an expected changing climate, we assumed that the maximum g_s value of each tree species decreased linearly to 50%. (T_c–T_a) was very sensitive to all simulated meteorological elements (Figure 5; $p < 0.01$) and increased linearly with the increase in direct radiation (DR) and relative humidity (RH) (Figure 5a,c). (T_c–T_a) decreased non-linearly with the increase in air temperature (T_a) and wind speed (Ws) (Figure 5b,d).

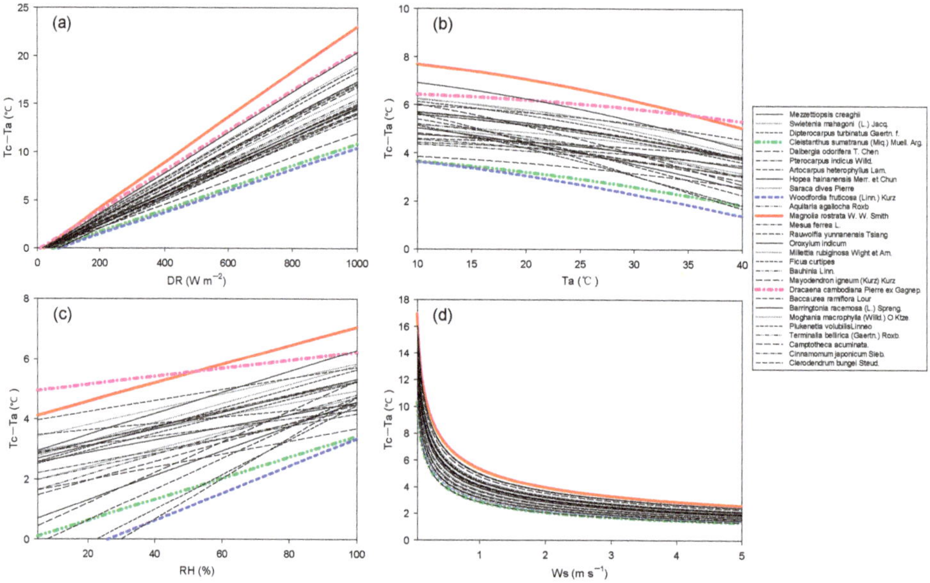

Figure 5. Response of canopy temperature difference (T_c–T_a) to climatic factors, including direct radiation (DR) (**a**), air temperature (T_a) (**b**), relative humidity (RH) (**c**), and wind speed (Ws) (**d**). Red line: *Magnolia rostrata W. W. Smith*; pink line: *Dracaena cambodiana Pierre ex Gagnep*; blue line: *Woodfordia fruticose* (Linn.) *Kurz*; green line: *Cleistanthus sumatranus* (Miq.) *Muell. Arg*.

With the increase in direct radiation (DR), the canopy temperature difference (CTD) of large-leaved trees increased more than that of small-leaved trees (Figure 5a). Species-specific differences in (T_c–T_a) became larger with the increasing DR.

The decrease amplitude of (T_c–T_a) with the increasing T_a of the tree species with the largest leaves (*Magnolia rostrata W. W. Smith*) was similar to the tree species with the smallest leaves (*Woodfordia fruticose (Linn.) Kurz*). (T_c–T_a) was negatively correlated with air temperature (T_a) and relative humidity (RH) (Figure 5b,c), indicating that (T_c–T_a) decreased with increasing water vapor pressure deficit (VPD). Species-specific differences in (T_c–T_a) became smaller with the increasing Ws (Figure 5d).

4. Discussion

Our results suggest that the species composition of trees has an important impact on the local climate of the primary tropical rain forest and are consistent with the study in a mixed deciduous forest [14]. The temperature changes found in this study may be critical in determining the diversity of canopy-dwelling plant communities. The patterns of temperature change identified by the study may be critical in determining canopy plant diversity [14].

The mean leaf temperatures of 28 tree species have a highly species-specific manner. Mean leaf temperature cannot be fully or solely explained by stomatal conductance or leaf area [21]. This is because of the complex feedback effect between leaf temperature and climatic conditions or leaf function [22]. Variations in leaf temperature play a major role in determining rates of photosynthetic CO_2 uptake and transpiration [23]. On the other hand, changes in photosynthesis and transpiration resulting from changes in leaf temperature may have an impact on the efficiency of nutrient use [7,20].

On the other hand, the response of canopy temperature difference of 28 tree species to climate change was also diverse. (T_c–T_a) of tree species with smaller leaves and larger stomatal conductance was lowly sensitive to climate factors, such as Woodfordia fruticose (Linn.) Kurz and Cleistanthus sumatranus (Miq.) Muell. Arg. In contrast, (T_c–T_a) of tree species (Dracaena cambodiana Pierre ex Gagnep.) with the smallest stomatal conductance was highly sensitive to climate factors. (T_c–T_a) of tree species (Magnolia rostrata W. W. Smith) with the largest leaves was also highly sensitive to climate factors, even if the stomatal conductance was not small. In other words, different tree species have their own strategies for climate change. With the maintenance of tree species diversity, the ability of ecosystems to resist climate change will be stronger. Conversely, loss of tree species diversity will make forests more vulnerable to climate change [24]. Climate change research for the future will involve a series of uncertainties, and it is still not sure what climate scenario will best reflect reality. What are the differences of short-term response and long-term adaptation of different tree species to climate change? [25]. Thus, tropical rainforest management for climate change has to deal with a range of uncertainties in the future.

5. Conclusions and Implications

Our study showed that the leaf size and stomatal conductance together decided the effect of climate change on (T_c–T_a) of the different tree species in the tropical rainforest. (T_c–T_a) of tree species with smaller leaves and larger stomatal conductance was lowly sensitive to climate factors. These results indicate species-specific functional traits which are needed to explicitly explore and model the interactions of individuals for improving the understanding and prediction of climate change impacts on vegetation.

Author Contributions: Q.S., C.S., and Y.Z. wrote the main manuscript text. H.B., H.Y., J.Z., W.Z., Y.L., Y.D., and L.S. performed the experiments. All authors have read and agreed to the published version of the manuscript.

Funding: The National Natural Science Foundation of China (41671209, U1602234, 31290221, 31770528), the National Natural Science Foundation of China and Thailand Research Fund (NSFC-TRF) (41961144017), the National Key Research and Development Program of China (2016YFC0502105), and the CAS 135project (2017XTBG-T01, 2017XTBG-F01).

Acknowledgments: We are grateful to Donghai Yang, Taoxiang Yi for their assistance in the experimental field. This work was supported by Xishuangbanna Ecological Station for Tropical rain forest. This research was funded by the National Natural Science Foundation of China (41671209, U1602234, 31290221, 41961144017, 31770528), the National Key Research and Development Program of China (2016YFC0502105), and the CAS 135project (2017XTBG-T01, 2017XTBG-F01).

Conflicts of Interest: The authors declare no conflict of interests.

References

1. Swift, M.J.; Heal, O.W.; Anderson, J.M. *Decomposition in Terrestrial Ecosystems*; Studies in Ecology; Univ of California Press: Berkeley, CA, USA, 1979; Volume 5.
2. Raich, J.W.; Russell, A.E.; Kitayama, K.; Parton, W.J.; Vitousek, P.M. Temperature Influences Carbon Accumulation in Moist Tropical Forests. *Ecology* **2006**, *87*, 76–87. [CrossRef]
3. Tan, Z.; Zhao, J.; Wang, G.; Chen, M.; Yang, L.; He, C.; Restrepocoupe, N.; Peng, S.; Liu, X.; Ribeiro Da Rocha, H.; et al. Surface conductance for evapotranspiration of tropical forests: Calculations, variations, and controls. *Agric. For. Meteorol.* **2019**, *275*, 317–328. [CrossRef]
4. Lin, H.; Chen, Y.J.; Zhang, H.L.; Fu, P.L.; Fan, Z.X. Stronger cooling effects of transpiration and leaf physical traits of the plants from a hot dry habitat than from a hot wet habitat. *Funct. Ecol.* **2017**, *31*, 2202–2211. [CrossRef]

5. Kumar, M.; Govindasamy, V.; Rane, J.; Singh, A.K.; Choudhary, R.L.; Raina, S.K.; George, P.; Aher, L.K.; Singh, N.P. Canopy temperature depression (CTD) and canopy greenness associated with variation in seed yield of soybean genotypes grown in semi-arid environment. *S. Afr. J. Bot.* **2017**, *113*, 230–238. [CrossRef]
6. Rahman, M.A.; Moser, A.; Gold, A.; Rotzer, T.; Pauleit, S. Vertical air temperature gradients under the shade of two contrasting urban tree species during different types of summer days. *Sci. Total Environ.* **2018**, *633*, 100–111. [CrossRef] [PubMed]
7. Page, G.F.; Lienard, J.F.; Pruett, M.J.; Moffett, K.B. Spatiotemporal dynamics of leaf transpiration quantified with time-series thermal imaging. *Agric. For. Meteorol.* **2018**, *256*, 304–314. [CrossRef]
8. Schimel, D.S.; Braswell, B.H.; Holland, E.A.; McKeown, R.; Ojima, D.S.; Painter, T.H.; Parton, W.J.; Townsend, A.R. Climatic, edaphic, and biotic controls over storage and turnover of carbon in soils. *Glob. Biogeochem. Cycles* **1994**, *8*, 279–293. [CrossRef]
9. Connell, J.H. Diversity in tropical rain forests and coral reefs. *Science* **1978**, *199*, 1302–1310. [CrossRef]
10. Dixon, R.K.; Brown, S.; Houghton, R.A.; Solomon, A.M.; Trexler, M.C.; Wisniewski, J. Carbon pools and flux of global forest ecosystems. *Science* **1994**, *263*, 185–190. [CrossRef]
11. Jobbàgy, E.G.; Jackson, R.B. The vertical distribution of soil organic carbon and its relation to climate and vegetation. *Ecol. Appl.* **2000**, *10*, 423–436. [CrossRef]
12. Van der Molen, M.K.; Dolman, A.J.; Ciais, P.; Eglin, T.; Gobron, N.; Meir, P.; Peters, W.; Phillips, O.L.; Reichstein, M.; Chen, T.; et al. Drought and ecosystem carbon cycling. *Agric. For. Meteorol.* **2011**, *151*, 765–773. [CrossRef]
13. Daniel, S.; Körner, C. Infrared thermometry of alpine landscapes challenges climatic warming projections. *Glob. Chang. Biol.* **2010**, *16*, 2602–2613.
14. Leuzinger, S.; Körner, C. Tree species diversity affects canopy leaf temperatures in a mature temperate forest. *Agric. For. Meteorol.* **2007**, *146*, 29–37. [CrossRef]
15. Leuzinger, S.; Roland, V.; Körner, C. Tree surface temperature in an urban environment. *Agric. For. Meteorol.* **2010**, *150*, 56–62. [CrossRef]
16. Solomon, S.; Qin, D. *Climate Change 2007: The Physical Science Basis, Contribution of Working Group I to the Third Assessment Report of the IPCC.*; Cambridge University Press: Cambridge, UK, 2007; p. 996.
17. Hu, Y.H.; Cao, M.; Lin, L.X. Dynamics of tree species composition and community structure of a tropical seasonal rain forest in Xishuangbanna, Southwest China. *Acta Ecol. Sin.* **2010**, *30*, 949–957.
18. Zhu, H. Forest vegetation of Xishuangbanna, south China. *For. Stud. China* **2006**, *8*, 1–58.
19. Maes, W.H.; Steppe, K. Estimating evapotranspiration and drought stress with ground-based thermal remote sensing in agriculture: A review. *J. Exp. Bot.* **2012**, *63*, 4671–4712. [CrossRef]
20. Atkinson, L.J.; Campbell, C.; Zaragozacastells, J.; Hurry, V.; Atkin, O.K. Impact of growth temperature on scaling relationships linking photosynthetic metabolism to leaf functional traits. *Funct. Ecol.* **2010**, *24*, 1181–1191. [CrossRef]
21. Jackson, R.D.; Idso, S.B.; Reginato, R.J.; Pinter, P.J. Canopy temperature as a crop water stress index. *Water Resour. Res.* **1981**, *17*, 1133–1138. [CrossRef]
22. Keenan, T.F.; Niinemets, U. Global leaf trait estimates biased due to plasticity in the shade. *Nat. Plants* **2017**, *3*, 1–6. [CrossRef]
23. Slot, M.; Reysanchez, C.; Winter, K.; Kitajima, K. Trait-based scaling of temperature-dependent foliar respiration in a species-rich tropical forest canopy. *Funct. Ecol.* **2014**, *28*, 1074–1086. [CrossRef]
24. Zhang, T.; Niinemets, U.; Sheffield, J.; Lichstein, J.W. Shifts in tree functional composition amplify the response of forest biomass to climate. *Nature* **2018**, *556*, 99–102. [CrossRef] [PubMed]
25. Slot, M.; Winter, K. In situ temperature response of photosynthesis of 42 tree and liana species in the canopy of two Panamanian lowland tropical forests with contrasting rainfall regimes. *New Phytol.* **2017**, *214*, 1103–1117. [CrossRef] [PubMed]

atmosphere

MDPI

Article

Simulation of Climate Change Impacts on Phenology and Production of Winter Wheat in Northwestern China Using CERES-Wheat Model

Zhen Zheng [1,2], Huanjie Cai [2,3,*], Zikai Wang [4] and Xinkun Wang [1]

1 Research Center of Fluid Machinery Engineering and Technology, Jiangsu University, ZhenJiang 212013, China; zhengzhenzzj@163.com (Z.Z.); xjwxk@126.com (X.W.)
2 Key Laboratory for Agricultural Soil and Water Engineering in Arid Area of Ministry of Education, Northwest A&F University, Yangling 712100, China
3 Institute of Water Saving Agriculture in Arid Areas of China (IWSA), Northwest A&F University, Yangling 712100, China
4 Zhenjiang Engineering Survey & Design Institute Co., Ltd., Zhenjiang 212013, China; wangzikai1989@163.com
* Correspondence: huanjiec@yahoo.com; Tel.: +86-298-708-2133

Received: 8 May 2020; Accepted: 23 June 2020; Published: 28 June 2020

check for updates

Abstract: Wheat plays a very important role in China's agriculture. The wheat grain yields are affected by the growing period that is determined by temperature, precipitation, and field management, such as planting date and cultivar species. Here, we used the CSM-CERES-Wheat model along with different Representative Concentration Pathways (RCPs) and two global circulation models (GCMs) to simulate different impacts on the winter wheat that caused by changing climate for 2025 and 2050 projections for Guanzhong Plain in Northwest China. Our results showed that it is obvious that there is a warming trend in Guanzhong Plain; the mean temperature for the different scenarios increased up to 3.8 °C. Furthermore, the precipitation varied in the year; in general, the rainfall in February and August was increased, while it decreased in April, October and November. However, the solar radiation was found to be greatly reduced in the Guanzhong Plain. Compared to the reference year, the results showed that the number of days to maturity was shortened 3–24 days, and the main reason was the increased temperature during the winter wheat growing period. Moreover, five planting dates (from October 7 to 27 with five days per step) were applied to simulate the final yield and to select an appropriate planting date for the study area. The yield changed smallest based on Geophysical Fluid Dynamics Laboratory (GFDL)-CM3 (−6.5, −5.3, −4.2 based on RCP 4.5, RCP 6.0, and RCP 8.5) for 2025 when planting on October 27. Farmers might have to plant the crop before 27 October.

Keywords: anthesis and maturity date; crop yield; SimCLIM; DSSAT model; planting date

1. Introduction

The growing wheat (*Triticum aestivum* L.) in China makes up 21.9% of the whole crop area sown in 2011, leading to China producing the highest wheat grain yield in the world [1,2]. However, wheat production is facing future changes in rainfall patterns, temperature conditions, and other factors that restrict farmers' ability to plant this crop. Thus, the whole world, including China, is paying attention to the risk of wheat production [3–5]. Previous studies have shown that wheat productivity will be vulnerable to climate change in southeastern Asia and southern China [6–8]. Thus, the appropriate strategies should be analyzed for adoption by policymakers and farmers.

In 30 or 50 years, the world will change in an unimaginable way and it is difficult to imagine how the future climate will be changed and how the crops respond to those climate changes, which results in many uncertainties in these studies [9–11]. Therefore, determining the possibile future changes in climate may affect the wheat yield, therefore finding strategies to adapt to ensure the continuation of the wheat supply are necessary. Combining the outputs of GCMs under different RCPs with the models is an active way to learn the effects of changed climate on crops yield [12–15]. At present, crop models have been proved the ability to provide useful views into the design of decision making in the agricultural management by simulating how cropping systems respond to climate change, management, and variety selection [16–19]. One of their advantages is that they can deal with crop responses for climate changes, i.e., drought, waterlogging, high temperatures, atmospheric CO_2 concentration changes and precipitation [20–22]. Therefore, many studies have attempted to investigate how the future climate will affect wheat growing under different scenarios by using crop models [23–25].

Generally, the projected changes in final production have quite a wide range, depending on the crop simulation models, GCMs, and RCP scenarios that were selected. SimCLIM has been used with a large number of crop simulations [26,27]. For example, SimCLIM was used in Georgia, USA to study the response of soybean phenology, development and yield to the changing climate coupled with the CSM-CROPGRO-Soybean model [28,29], and it also has been applied to project the climate variability and its impact on cotton production in southern Punjab, Pakistan [29]. Moreover, SimCLIM has provided an easier way to learn climatic factors for different fields such as agriculture [30] and ecosystem resilience [31]. The CSM-CERES-Wheat model could analyze the influence of soil, field management (like irrigation, fertilization, planting date, cultivar) as well as climate on crop growth and grain yield [32–34]. The model can simulate wheat development, water balance, phosphorous, nitrogen balance, and aboveground biomass and grain yield in relation to weather, soil, phenotype factors and management practices [35–37].

In this research, we studied the future climate change in the two future projections of 2025 and 2050 compared with the baseline period (1961–1990), and the response of winter wheat production to it and compared with the reference years (19834–2013). The greenhouse gas CO_2 emissions of three RCPs were considered. The CERES-Wheat model was applied to study crop yield simulation in cooperation with the GCM climate. The main goals of this analysis were: (1) to identify the future climate change in Guanzhong Plain, (2) to study the future climate change impacts on winter wheat phenology and productivity in this region, and (3) to provide suggestions for potential adaptation strategies for winter wheat growth in Guanzhong Plain.

2. Materials and Methods

2.1. Study Location and Crop Management

Yangling, an arid area of Guanhzong Plain, China (34.38° N, 107.15° E), was selected as a case study (Figure 1) [38]. Guanzhong Plain, located in the southestern China, a winter wheat-summer corn double cropping system was applied in this area. The cultivar "Xiaoyan 22" was selected as the planting cultivar with the recommendation of local farmers. The data of growth and yield for "Xiaoyan 22" were validated with different irrigation levels by the CERES-Wheat model; the details were provided by Zheng et al. [39]. Previous results showed that the validated model could simulate winter wheat phenology, total biomass and final yield greatly, with a lower normalized root mean square error (RMSEn). However, the RMSEn was a bit high when simulating aboveground biomass in the treatments that had water stress. With the RMSEn less than 2% for phenology, 15% for total biomass, and 15% for the yield. The genetic coefficient for "Xiaoyan 22" is shown in Table 1.

Further detailed information about basic field conditions and management strategies was pursued by Zheng et al. [40]. The soil parameters are listed in Table 2, and the initial conditions of soil used in the simulation are shown in Table 3. The sowing density was 340 plants m^{-2}, and 130 kg ha^{-1} N was applied on the planting date and wintering time, independently. The simulation was set as a rainfed condition.

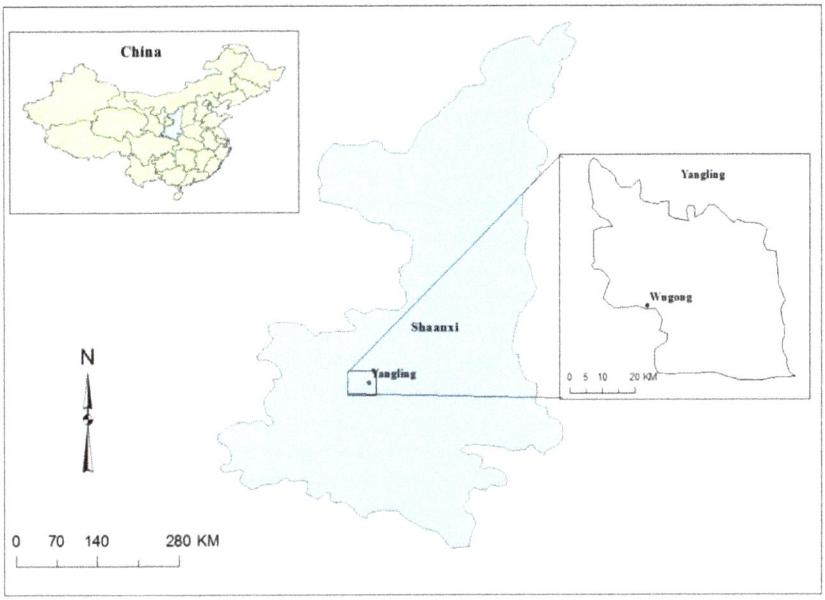

Figure 1. Location of the study area.

Table 1. Validated "Xiaoyan 22" wheat cultivar parameters.

Abbreviation	Definition	Unit	Value
P1V	Vernalization sensitivity coefficient	degree-days	6.62
P1D	Photoperiod parameter	-	81.37
P5	Grain filling phase duration	°C. d	572.10
G1	Kernel number per unit canopy weight at anthesis	#/g	23.30
G2	Potential kernel growth rate	mg	33.70
G3	Standard, non-stressed dry weight (total, including grain) of a single tiller at maturity	g	1.55
PHINT	Thermal time between the appearance of leaf tips	°C. d	97.20

Table 2. Soil physical parameters for the study area, Yangling.

Depth (cm)	Bulk Density ($g{\cdot}cm^{-3}$)	Field Capacity	Wilting Moisture	Soil Texture (%)		
				sand	silt	clay
0–23	1.3	0.28	0.12	26.7	40.8	32.1
23–35	1.4	0.28	0.13	25.0	42.8	32.1
35–74	1.4	0.27	0.15	24.1	44.8	31.0
74–95	1.4	0.28	0.19	22.7	38.8	38.5
95–163	1.4	0.27	0.14	21.3	38.6	40.1
163–196	1.3	0.26	0.13	24.3	36.9	38.9

Table 3. Initial conditions of soil profile and physical characteristics in the field, Yangling.

Soil Depth (cm)	Wilting Point ($cm^3 \cdot cm^{-3}$)	Field Capacity ($cm^3 \cdot cm^{-3}$)	Saturation ($cm^3 \cdot cm^{-3}$)	Initial Water Content ($cm^3 \cdot cm^{-3}$)	NH_4-N Conc. ($g \cdot Mg^{-1}$)	NO_3-N Conc. ($g \cdot Mg^{-1}$)
0–5	0.10	0.28	0.45	0.28	1.90	12.90
5–35	0.11	0.28	0.46	0.24	0.50	11.20
35–70	0.12	0.28	0.46	0.22	0.40	12.60
70–90	0.14	0.28	0.49	0.22	0.60	11.80
90–100	0.14	0.28	0.50	0.23	0.60	10.50

2.2. Climate Models

The SimCLIM [41] as initially developed to enable integrated estimate of future climate on different regions in New Zealand [42,43]. SimCLIM 2013 [44] mainly relies on the IPCC CMIP5 datasets. Generally, 1986 to 2005 was used as the baseline period for the SimCLIM 2013; the previous standard 1961 to 1990 can also be used. Thus, we used 1961–1990 as the baseline period and chose 1984–2013 as the reference year in our study. The climate projections from ranged from 1991 to 2100 around the world.

2.3. Yield Simulation with the Crop Model

Here, DSSAT Version 4.6 [35–37] was used to simulate the wheat phenology, as well as the winter wheat grain yield for 2025 and 2050 projections. The inputs of daily weather data for simulations from future projections were modified from SimCLIM based on the reference years weather. The daily weather inputs included sunshine hours, rainfall, and maximum and minimum temperatures. These data for the reference time 1984–2013 and baseline 1961–1990 at the study area were downloaded from the China Meteorological Data Service Center (CMDC) [45]. The RCPs (RCP 2.6, RCP 4.5, RCP 6.0, and RCP 8.5) are named after a possible range of radiative forcing values in the year 2100 (of 2.6, 4.5, 6.0, and 8.5 W/m^2, respectively) [46]. Three scenarios (RCP 4.5, RCP 6.0, and RCP 8.5) were selected in this study. RCP 6.0 represents the median value of medium climate prediction sensitivity, while RCPs 4.5 and 8.5 with low and high climate sensitivity, respectively. Furthermore, two GCM models (GFDL-CM3 and MRI-CGCM3) were selected from SimCLIM; both of these GCMs provided all the climate variables including temperature, precipitation, SRAD, wind speed, relative humidity, and sea level. These two GCMs can project future climate change accurately, so their prediction for future temperature, SRAD and rainfall have been accepted [28,47]. The present average planting date was around October 15 in the study area. Five planting dates, about 10 days in advance of and 10 days after (October 7, 12, 17, 22 and 27), were set to simulate the anthesis date, maturity date, and yield in the 2025 and 2050 projections. The simulated phenology and final yield in projections 2025 and 2050 were compared with 1984–2013.

3. Results

3.1. Climate Projections for 2025 and 2050

The predicted monthly change of solar radiation (SRAD) (Figure 2), percentage of precipitation (Figure 3) and mean temperature (T_{mean}) (Figure 4) for 2025 and 2050 were modified with SimCLIM based on two GCMs (i.e., GFDL and MRI) and three RCPs. The mean daily radiation for 1961–1990 in the study area was 15.2 MJ m^{-2}; the results showed that the average SRAD decreased for all three RCPs compared with the baseline. The SRAD showed difference by year and month for the GFDL model for the future projections in 2025 and 2050. The predicted SRAD change (based on GFDL) as 2025 was the same for the given months compared to 2050 projection, with a slight decreasing trend in January, February, and March and a slight increasing trend in the rest of the months. Among the three RCPs, the projection for SRAD for 2025 only showed a slight difference, but the differences among the three scenarios in 2050 were greater than in 2025. The predicted trends for the change in solar radiation based on MRI were similar to GFDL, and the differences among the three scenarios were smaller than the GFDL.

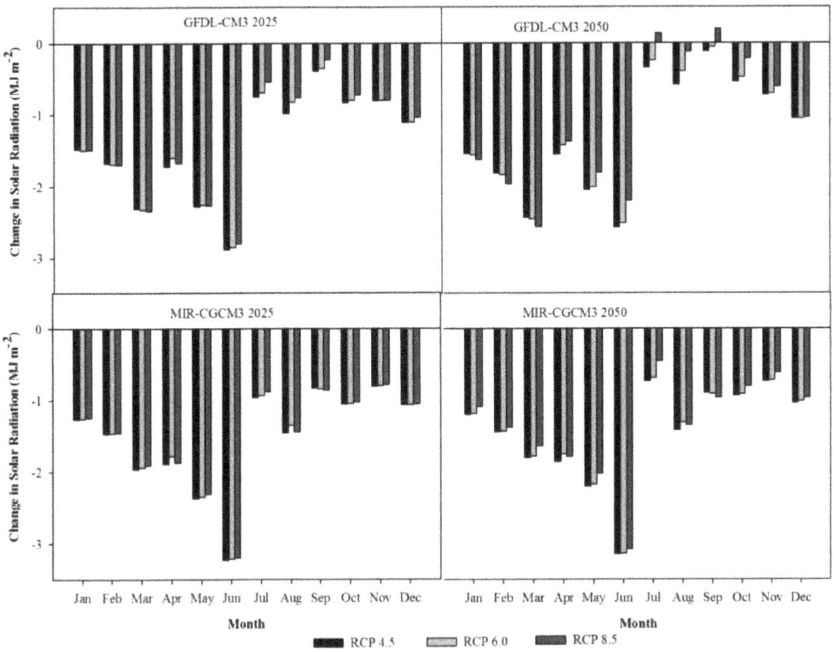

Figure 2. Changes in monthly SRAD (MJ m^{-2}) as projected for 2025 and 2050 based on three RCPs for two GCMs compared with 1961–1990 for Guanzhong Plain.

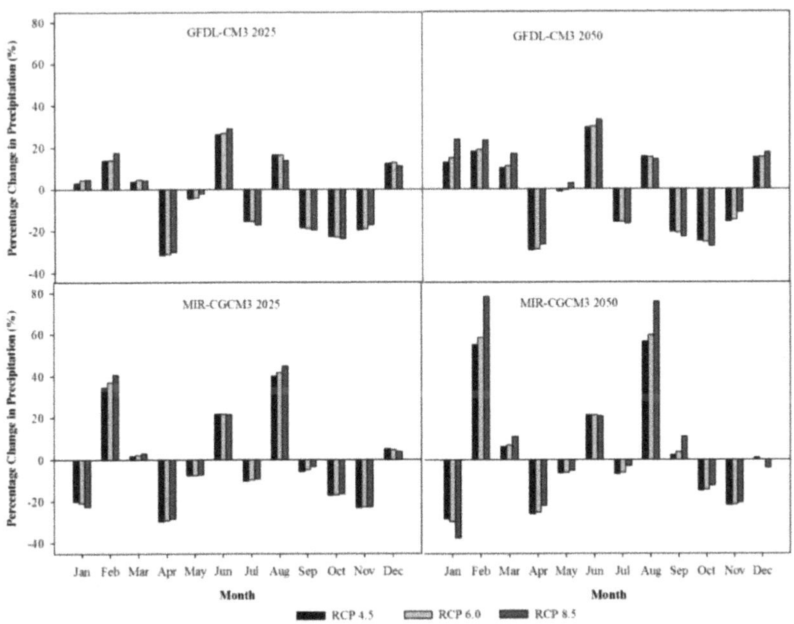

Figure 3. Changes in precipitation (%) as projected for 2025 and 2050 based on three RCPs for two GCMs compared with 1961–1990 for Guanzhong Plain.

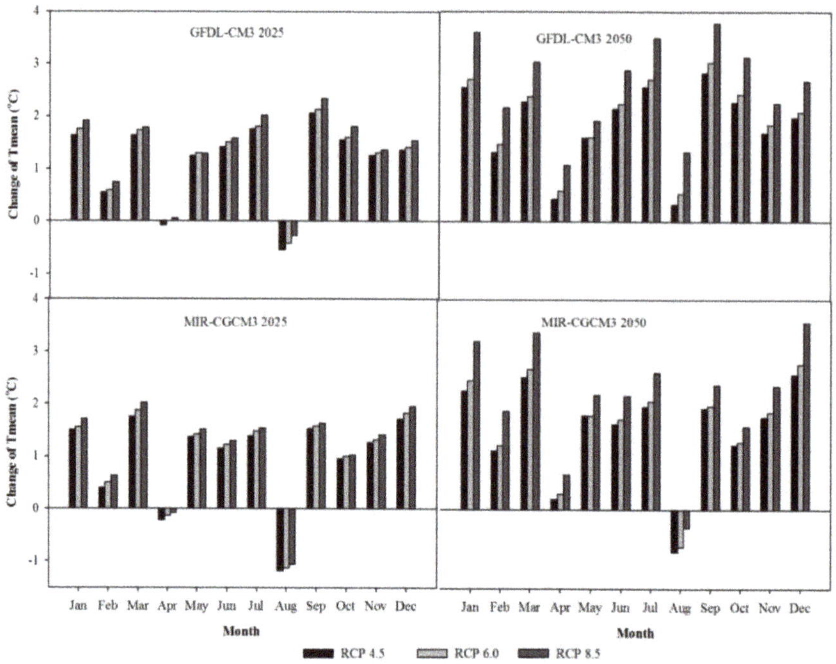

Figure 4. Changes in mean temperature (°C) projected for 2025 and 2050 based on three RCPs, and RCP 8.5 for two GCMs compared with 1961–1990 for Guanzhong Plain.

The average yearly rainfall for 1961–1990 at the experiment site was 623.5 mm. Projected rainfall showed a difference between two GCMs, with one GCM simulating an increase in rainfall and another projecting a decrease (Figure 3). Projected rainfall usually increased in February, June, August, while it decreased in April, July, October, and November compared with the baseline. The rainfall was projected to increase by 13–40% for February and 1.5–5% for March for the 2025 projection. The differences among RCPs were no more than 3% for 2025. While the yearly difference, between the projections of 2025 and 2050, was larger based on MIR compared with GFDL; it was approximately 61% between 2025 and 2050.

The average daily T_{mean} for 1961–1990 at the study area was 13.9 °C. Figure 4 shows changes in T_{mean} projected by two GCMs and three RCPs compared with the baseline. Overall, the T_{mean} had an increasing trend in the 2025 and 2050 projections. The mean temperature increases were 1.2, 2.1, 1.0 and 1.8 °C for the GFDL 2025, GFDL 2050, MRI 2025, and MRI 2050 projections. A small decreasing trend was found in the projections except for the GFDL 2050 projection.

3.2. Projected Phenology Changes

To know the impact of changing climate on winter wheat growing period, we simulated the number of days from planting to anthesis (ADAPS) and the number of days from planting to maturity (MDAPS) for different planting dates based on the two GCMs and three RCPs, the results of which are illustrated in Figure 5. Obvious decreases were projected for both two future periods compared with the reference year. The ADAPS decreased from 4.8 to 5.9 days on average and from 8.3 to 12.7 days based on GFDL for the 2025 and 2050 projections, respectively. The largest change of ADAPs was observed in 2050, according to GFDL model in RCP 8.5, with a decrease of 17.9 days, and with a decrease of 18.1 days based on MIR under RCP 8.5.

Similarly, the MDAPS was shortened compared with the reference years for both GCMs; the predictions showed a difference in planting date, GCMs, scenarios, and projected years (Figure 6).

The MDAPS had a decreasing trend under different scenarios and different planting dates. Among the three RCPs, the largest decrease was occurred for the RCP 8.5, followed by RCP 6.0, while RCP 4.5 showed similar changes based on both GCMs. The greatest shortening of MDAPS was projected by MIR on October 27 during 2050 under RCP 8.5, reaching 24.3 days. The MDAPS decreased from 6.3 to 7.1 days on average and from 9.7 to 14.5 days based on GFDL for the 2025 and 2050 projections, respectively.

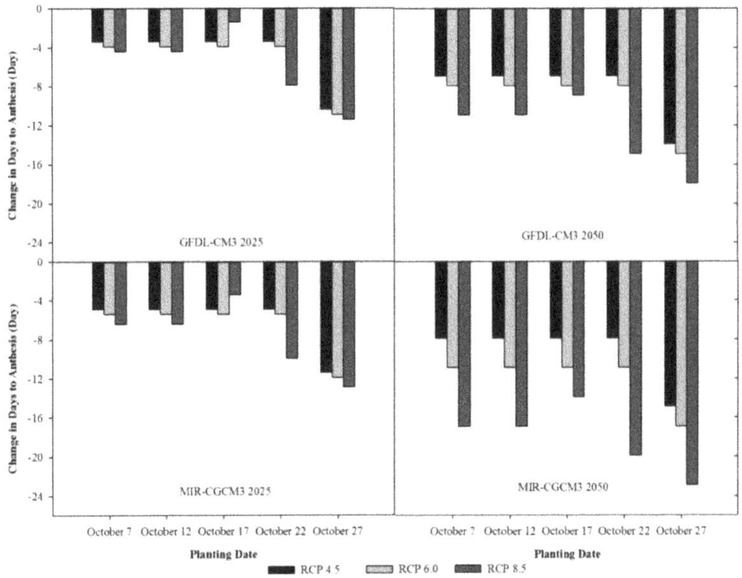

Figure 5. Simulated ADAPS of winter wheat for Guanzhong Plain based on three RCPs under different planting dates based on two GCMs in the 2025 and 2050 projections.

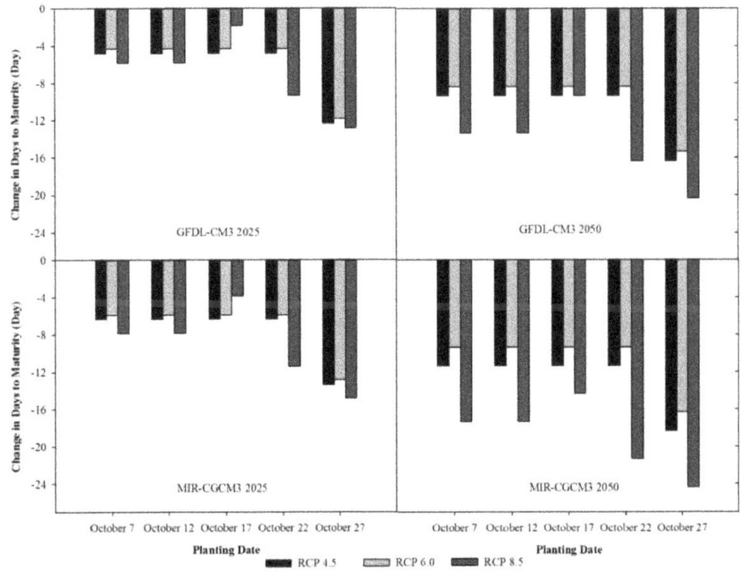

Figure 6. Predicted MDAPS of winter wheat for Guanzhong Plain based on three RCPs under different planting date based on two GCMs in the 2025 and 2050 projections.

3.3. *Projected Changes in Winter Wheat Yields*

As mentioned before, the historical weather data based on the reference years were modified through two GCMs and three RCP scenarios to predict the yield productions in the future projections under changed sowing windows by using crop model. The predicted grain yields for the 2025 and 2050 projections based on two GCMs are shown in Table 4. For all the scenarios, the winter got a higher yield when planting on October 17, and the yield decreased largely when planting date shifted to October 27. In our study, we compared the simulations of winter wheat for the reference years with the future projections based on two GCMs instead of analyzing the absolute wheat yield prediction (Figure 7).

Table 4. Simulated yields for the 2025 and 2050 projections based on GFDL and MRI GCM.

Planting Date	Projections											
	GFDL-CM3 2025			GFDL-CM3 2050			MRI-CGCM3 2025			MRI-CGCM3 2050		
	RCP 4.5	RCP 6.0	RCP 8.5	RCP 4.5	RCP 6.0	RCP 8.5	RCP 4.5	RCP 6.0	RCP 8.5	RCP 4.5	RCP 6.0	RCP 8.5
10.7	4216	4282.5	4271	4552	4643	5034	4157	4272	4178.5	4247	4542	4121
10.12	4216	4282.5	4271	4552	4643	5034	4157	4272	4178.5	4247	4542	4121
10.17	4216	4282.5	4340	4552	4643	5002	4157	4272	4243	4247	4542	3982
10.22	4216	4282.5	4068.5	4552	4643	4870	4157	4272	4137.5	4247	4542	4466
10.27	3734	3778.5	3824	4010	4235	4603	3814	3942	3979	4160	4477	4582

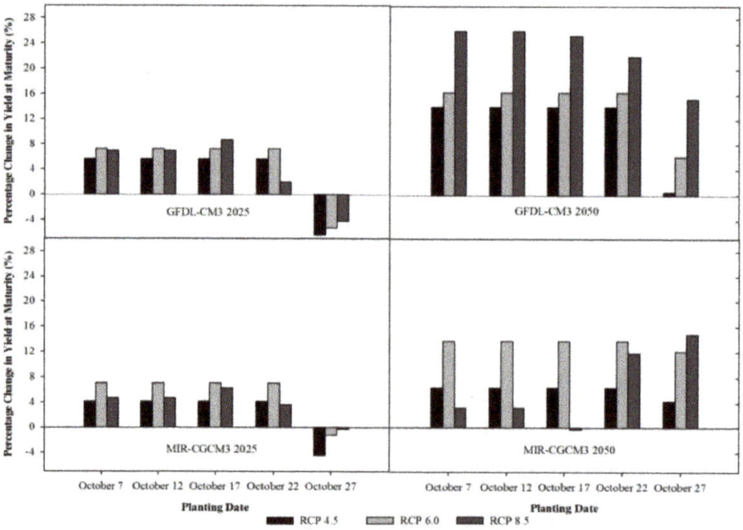

Figure 7. Predicted winter wheat yield for Guanzhong Plain based on three RCPs under changed planting dates based on two GCMs in the 2025 and 2050 projections.

Among the three RCPs, the increases in grain yield between the scenarios were different and they depended on the sowing date. The yield increased higher for RCP 6.0, followed by RCPs 8.5 and 4.5 in the 2025 projection before October 12, while in the 2050 projection, the increase in yield for RCP 8.5 was higher, followed by RCPs 6.0 and 4.5 based on GFDL. For MRI GCM, the yield increased higher for RCP 6.0 and followed by RCP 4.5 and RCP 8.5. Due to the large increase in rainfall for the 2050 projection, the yield rose larger than for the 2025 projection. The grain yield at maturity had a deceasing trend when planting on October 27 based on all the RCPs and both GCMs for the 2025 projection, and except for the MRI 2050 projection, the grain yield had a declining trend based on all three RCPs and two GCMs when the planting date was delayed to 17 October. The largest increases in grain yield were 26.1%, 16.3%, and 14% based on the GFDL 2050 projection for the RCP 8.5, RCP 6.0, and RCP4.5, respectively, when planting on October 7 and 12.

4. Discussion

In our study, by using the crop simulation model, the wheat grain yield in Guanzhong Plain would increase by 2.8% and 8.6% under RCP 4.5, 5.1% and 13.9% under RCP 6.0, and 3.9% and 14.8% under RCP 8.5 for the 2025 and 2050 projections. The results were consistent with a previous study which found that the warming climate in the last 30 years increased wheat yield by 0.9–12.9% in north part of China but decreased 1.2–10.2% in south part of China, differed in location, and the reason was due to the final impacts depends on the combined effect of changes in all climate variables. One zone was sensitive to mean temperature and the other was most sensitive to solar radiation during the growing period [6]. The adverse effects of changed climate can be reduced by choosing optimum sowing dates [48,49], and increasing rainfall during this time is also beneficial [50]. Our study illustrated that the winter wheat planted after October 17 would decrease the grain yield by 0.3–6.5%. For the 2025 projection, the average yield increased less for RCP 8.5 compared with the other RCPs based on GFDL and MRI GCM. The reason for this may due to the larger decrease in MDPAS.

Physiologically, wheat is a C3 plant, which greatly benefits from an increase in CO_2 concentration; that is, the increase in CO_2 concentration has a fertilization effect that can increase in the photosynthetic rate and it also has a water-saving effect by decreasing transpiration [51,52]. Generally, increases in CO_2, high mean temperature, and SRAD can improve photosynthesis leading to a final yield increase. Therefore, changes in CO_2, T_{mean}, and SRAD would affect the crop production significantly [53]. Parry et al. [54] illustrated that, because of the "CO_2-fertilization effect", increasing in CO_2 concentration would counteract the passive influences (such as yield reduction) of climate change in the future projections. The yield gains for RCP 8.5 were larger based on GFDL. The reason for this may due to the CO_2 fertilization offsetting the interactions, such as higher temperature [55]. Semenov and Shewry [56] found that, although earlier flowering with increasing temperatures allowed crops to escape increasing terminal drought, compared to RCP 4.5, the RCP 8.5 with higher CO_2 concentrations can also counteract the increased negative impacts of rainfall reduction and shorter growth period. Thus, an appropriate decision to support the arid area could be to plant a cultivar that flowers early [52].

Obviously, there were some uncertainties and limitations in the method of combining different scenarios and crop models in our study. The crop models are useful tools in predicting the impacts of different weather conditions on crop development and final productivity, but they have limitations regarding extreme weather events and soil conditions, and the soils used for simulation were also sources of uncertainty, as different calibration results could lead to different simulation results. Our results showed that the phenology of winter wheat totally decreased in the future and the yield increased in Guanzhong Plain by the midcentury. Hernandez-Ochoa et al. [55] indicated that applying the wheat-crop-climate multi-model ensemble may counteract the negative impact of climate change on wheat yield in Mexico. Parry et al. [57] suggested about 5% to 10% wheat yield may decline around the world by midcentury, even changing the sowing dates, choosing the different varieties, applying the appropriate fertilizer and irrigation amount or other adaptation strategies applied. In our further studies, we will take into account other wheat cultivars that may be more heat tolerant and drought resistant, as well as other potential adaptation scenarios such as irrigation and fertilizer management.

5. Conclusions

The present study indicated that the solar radiation mainly reduced from 0.3 to 3.3 MJ m^{-2} in the future projection and decreased most in June. Rainfall normally raised in February, June and August, but reduced in April, October and November in the study area. The precipitation change for the RCP 8.5 scenario was the largest, followed by RCPs 6.0 and 4.5. The mean temperature in most months rose compared with the baseline, among which the temperature in January, March, and December increased the most. The winter wheat anthesis date was shortened 3–23 days, the maturity date was shortened 4–24 days under different projections, and the winter wheat yield increased up to 28% among all scenarios.

Overall, the effect of the future climate on winter wheat production in Guanhzong Plain is positive, and the negative impact of climate change depends on the climate projections considered, as some of the GCMs showed an increase in grain yield and some showed a reducing trend. For the planting date, October 7–17 is the optimum choice, and the winter wheat yield would have a declining trend when planting after October 17. However, the simulated results were based on the rainfed scenario; the grain yield of rainfed wheat is very sensitive to climate change. Due to the great uncertainty in the future change of rainfed wheat yield in the Guanzhong area, irrigation management should be considered.

Author Contributions: Z.Z. and Z.W., methodology, software, investigation, and data curation. X.W., formal analysis. Z.Z., conceptualization, writing-original draft preparation. H.C., validation, resources, writing-review and editing, supervision, project administration, and funding acquisition. All authors have read and agreed to the published version of the manuscript.

Funding: This research was supported by the National Key Research and Development Program of China (grant number 2016YFC0400202), the Project of the Faculty of Agricultural Equipment of Jiangsu University, and a Project Funded by the Priority Academic Program Development of Jiangsu Higher Eeucation Institutions (No. PAPD-2018-87).

Conflicts of Interest: The authors declare no conflict of interest.

References

1. National Bureau of Statistics of China. *China Statistical Yearbook 2011*; China Statistics Press: Beijing, China, 2011.
2. FAO. *Statistical Yearbook: World Food and Agriculture*; FAO (Food and Agriculture Organization of the United Nations): Rome, Italy, 2012; p. 184.
3. Chen, Y.; Zhang, Z.; Wang, P.; Song, X.; Wei, X.; Tao, F. Identifying the impact of multi-hazards on crop yield—A case for heat stress and dry stress on winter wheat yield in northern China. *Eur. J. Agron.* **2016**, *73*, 55–63. [CrossRef]
4. Liang, S.; Li, Y.; Zhang, X.; Sun, Z.; Sun, N.; Duan, Y.; Xu, M.; Wu, L. Response of crop yield and nitrogen use efficiency for wheat-maize cropping system to future climate change in northern China. *Agric. For. Meteorol.* **2018**, *262*, 310–321. [CrossRef]
5. Hernadez-Ochoa, I.M.; Asseng, S.; Kassie, B.T.; Xiong, W.; Robertson, R.; Pequeno, D.N.L.; Sonder, K.; Reynolds, M.; Babar, M.A.; Milan, A.M.; et al. Climate change impact on Mexico wheat production. *Agric. For. Meteorol.* **2018**, *263*, 373–387. [CrossRef]
6. Tao, F.; Zhang, Z.; Xiao, D.; Zhang, S.; Rotter, R.P.; Shi, W.; Liu, Y.; Wang, M.; Liu, F.; Zhang, H. Response of wheat growth and yield to climate change in different climate zones of China, 1981–2009. *Agric. For. Meteorol.* **2014**, *189*, 91–104. [CrossRef]
7. Xiao, D.; Tao, F. Contributions of cultivars, management and climate change to winter wheat yield in the North China Plain in the past three decades. *Eur. J. Agron.* **2014**, *52*, 112–122. [CrossRef]
8. Kaushika, G.S.; Himanshu Arora, H.; KS, H.P. Analysis of climate change effects on crop water availability for paddy, wheat and berseem. *Agric. Water Manag.* **2019**, *225*, 105734.
9. Zhang, H.; Zhou, G.; Liu, D.; Wang, B.; Xiao, D.; He, L. Climate-associated rice yield change in the Northeast China Plain: A simulation analysis based on CMIP5 multi-model ensemble projection. *Sci. Total Environ.* **2019**, *666*, 126–138. [CrossRef]
10. IPCC. *Climate Change 2001: The Scientific Basis. Contribution of Working Group I to the Third Assessment Report of the Intergovernmental Panel on Climate Change*; Houghton, J.T., Ding, Y., Griggs, D.J., Noguer, M., van der Linden, P.L.J., Dai, X., Maskell, K., Johnson, C.A., Eds.; Cambridge University Press: Cambridge, UK; New York, NY, USA, 2001; p. 881.
11. Lobell, D.B.; Field, C.B.; Cahill, K.N.; Bonfils, C. Impacts of future climate change on California perennial crop yields: Model projections with climate and crop uncertainties. *Agric. For. Meteorol.* **2006**, *141*, 208–218. [CrossRef]
12. Xiong, W.; Holman, I.; Conway, D.; Lin, E.; Li, Y. A crop model cross calibration for use in regional climate impacts studies. *Ecol. Model.* **2008**, *213*, 365–380. [CrossRef]

13. Rosenzweig, C.; Elliott, J.; Deryng, D.; Ruane, A.C.; Müller, C.; Arneth, A.; Boote, K.J.; Folberth, C.; Glotter, M.; Khabarov, N.; et al. Assessing agricultural risks of climate change in the 21st century in a global gridded crop model inter comparison. *Proc. Natl. Acad. Sci. USA* **2014**, *111*, 3268–3272. [CrossRef]
14. Wang, W.; Yu, Z.; Zhang, W.; Shao, Q.; Zhang, Y.; Luo, Y.; Jiao, X.; Xu, J. Response of rice yield, irrigation water requirement and water use efficiency to climate change in China: Historical simulation and future projections. *Agric. Water Manag.* **2014**, *146*, 249–261. [CrossRef]
15. Yan, C.R.; Liu, L.; Huang, G.H. Multi-model projections of future climate change under different RCP scenarios in arid inland region of north China. *J. Drain. Irrig. Mach. Eng.* **2018**, *36*, 1193–1199. (In Chinese with English abstract)
16. Rotter, R.P.; Carter, T.R.; Olesen, J.E.; Porter, J.R. Crop-climate models need an overhaul. *Nat. Clim. Chang.* **2011**, *1*, 175–177. [CrossRef]
17. Chenu, K.; Porter, J.R.; Martre, P.; Basso, B.; Chapman, S.C.; Ewert, F.; Bindi, M.; Asseng, S. Contribution of crop models to adaptation in wheat. *Trends Plant Sci.* **2017**, *22*, 472–490. [CrossRef] [PubMed]
18. Martre, P.; Wallach, D.; Asseng, S.; Ewert, F.; Jones, J.W.; Rötter, R.P.; Boote, K.J.; Ruane, A.C.; Thorburn, P.J.; Cammarano, D.; et al. Multi-model ensembles of wheat growth: Many models are better than one. *Glob. Chang. Biol.* **2015**, *21*, 911–925. [CrossRef] [PubMed]
19. Gu, Z.; Qi, Z.; Ma, L.; Gui, D.; Xu, J.; Fang, Q.; Yuan, S.; Feng, G. Development of an irrigation scheduling software based on model predicted crop water stress. *Comput. Electron. Agric.* **2017**, *143*, 208–221. [CrossRef]
20. Asseng, S.; Ewert, F.; Martre, P.; Rötter, R.P.; Lobell, D.B.; Cammarano, D.; Kimball, B.A.; Ottman, M.J.; Wall, G.W.; White, J.W.; et al. Rising temperatures reduce global wheat production. *Nat. Clim. Chang.* **2015**, *5*, 143–147. [CrossRef]
21. Bai, H.; Tao, F. Sustainable intensification options to improve yield potential and co-efficiency for rice-wheat rotation system in China. *Field Crop Res.* **2017**, *211*, 89–105. [CrossRef]
22. Yang, W.C.; Mao, X.M. Uncertainty of crop models under influence of climate change. *J. Drain. Irrig. Mach. Eng.* **2018**, 36, 874–879, 902. (In Chinese with English abstract)
23. Liu, Y.; Tao, F. Probabilistic change of wheat productivity and water use in China for global mean temperature changes of 1, 2, and 3 oC. *J. Appl. Meteorol. Climatol.* **2013**, *52*, 114–129. [CrossRef]
24. Gennady, B.M.; Peter, T.H.; Bertram, O. Modelling long-term risk profiles of wheat grain yield with limited climate data. *Agric. Syst.* **2019**, *173*, 393–402.
25. Rashid, M.A.; Jabloun, M.; Andersen, M.N.; Zhang, X.; Olesen, J.E. Climate change is expected to increase yield and water use efficiency of wheat in the North China Plain. *Agric. Water Manag.* **2019**, *222*, 193–203. [CrossRef]
26. Warric, R.A.; Kenny, G.J.; Harman, J.J. *The Effects of Climate Change and Variation in New Zealand: An Assessment Using the CLIMPACTS System*; The International Global Change Institute (IGCI), University of Waikato: Hamilton, New Zealand, 2001. Available online: http://hdl.handle.net/10289/897 (accessed on 28 June 2020).
27. Warrick, R.A. Using SimCLIM for modelling the impacts of climate extremes in a changing climate: A preliminary case study of household water harvesting in Southeast Queensland. In Proceedings of the 18th World IMACS. In MODSIM Congress, Cairns, Australia, 13–17 July 2009; pp. 2583–2589.
28. Bao, Y.; Hoogenboom, G.; McClendon, R.; Urich, P. Soybean production in 2025 and 2050 in the southeastern USA based on the SimCLIM and the CSM-CROPGRO-Soybean models. *Clim. Res.* **2015**, *63*, 73–89. [CrossRef]
29. Amin, A.; Nasim, W.; Mubeen, M.; Ahmad, A.; Nadeem, M.; Urich, P.; Fahad, S.; Ahmad, S.; Wajid, A.; Tabassum, F.; et al. Simulated CSM-CROPGRO-cotton yield under projected future climate by SimCLIM for southern Punjab, Pakistan. *Agric. Syst.* **2018**, *167*, 213–222. [CrossRef]
30. Kenny, G.J.; Harman, J.J.; Flux, T.L.; Warrick, R.A.; Ye, W. The Impact of Climate Change on Regional Resources: A Case Study for Canterbury and Waikato Regions. In *The Effects of Climate Change and Variation in New Zealand: An Assessment Using the CLIMPACTS System*; Warrick, R.A., Kenny, G.J., Harman, J.J., Eds.; The International Global Change Institute (IGCI), University of Waikato: Hamilton, New Zealand, 2001. Available online: http://hdl.handle.net/10289/897 (accessed on 28 June 2020).
31. Storey, L.P. Effect of climate and land use change on invasive species: A case study of Tradescantiafluminensis (Vell.) in New Zealand. Ph.D. Thesis, University of Waikato, Hamilton, New Zealand, 2009. Available online: http://hdl.handle.net/10289/2634 (accessed on 28 June 2020).

32. Li, Z.; Song, M.; Feng, H.; Zhao, Y. Within-season yield prediction with different nitrogen inputs under rain-fed condition using CERES-Wheat model in the northwest of China. *J. Sci. Food Agric.* **2016**, *96*, 2906–2916. [CrossRef]
33. Liu, J.; Feng, H.; He, J.; Chen, H.; Ding, D.; Luo, X.; Dong, Q. Modeling wheat nutritional quality with a modified CERES-Wheat model. *Eur. J. Agron.* **2019**, *109*, 125901. [CrossRef]
34. Dar, E.A.; Brar, A.S.; Mishra, S.K.; Singh, K.B. Simulating response of wheat to timing and depth of irrigation water in drip irrigation system using CERES-Wheat model. *Field Crop Res.* **2017**, *214*, 149–163. [CrossRef]
35. Hoogenboom, G.; Jones, J.W.; Wilkens, P.W.; Porter, C.H.; Boote, K.J.; Hunt, L.A.; Singh, U.; Lizaso, J.L.; Whiht, J.W.; Uryasev, O.; et al. *Decision Support System for Agrotechnology Transfer, Version 4.5*; University of Hawaii: Honolulu, HI, USA, 2011.
36. Jones, J.W.; Hoogenboom, G.; Porter, C.H.; Boote, K.J.; Batchelor, W.D.; Hunt, L.A.; Wilkens, P.W.; Singh, U.; Gijsman, A.J.; Ritchie, J.T. The DSSAT cropping system model. *Eur. J. Agron.* **2003**, *18*, 235–265. [CrossRef]
37. Hoogenboom, G.; Porter, C.H.; Boote, K.J.; Shelia, V.; Wilkens, P.W.; Singh, U.; White, J.W.; Asseng, S.; Lizaso, J.I.; Moreno, L.P.; et al. The DSSAT crop modeling ecosystem. In *Advances in Crop Modeling for a Sustainable Agriculture*; Boote, K.J., Ed.; Burleigh Dodds Science Publishing: Cambridge, UK, 2019; pp. 173–216.
38. Saddique, Q.; Cai, H.; Ishaque, W.; Chen, H.; Chau, H.W.; Chattha, M.U.; Hassan, M.U.; Khan, M.I.; He, J. Optimizing the sowing date and irrigation strategy to improve maize yield by using CERES (Crops Estimation through Resource and Environment Systhesis)-Maize model. *Agronomy* **2019**, *9*, 109. [CrossRef]
39. Zheng, Z.; Cai, H.; Yu, L.; Hoogenboom, G. Application of the CSM–CERES–Wheat Model for Yield Prediction and Planting Date Evaluation at Guanzhong Plain in Northwest China. *Agron. J.* **2017**, *109*, 204. [CrossRef]
40. Zheng, Z.; Cai, H.; Hoogenboom, G.; Chaves, B.; Yu, L. Limited Irrigation for Improving Water Use Efficiency of Winter Wheat in the Guanzhong Plain of Northwest China. *Trans. ASABE* **2016**, *59*, 1841–1852.
41. What is SimCLIM? Available online: http://www.climsystems.com/simclim (accessed on 28 June 2020).
42. Warrick, R.A.; Ye, W.; Kouwenhoven, P.; Hay, J.E.; Cheatham, C. New Developments of the SimCLIM Model for Simulating Adaptation to Risks Arising from Climate Variability and Change. *MODSIM 2005. International Cogress on Modelling and Simulation*; Zerger, A., Argent, R.M., Eds.; Modelling and Simulation Society of Australia and New Zealand, 2005. Available online: https://hdl.handle.net/10289/5486 (accessed on 28 June 2020).
43. Kenny, G.J.; Warrick, R.A.; Campbell, B.D.; Sing, G.C.; Camilleri, M.; Jamieson, P.D.; Mitchell, N.D.; Mcpherson, H.G.; Salinger, M.J. Investigating climate change impacts and thresholds: An application of the CLIMPACTS integrated assessment model for New Zealand agriculture. *Clim. Chang.* **2000**, *46*, 91–113. [CrossRef]
44. Yin, C.; Li, Y.; Urich, P. *SimCLIM 2013 Data Manual*; CLIMsystems Ltd.: Hamilton, New Zealand, 2013. Available online: http://documents.climsystems.com/news/6-11-2013/SimCLIM_2013_AR5_data_manual.pdf (accessed on 28 June 2020).
45. China Meteorological Data Service Center. Available online: http://data.cma.cn/ (accessed on 28 June 2020).
46. IPCC. *Climate Change 2014: Synthesis Report. Contribution of Working Groups I, II and III to the Fifth Assessment Report to the Intergovernmental Panel o Climate Change*; Core Writing Team, Pachauri, R.K., Meyer, L.A., Eds.; IPCC: Geneva, Swizerland, 2014; p. 151, In IPCC AR5 Synthesis Report website.
47. Bao, Y.; Hoogenboom, G.; McClendon, R.W.; Paz, J.O. Potential adaptation strategies for rainfed soybean production in the south-eastern USA under climate change based on the CSM-CROPGRO-Soybean model. *J. Agric. Sci.* **2015**, *153*, 798–824. [CrossRef]
48. Carbone, G.J.; Kiechle, W.; Locke, C.; Mearns, L.O.; McDaniels, L.; Downton, M.W. Response of soybean and sorghum to varying spatial scales of climate change scenarios in the southeastern United States. *Clim. Chang.* **2003**, *60*, 73–98. [CrossRef]
49. Nasim, W.; Belhouchette, H.; Ahaman, M.H.; Jabran, K.; Ulah, K.; Fahad, S.; Shakee, M.; Hoogenboom, G. Modelling climate change impacts and adaptation strategies for sunflower in Punjab-Pakistan. *Outlook Agric.* **2016**, *45*, 39–45. [CrossRef]
50. IPCC. Climate Change 2007: Impacts, Adaptation and Bulnerability. In *Contribution of Working Group II to the Fourth Assessment Report of the Inter Governmental Panel on Climate Change*; Parry, M.L., Canziani, O.F., Palutikof, J.P., van der Linden, P.J., Hanson, C.E., Eds.; Cambridge University Press: Cambridge, UK, 2007; pp. 589–662.

51. Dettori, M.; Cesaraccio, C.; Duce, P. Simulation of climate change impacts on production and phenology of durum wheat in Mediterranean environments using CERES-Wheat model. *Field Crops Res.* **2017**, *206*, 43–53. [CrossRef]
52. Qu, C.; Li, X.; Ju, H.; Liu, Q. The impacts of climate change on wheat yield in the Huang-Huai-Hai Plain of China using DSSAT-CERES-Wheat model under different climate scenarios. *J. Integr. Agric.* **2019**, *18*, 1379–1391. [CrossRef]
53. Wang, B.; Liu, D.; Asseng, S.; Macadam, I.; Yu, Q. Modelling wheat yield change under CO_2 increase, heat and water stress in relation to plant available water capacity in eastern Australia. *Eur. J. Agron.* **2017**, *90*, 152–161. [CrossRef]
54. Parry, M.L.; Rosenzweig, C.; Iglesias, A.; Livermore, M.; Fischer, G. Effects of climate change on global food production under SRES emissions and social-economic scenarios. *Global Environ. Change* **2004**, *14*, 53–67. [CrossRef]
55. Araya, A.; Hoogenboom, G.; Luedeling, E.; Hadgu, K.M.; Kisekka, I.; Martorano, L.G. Assessment of maize growth and yield using crop models under present and future climate in southwestern Ethiopia. *Agric. For. Meteorol.* **2015**, *214*, 252–265. [CrossRef]
56. Semenov, M.A.; Shewry, P.R. Modelling predicts that heat stress, not drought, will increase vulnerability of wheat in Europe. *Sci. Rep.* **2011**, *1*, 66. [CrossRef]
57. Hernandez-Ochoa, I.M.; Pequeno, D.N.; Reynolds, M.; Babar, M.A.; Sonder, K.; Milan, A.M.; Hoogenboom, G.; Robertson, R.; Gerber, S.; Rowland, D.L.; et al. Adapting irrigated and rainfed wheat to climate change in semi-arid environments: Management, breeding options and land use change. *Eur. J. Agron.* **2019**, *109*, 125915. [CrossRef]

atmosphere

Article

Modelling Canopy Actual Transpiration in the Boreal Forest with Reduced Error Propagation

M. Rebeca Quiñonez-Piñón [1] and Caterina Valeo [2,*]

[1] Geomatics Engineering, Schulich School of Engineering, University of Calgary, 2500 University Drive NW, Calgary, AB T2N 1N4, Canada; mrquinon@gmail.com
[2] Mechanical Engineering, University of Victoria, P.O. Box 1700 STN CSC, Victoria, BC V8P 5C2, Canada
* Correspondence: valeo@uvic.ca

Received: 5 September 2020; Accepted: 22 October 2020; Published: 27 October 2020

Abstract: The authors have developed a scaling approach to aggregate tree sap flux with reduced error propagation in modeled estimates of actual transpiration ($\overline{T}_{plot}$) of three boreal species. The approach covers three scales: tree point, single tree trunk, and plot scale. Throughout the development of this approach the error propagated from one scale to the next was reduced by analyzing the main sources of error and exploring how some field and lab techniques, and mathematical modeling can potentially reduce the error on measured or estimated parameters. Field measurements of tree sap flux at the tree point scale are used to obtain canopy transpiration estimates at the plot scale in combination with allometric correlations of sapwood depth (measured microscopically and scaled to plots), sapwood area, and leaf area index. We compared the final estimates to actual evapotranspiration and actual transpiration calculated with the Penman–Monteith equation, and the modified Penman–Monteith equation, respectively, at the plot scale. The scaled canopy transpiration represented a significant fraction of the forest evapotranspiration, which was always greater than 70%. To understand climate change impacts in forested areas, more accurate actual transpiration estimates are necessary. We suggest our model as a suitable approach to obtain reliable $\overline{T}_{plot}$ estimates in forested areas with low tree diversity.

Keywords: climate change; actual evapotranspiration; modified Penman–Monteith; sap flow; scaling methods; allometric correlations; sapwood depth; sapwood area; leaf area index

1. Introduction

Climate change is reflected in almost every ecosystem by rising temperatures and changing precipitation patterns [1]. Higher seasonal temperatures in the boreal forest will increase evapotranspiration rates, decrease groundwater recharge, and affect water runoff levels [2–4]. Recent findings project a decrease in boreal forest biomass due to climate change, especially in the dominant conifer species [5], and these factors will likely change the boreal forest composition [6] and therefore change overall forest evapotranspiration. In vegetated areas, reliable water balance component estimates are of great importance for water management, sustainability, wildlife conservation, and nowadays, to understand and define plant species' challenges to climate adaptation [7]. Evaporation and transpiration are two water balance components whose estimates are well known to carry uncertainty due to the use of equations and models that (1) lump the components into a single estimate of evapotranspiration (ET), and (2) focus on estimating ET under ideal atmospheric conditions and with an unlimited source of water (i.e., potential ET) [8,9]. This paper focuses on estimating actual transpiration ($\overline{T}_{plot}$) and actual ET (E_a) as a first step to reduce the uncertainty introduced by assumptions of ideal conditions. Recent studies have proven that modeling actual evapotranspiration and using site-specific calibrated variables greatly reduces the error and improves estimates of actual evapotranspiration [10,11].

Scaling a single tree's point sap flow measurements to the watershed level is a complex and challenging task, not to mention the error propagated during the scaling process. Thus, most studies tend to focus on refining single components of the scaling process. For instance, direct ways to estimate actual transpiration of a single tree include the heat thermal dissipation method developed by Granier [12]. Still, sap flow radial variability should be accounted for and several researchers focus their effort on modelling and reducing the error associated with tree sap flow fluctuations [13–16].

Initial efforts to account for radial variation was to measure sap flow at different depths [17–20] of the tree trunk. The problem with measuring sap flow at different depths is that none of the thermal techniques are sensitive enough to determine the boundary between the sapwood and heartwood (i.e., sapwood depth) of a tree, and some radial flow and moisture transfer between sapwood and heartwood can be confounded with transversal sap flow. These radial variations in sap flow are also related to the species vascular structure, specifically the radial distribution and length of sapwood depth around the tree circumference [21–24]. Thus, the influence of a single tree's sapwood depth variability on estimating sap flow velocity should be acknowledged. The key point for most of the thermal techniques is that accurate measures of sapwood depth or sapwood area are required to take into account the sap flow radial pattern. Most studies measure sapwood depth using visual methods, which carry uncertainty [25], and studies suggest that more accurate methods to measure sapwood depth and sapwood area should be included in the scaling process in order to reduce error propagation at the tree scale [25–27]. The use of more accurate sapwood depth values to estimate sapwood area and tree sap mass flux should be a first step towards reduction of error propagation to larger scales (i.e., canopy).

Based on our previous work [25,26], the importance of accurate estimates of sapwood area and sapwood depth to model robust allometric correlations [28] cannot be underestimated. Allometric correlations to scale up sap flow to sap flux and to canopy transpiration normally require estimates or measurements of the structural characteristics of trees, such as diameter at the breast height, leaf area, and leaf area index. These findings indicate that there is not always a positive, direct correlation between the structural characteristics [29–38], and it is necessary to create species specific allometric correlations in the area of study.

Canopy heterogeneity is another source of uncertainty when estimating canopy and basin transpiration estimates [20,37–44]. Due to the differences in vegetation structure, each type of plant differs in its physiological process and therefore, in the amount of water required for transpiration [45]. For instance, coniferous trees require less water than deciduous trees because of their more conservative vascular structure and their tolerance to growing in xeric–mesic environments [46]. At the same time, trees' transpiration rates change according to micrometeorological conditions, solar energy, and water availability [47]. This combination of physiological, micrometeorological and energy factors generate spatial and temporal heterogeneity. Indeed, it is expected that the total transpiration of a forested area with mixed vegetation will be the aggregation of each tree's transpiration at a given time and under specific conditions.

In this work, we hypothesize that an accurate scaling approach can be reached by using the current, and most robust techniques to measure the necessary scaling parameters and in situ sap flow. We also consider that the main sources of error while scaling transpiration are (1) leaf area and sapwood area estimates, (2) sap flow radial variability, (3) canopy heterogeneity and density (interspecific and intraspecific variability), and (4) forest fragmentation. In addition to these, it is necessary to validate the final estimates, and also to estimate the propagated error.

Our previous work demonstrates that accurate estimates of canopy sapwood area [25,26,28] are predicated on obtaining sapwood depth values with a negligible error. Hence, sapwood areas of single trees can produce accurate regression models for estimating canopy sapwood area while considering the canopy's heterogeneity. This paper presents the computation of sap flow radial variations for each tree species included into the scaling approach. We use the allometric models—sapwood area and leaf area—reported in [28] to scale tree sap flow to canopy actual transpiration in a five-day period. These canopy actual transpiration estimates are compared to actual evapotranspiration (E_a) and actual transpiration values calculated with the Penman–Monteith equation, and the modified Penman–Monteith equation, respectively.

2. Methodology

2.1. Scaling Approach Concept

Our scaling approach is based on the concept that calibrated values of tree sap flow that are aggregated to the tree scale by using allometric correlations with low uncertainty, will provide estimates of tree sap flux that can be used to compute low error estimates of canopy transpiration when plot's vegetation heterogeneity is integrated into the mathematical model. For a more accurate estimate of canopy transpiration, we included canopy heterogeneity into our model by counting the total number of individuals of each species within our 60 × 60 m^2 delimited plots, and we measured each tree's circumference and estimated outer bark diameter, D_{OB}. Based on allometric correlations between D_{OB} and sapwood depth, we computed each plot's total sapwood area. These allometric correlations are reported in [26,28]. Figures 1 and 2 explain graphically the scaling approach to estimate canopy actual transpiration.

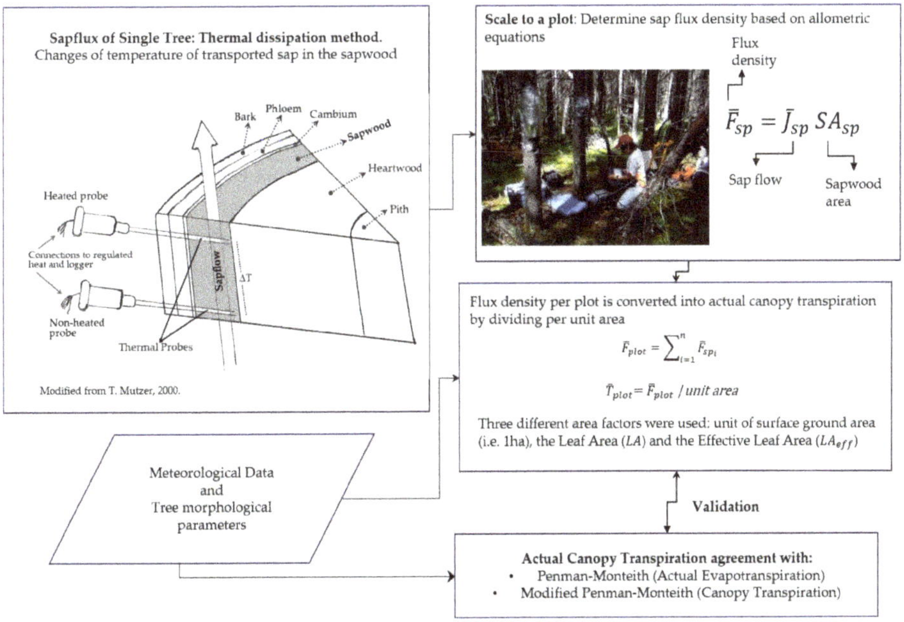

Figure 1. Sap flow measured at a single tree point, can provide actual tree sap flux estimates when the tree's sapwood area is known or accurately estimated. Each tree species sap flux can provide accurate estimates of a plot's actual transpiration by using reduced error allometric correlations of canopy structure parameters such as Leaf Area or Sapwood Area. The plot scale can provide estimates of ET and T comparable to estimates from Penman–Monteith (P-M) and Modified P-M method, respectively. Appendix A.4 details the calibration of tree sap flow measurements.

Figure 2. Thermal Dissipation Probes (TDPs) installed in a boreal, coniferous tree with the isolation material (upper part of the picture) ready to cover the sensors and protect them from direct exposure to solar radiation that could affect the sensors' temperature.

In summary, the expected outcomes are:

1. To aggregate mass sap flow from single trees to the plot scale;
2. To estimate the transpiration rates of a single plot (i.e., canopy transpiration);
3. To obtain estimates of canopy transpiration and validate these results through their comparison with other well-known and reliable methods (i.e., Penman–Monteith).

2.2. Study Site

A forest region located in the Sibbald Areas of Kananaskis Valley in Alberta, Canada, was the study site for field measurements to support this work. The Kananaskis Valley is a Montane closed forest formation [48] within the Rocky Mountains [49,50]. This type of forest has ridged foothills and a marked rolling topography. The Montane forest is classified as an ecoregion within the Cordilleran eco-province and experiences unique climatic conditions arising from the combination of physiography and air masses [48]. Within the province of Alberta, the Montane forest maintains the warmest temperatures during the winter than any other forested ecosystem.

Two plots of 60 × 60 m were delimited and used to scale up sap mass flow and to calculate the total rate of transpiration per plot. One plot was a pure coniferous site of *Pinus contorta* (Lodgpole pine) mixed with *Picea glauca* (White spruce), while the other was a pure deciduous site composed of *Populus tremuloides* (Trembling aspen). They are henceforth to be referred to as Coniferous site and Deciduous site, respectively. These two plots were part of the plot samples used to create allometric regression models to scale up five boreal species sapwood area to the plot level (SA_{plot}) [28].

Field campaigns conducted in 2004 and 2005 were used to collect all the material and biometrics required for the entire scaling process. Field data collected at each plot included: each plot's number of trees per species, each tree's outside bark diameter at breast height (D_{OB}), leaf area index (LAI_{plot}), soil moisture, and sap flow velocity (J_i). We measured LAI_{plot} within the 60 × 60 m plots using the Tracing Radiation and Architecture Canopies device (TRAC, 3rd Wave Engineering Co., Nepean, ON, Canada).

2.3. Measuring Single Trees' Sap Flow

Sap flow in a single tree was measured using the heat dissipation technique [51]. At each plot, a group of four trees (for each species inside the plot) were set up with thermal dissipation probes (TDP-30, Dinamax, Inc., Houston, TX, USA) for periods of 48 h. The thermal dissipation probes (TDP's) were installed in the North side of the trees and covered with a special insulating material (Figure 2) to avoid direct solar incidence and overheating of the sensors that might alter the logger readings. At the same time, a set of soil moisture sensors (six sensors) was placed in the soil (below the litter) to observe the changes in soil moisture content and to later compare with the trees' water uptake. The soil moisture values are also used in the empirical calculation of E_a. After 48 h, another group of four trees was set up with the TDP's and soil moisture sensors. This task was performed at least four times within each plot. We were interested in capturing the sap flow patterns of trees with different diameters (i.e., interspecific heterogeneity); therefore, the trees selected for measuring daily sap mass flow were chosen in order to cover the range of trees' outside bark diameter at breast height (D_{OB}) in each plot (i.e., the largest, the smallest, the mean, and other intermediate D_{OB} values of each species found inside the plot).

Sap mass flow measurements were corrected by applying the original calibration presented by Granier [52], and presented here in Appendix A. The canopy transpiration estimates were computed after sap mass flow data were corrected for radial patterns of sap flow. Trembling aspen individuals were excluded from the radial correction, since it had been proven that diffuse-porous tree radial sap flow does not vary significantly [18,53,54]. The method to compute canopy transpiration from sap mass flow data is detailed in Section 2.2.

In order to validate the scaled transpiration values, the actual forest evapotranspiration (E_a) and plant actual transpiration (T_{plant}) were estimated for both sites. The former estimate was computed using the Penman–Monteith equation, while the latter was computed using a modified version of the Penman–Monteith equation [47]. The mathematical theory behind the three models' computations is detailed in Sections 2.3 and 2.4.

The meteorological data required to compute E_a and T_{plant} were collected using a HOBO meteorological station (Onset Computer Corporation, Bourne, MA, USA), which was set up in a 25 m radius clearing located inside the Barrier Lake forestry trails nearby. The installed sensors measured temperature, relative humidity, dew point, rainfall, atmospheric pressure, wind speed, gust speed, wind direction, solar radiation, and photosynthetically active radiation. The sensors were placed at height of about 3 metres above the ground level. All the variable data were collected every minute.

The sap flow values were assessed by observing the order of magnitude and their agreement between some meteorological variables and the sap flow trends. It was expected that sap flow rates would be greater in sunny, calm days than in rainy, cold, cloudy days, with a plateau at night. There are periods of the day when sap flow decreases to avoid desiccation, and some other periods in which it is known that all trees reach their maximum sap flow rates.

2.4. Scaling Actual Canopy Transpiration

2.4.1. Radial Patterns of Sap Flow

The acropetal sap transport rate has a radial gradient that decreases from the outermost part of the sapwood towards the pith. Since there is enough evidence of the significance of the sap flow radial gradient while scaling up sap flux density from a single point to the entire tree [17,18,24,55,56] a sap flow radial profile function developed by [57] was used to calculate the sap flow velocity along the entire sapwood depth of each tree. The radial profile function accounts for the fractional changes in sap flow as a function of the maximum sap flow rate, the sapwood depth at which this rate occurs, the total sapwood depth and the rate at which the sap flow velocity decreases from the outer to the inner sapwood:

$$f(x) = \varpi exp(-0.5\,[\frac{x - x_0}{\beta}]^2) \quad (1)$$

where $f(x)$ is the sap flow rate index (expressed as a fraction), ϖ is the maximum sap flow rate (equal to 1.0) occurring at the x_o sapwood depth, and $1/\beta$ is the rate at which the sap flow radially decreases towards the pith's trunk. In order to calculate sap flow velocity changes instead of fractional changes, Equation (1) was modified to the following form:

$$V_{0-3}/V_{max} = \frac{1}{\beta\sqrt{2\pi}} \int_0^3 exp(-0.5\left[\frac{x - x_0}{\beta}\right]^2)\,dx \quad (2)$$

where V_{0-3} is the sap flow velocity in the first three centimeters of sapwood, and the maximum sap flow velocity is V_{max}. Studies in variations of radial sap flow have found that in conifers, the maximum velocity or the largest portion of sap flow occurs in the first centimetre [17], the first 2 cm and 3 cm [58] of sapwood depth (from cambium to pith). Research outcomes [56] reported graphs showing that maximum sap flow occurs at 20% of the depth (from cambium to pith as well). It seems that the depth at which the maximum sap flow occurs is a standard pattern independent of the tree size. Based on these previous results, here it is assumed that V_{max} occurs somewhere between the first two centimeters; thus, x_0= 2 cm. Other studies have reported that the rate of decrement in radial sap flow is about 20–24% in conifers; thus, β has been assumed to equal 4 (i.e., a 25% of decrement). As V_{0-3} is known; that is, it is calculated from the field measurements, V_{max} can be estimated from Equation (2). Finally, V_{max} is used to estimate the sap flow velocity along the entire sapwood depth ($\overline{sd}$) at a specific time:

$$V_{0-\overline{sd}} = V_{max}\frac{1}{4\sqrt{2\pi}}\int_{0}^{\overline{sd}} exp(-0.5\left[\frac{x-2}{4}\right]^2)\,dx \quad (3)$$

Note that $V_{0-\overline{sd}}$ is J_i, the original symbol used by [53] to define sap flow velocity; thus,

$$F_s = SAJ_i = SA_{tree}V_{0-\overline{sd}_i} \quad (4)$$

where the sap flow velocity, $V_{0-\overline{sd}i}$ (cm s^{-1}), is converted into the total trunk's mass flow, F_s, (cm^3 s^{-1}), by scaling from the point of measurement, to the total sapwood cross-sectional area, SA_{tree} (cm^2). Values of $V_{0-\overline{sd}i}$ were computed with Equation (3) at each time step (5 min) and then used to estimate each tree's daily F_s.

To estimate an average canopy transpiration rate, $\overline{T}_{plot}$, single tree transpiration values where scaled up to the whole plot. First, a diurnal average sap flow per species was estimated, and multiplied by the total sapwood area of that species inside the plot, obtaining the total average canopy water mass flow:

$$\overline{J}_{sp} = \frac{1}{m}\sum_{i=1}^{m}\overline{V}_{0-\overline{sd}} \quad (5)$$

and

$$\overline{F}_{sp} = \overline{J}_{sp}\,SA_{sp} \quad (6)$$

where $\overline{J}_{sp}$ is the average sap flow velocity of the species *sp* obtained by the summation of the diurnal average sap flow velocity of each *i*th individual and divided by the total *m* individuals of the same species whose sap flow was measured. $\overline{F}_{sp}$ is the average of the species *sp* total mass flow, and SA_{sp} is the total sapwood area of the species *sp*. present in a specific plot The calculation of the average canopy water mass flow ($\overline{F}_{plot}$) is through the summation of each plot's species total water mass flow:

$$\overline{F}_{plot} = \sum_{i=1}^{n}\overline{F}_{sp_i} \quad (7)$$

The average sap flow within the plot is:

$$\overline{J}_{plot} = \sum_{i=1}^{n}\overline{J}_{sp_i} \quad (8)$$

To estimate $\overline{T}_{plot}$, the $\overline{F}_{plot}$ is normally divided by a unit area of ground (i.e., 1 ha). This division allows one to observe the agreement between canopy transpiration and actual forest evapotranspiration (E_a), or actual forest transpiration (T_{plant}). Here, we assessed three different units of surface ground area: Plot sapwood area (SA_{plot}), the actual leaf area (LA), and the effective leaf area (LA_{eff}). Finally, $\overline{T}_{plot}$ values were compared with an average of E_a and T_{plant}.

2.4.2. Azimuthal Sap Flow Variation

Previous work [27] where we reported that we measured sapwood depth in four different sides of the tree to account for the sapwood depth variation around the tree trunk, and therefore the azimuthal variation in sap flow. The sapwood depth was measured under the microscope, and for each tree, an average sapwood depth was computed. We consider that this is sufficient to account for azimuthal sap flow variation.

2.4.3. Water Storage Capacity Estimates

It is assumed that the water stored in the tree trunk equals the amount of water replenished at night. The assumption is based on previous research focused on the contribution of a tree trunk's stored water to transpiration, under dry and wet conditions. Reported results show that on average, of the daily amount of water transpired by a tree, 14.8–20.0% corresponds to the trunk's stored water [59–62]. In Trembling aspen, the water trunk provided 11.6% of the mean daily transpiration [61] and most of the time, full replenishment for the tree trunk occurs at night [60], which creates a water balance between the tree water lost during the day and the water recharged at night. In addition, it has been determined that for scaling purposes, the error associated with water storage capacity is practically null if between individuals, the sap flux variability is low [60].

2.5. Estimating Actual Forest Evapotranspiration

The Penman–Monteith equation estimates the actual evapotranspiration of vegetated surfaces by accounting for all the micrometeorological factors that influence evapotranspiration as well as the influence of the canopy conductance and aerodynamic resistance in the rates of vegetation transpiration:

$$\lambda E_a = \frac{\Delta(R_n - G) + \rho_a c_p (e^o - e_a)/r_a}{\Delta + \gamma\left[1 + \frac{r_c}{r_a}\right]} \tag{9}$$

where λE_a is the latent heat of actual evapotranspiration, Δ is the slope of the saturation vapor pressure curve (kPa°C^{-1}), R_n is the net solar radiation, and G is the soil heat flux (all these terms in units of (J m^{-2} s^{-1})). The air density is denoted by ρ_a (kg m^{-3}) and c_p is the specific heat of air at constant pressure (i.e., 1010 Jkg^{-1}°C^{-1}). The term $(e^o - e_a)$ is the vapor pressure deficit (VPD) calculated by the difference between the saturation vapor pressure (e^o, (kPa)) and the actual vapor pressure (e_a, (kPa)). The psychrometric constant γ is in units of (kPa°C^{-1}). The aerodynamic terms, r_a and r_c are the aerodynamic resistance to vapor and heat transfer, and the bulk canopy resistance (both expressed in sm^{-1}). To convert the latent heat of evapotranspiration to actual evapotranspiration (E_a), $E_a = \frac{\lambda E_a}{\lambda}$ in units of mms^{-1} was employed. The equations used to solve the aerodynamic and energy parameters of the Penman–Monteith equation are detailed in Appendix A.

2.6. Estimating Actual Canopy Transpiration

Liu et al. [47] presented a modified version of the Penman–Monteith equation in order to estimate actual canopy transpiration at large scales. According to Liu et al. [47], a model such as Penman–Monteith should be adjusted by separately estimating the transpiration of shaded and sunlit leaves as follows (stratified model):

$$T_{plant} = T_{sun} LAI_{sun} + T_{shade} LAI_{shade} \tag{10}$$

where T_{sun} and T_{shade} are the actual transpiration of sunlit and shaded leaves, respectively; LAI_{sun} and LAI_{shade} are the leaf area indices for sunlit and shaded leaves as well. The Penman–Monteith equation is then used by Liu et al. [47] to estimate T_{sun} and T_{shade}:

$$\lambda T_{sun} = \frac{\Delta(R_{n,sun}) + \rho_a c_p (e^o - e_a)/r_a}{\Delta + \gamma[1 + r_s/r_a]} \tag{11}$$

and

$$\lambda T_{shade} = \frac{\Delta\left(R_{n,shade}\right) + \rho_a c_p (e^o - e_a)/r_a}{\Delta + \gamma[1 + r_s/r_a]} \tag{12}$$

where $R_{n,sun}$ and $R_{n,shade}$ are the net solar radiation available for sunlit and shaded leaves ($Jm^{-2}s^{-1}$), respectively, and r_s is the stomatal resistance (sm^{-1}). The rest of the parameters and units remain the same as in Equation (9).

The boreal ecosystem productivity simulator (BEPS) provides a set of equations to calculate $R_{n,sun}$ and $R_{n,shade}$ (Liu et al., [47,63]). The equations compute the shortwave solar radiation for sunlit and shaded leaves as well. The net longwave solar radiation is assumed to behave equally for sunlit and shaded leaves; therefore, a single equation is used to calculate net longwave solar radiation. Thus, $R_{n,sun}$ and $R_{n,shade}$ are respectively given by:

$$R_{n,sun} = R_{s,sun} + R_{nl,sun} \tag{13}$$

and

$$R_{n,shade} = R_{s,shade} + R_{nl,shade} \tag{14}$$

where $R_{s,sun}$ and $R_{s,shade}$ are the shortwave solar radiation for sunlit and shaded leaves, respectively, and $R_{nl,sun}$ and $R_{nl,shade}$ are the net longwave solar radiation for sunlit and shaded leaves, respectively. The solution to these equations is formulated in Appendix A, and also provided by Liu et al. [47].

3. Results

Not all of the instrumented trees provided credible data and after all the sap flow data collection were checked for quality, only four Trembling aspen, five Lodgepole pine, and four White spruce trees provided credible sap flow measurements adequate for scaling up to the plot scale. In the case of the Coniferous site, eight days of sap flow measurements were used to calculate $\overline{F}_{plot}$ and $\overline{T}_{plot}$. The Deciduous site provided four days of sap flow measurements and meteorological data. For the same dates, E_a and T_{plot} were computed. Each site's daily values were averaged and compared with the average $\overline{T}_{plot}$ obtained for their respective time periods.

3.1. Scaling Canopy Transpiration

The Deciduous plot's ratio of SA_{plot} to the plot's basal area was 0.57, while in the Conifer site, the ratio was 0.54 for the Lodgepole pine trees and 0.38 for the White spruce trees. Thus, the Trembling aspen showed a larger sapwood area per unit basal area at the plot scale than the conifer species. That was expected since diffuse-porous trees have larger sapwood areas in order to meet their water demand (i.e., they are less efficient at transporting water). As it is shown in the following sections, the Deciduous site drew larger mass flow per plot than the Coniferous site.

Figures 3–5 exemplify the diurnal sap flow patterns in Lodgepole pine, White spruce, and Trembling aspen. In each plot, the dashed line is R_s and the solid line is J_i. Two individuals of different D_{OB} are presented in order to exemplify the differences in J_i due to the tree size. Notice that the Lodgepole pine J_i is somewhat tempered in comparison to R_s.

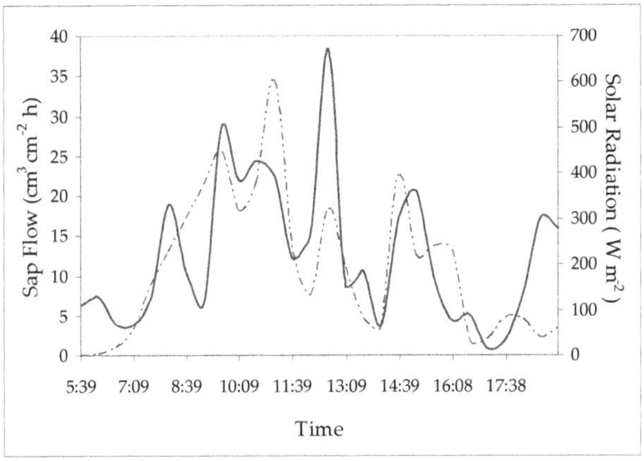

Figure 3. An example of a Lodgepole pine tree diurnal sap flow. Tree's diameter at breast height was 17 cm. Day of the year: 216, in 2004. Dashed line is R_s and the solid line is J_i.

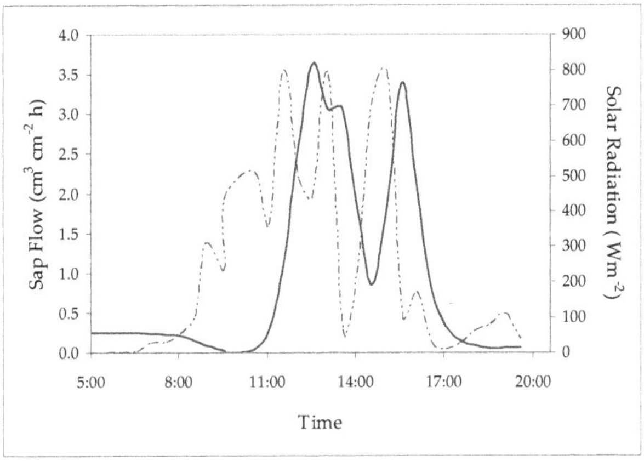

Figure 4. An example of a White spruce tree diurnal sap flow. Tree's diameter at breast height was 18 cm. Day of the year: 232, in 2004. Dashed line is R_s and the solid line is J_i.

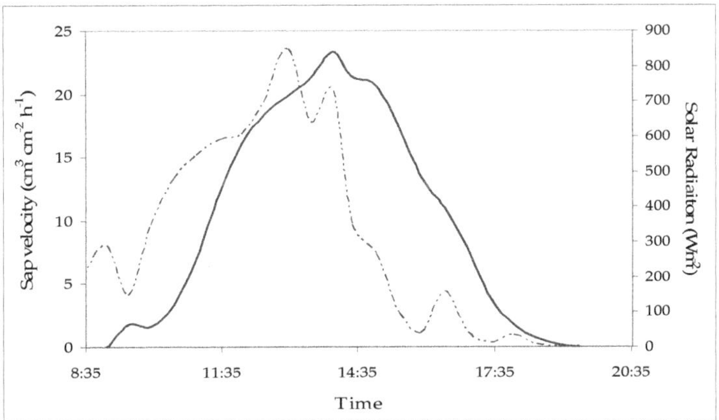

Figure 5. An example of a Trembling aspen tree diurnal sap flow. Tree's diameter at breast height was 31 cm. Day of the year: 228, in 2004. Dashed line is R_s and the solid line is J_i.

Each tree's sap flow pattern was analyzed in order to determine the times of initial and final daily transpiration activity. The transpiration patterns of the sampled trees showed activity starting early in the morning (around 500 and 545 h) and finishing between 1700 and 1900 h. Variations in the time at which the tree stopped transpiring and starting transpiring again were related to the meteorological changes.

The radial profile function to correct the sap flow velocity showed that the sap flow velocity values could have an underestimation of 12.5% in trees with a relatively small sapwood depth (3.5 ± 1.5 cm). The average $\overline{sd}$ in conifers ranged between 3.10 cm and 3.50 cm. In this particular case, if the radial profile correction could not be applied, the sap velocity will be underestimated when scaled to the entire tree. Each species, $\overline{F}_{sp}$ and $\overline{F}_{plot}$ are reported in Table 1. The Coniferous site total mass flow is the summation of the two species populating the site.

Table 1. $\overline{F}_{sp}$ and $\overline{F}_{plot}$ (m^3day^{-1}) at each site (n is the number of individuals used per plot).

Site	Tree Type	n	$\overline{F}_{sp}$	$\overline{F}_{plot}$	Days Averaged
Conifer-4	Lodgepole pine	5	12.64		8
	White spruce	4	2.57	15.21 [1]	8
Deciduous-6	Trembling aspen	4	32.44	32.44	4

[1] Conifer-4 $\overline{F}_{plot}$ is the sum of $\overline{F}_{sp}$ of both species within the plot.

3.2. Actual Evapotranspiration Results

All the mathematical models to calculate E_a are detailed in Appendix A. The most complex parameter to obtain is r_c. A series of reduction functions were used, and the assumptions made provided half-hourly r_c values that are in reasonable agreement with the values listed by [64–66]. The other parameter that was estimated in an uncommon way was R_n. This was done by integrating parameters that take into account the influence of LAI, gap fraction and emmisivity of understory and overstory. Since the determination of LAI_u and Ω_u was essentially based on previous reports, which at the same time are based on a few assumptions, it was necessary to observe the influence of LAI_u and Ω_u values on the calculation of E_a. Thus, a sensitivity analysis of E_a was performed by using different LAI_u and Ω_u values. The range of values to test LAI_u and Ω_u were 0.6–1.5 and 0.5–0.9, respectively. The obtained estimates of E_a with respect to the initial E_a differ in the range of -2.0×10^{-4} to 9.0×10^{-4} mmd^{-1}. When LAI_u and Ω_u are set up as 0.6 and 0.9, respectively, E_a estimates are practically the same as when LAI_u and Ω_u are set up as 1.0 and 0.5 (the values used here), respectively. The sensitivity analysis was also performed to see the impact on the average of E_a (i.e., $\overline{E}_a$) per day. The analysis showed differences among values in the range of 2.0×10^{-4} to 6.0×10^{-4}. In conclusion, the assumed LAI_u and Ω_u values were considered adequate. Final estimates of E_a are listed in Table 2. The E_a values are shown per date and sorted by the type of site that was set up for sap flow measurements in the same dates.

Table 2. Penman–Monteith E_a and $\overline{E}_a$ estimates during the same days that sap flow was measured at each site (in 2004). $\overline{E}_a$ is the daily E_a average.

Day of the Year	Conifer-4 E_a(mm/d)	Day of the Year	Deciduous-6 E_a(mm/d)
212	1.50	225	4.79
213	0.78	226	5.82
215	3.01	227	3.29
216	1.68	228	3.27
231	0.90		
232	0.87		
234	3.63		
235	0.07		
$\overline{E}_a$	1.56		4.29

Liu et al. [47] reported that Canadian boreal forest evapotranspiration values range between 100 and 300 mm year^{-1}. Additionally, Liu et al. [47] estimated that a coniferous land cover could have a yearly transpiration of 123 mm with an s = 55 mm; and deciduous and mixed forests land covers were reported with yearly transpiration values of 327 mm and 244 mm, respectively. On examination of the previous results, it would seem that there is an overestimation of E_a; however, 2004 had a particularly wet and hot summer, that exceeded reported rainfall normals [67] by a factor of 0.75 in July and 2.27 in August. Moreover, daily maximum temperatures during the months of July and August were greater than the daily maximum values reported in Environment Canada's Climatic Normals [68]. That is, July and August maximum temperatures varied between 24 and 29 °C, respectively, while the Climatic Normals reported maximum temperatures of 21.5 and 21.1 °C, respectively. Thus, the conditions for large evapotranspiration amounts that are greater than normal maximums could be considered reasonable for this wet and hot summer.

Variation in the Soil Field Capacity

The soils in the field plots were comprised of a sandy loam. The field capacity of a sandy loam soil varies between 0.16 and 0.22, and its wilting point is 0.073 [68]. The reported E_a was calculated using an average value of the soil field capacity. Calculations of θ_e, g_s, and E_a were made using the lower and upper bounds of the soil's θ_{fc}.

Results showed that in days when $\theta_e \leq 0.00$, the function limiting E_a was $g(\theta_{sm})$, causing g_s to become practically null, and making r_c reach its maximum value. In these days, there was no difference in the final E_a since the computation of θ_e will always be zero or negative, no matter the θ_{fc} value. Of course, in those days the factor limiting E_a was soil moisture to the point that observed E_a values were lower than 1 mmd^{-1} (e.g., days 213 and 235, Coniferous site).

When $0.16 \geq \theta_{sm} \geq 0.22$, soil moisture is not limiting at all, and other environmental factors drive E_a. In these cases, there was no variation in the final E_a estimate. It was noticed as well that the immediate limiting factor was VPD, and then R_s (e.g., days 231, Coniferous site).

Finally, if $\theta_{sm} \approx \theta_{wp}$, there is variation in the estimates of E_a. This was noticeable for just two days in the whole data set used here (days 215 and 216, set up in the Conifer site). When θ_{sm} varied from 0.0750 to 0.0795, the changes in θ_{fc} generated E_a to vary between 2.54 mmd^{-1} and 3.73 mmd^{-1}, when θ_{fc} was set up as 0.22 and 0.16 respectively (day 215). When θ_{sm} varied from 0.0735 to 0.0743, the changes in θ_{fc} caused E_a of 0.90 mmd^{-1}, either θ_{fc} was 0.22 or 0.16, respectively (day 216). The reported E_a values for these two days are 3.01 mmd^{-1} and 1.68 mmd^{-1}. In those two days, it could be said that there is a variation in the E_a estimates between 0.47 mmd^{-1} and 0.78 mmd^{-1}.

3.3. Actual Canopy Transpiration Results

Appendix A details the mathematical models, and the computation of T_{sun} and T_{shade} is very similar to the one applied for computing E_a. The main changes rely on substituting R_n by either R_{sun} or R_{shade} and the use of r_s instead of r_c. Tables 3 and 4 show the obtained transpiration estimates for shaded, sunlit leaves, and the total canopy transpiration in the Conifer and Deciduous sites, respectively. It is worth mentioning that for the Deciduous site, the T_{plant} estimates were based on the estimation of $g(\mathrm{VPD})$ computed with $K_{VPD} = 0.84$ kPa.

Table 3. Modified Penman–Monteith T_{plant} estimates during the same days that sap flow was measured at the Coniferous site. T_{plant} is the summation of T_{shade} and T_{sun}. $\overline{T}_{plant}$ is the average of the daily T_{plant}. All estimates are in mm/d.

Day of the Year	T_{shade}	T_{sun}	T_{plant}
212	0.38	1.08	1.46
213	0.19	0.56	0.75
215	1.80	0.73	2.53
216	1.26	0.41	1.67
231	0.56	0.66	1.22
232	0.52	0.64	1.16
234	2.59	0.95	3.54
235	0.11	0.04	
$\overline{T}_{plant}$			1.56

Table 4. Modified Penman–Monteith T_{plant} estimates during the same days that sap flow was measured at the Deciduous site (measured in 2004). T_{plant} is the summation of T_{shade} and T_{sun}. $\overline{T}_{plant}$ is the average of the daily T_{plant}. All estimates are in mm/d.

Day of the Year	T_{shade}	T_{sun}	T_{plant}
225	3.00	1.75	4.75
226	3.67	2.13	5.80
227	2.44	1.42	3.86
228	2.05	1.20	3.25
$\overline{T}_{plant}$			4.42

Variation of the Soil Field Capacity

As in the E_a estimates, there is variation in the estimates of T_{plant} if $\theta_{sm} \approx \theta_{wp}$. At the Coniferous site, days 215 and 216 showed the variations at the Coniferous site. When θ_{sm} varied from 0.0750 to 0.0795, the changes in θ_{fc} generated T_{plant} to vary between 2.10 mmd^{-1} and 3.22 mmd^{-1}, (keeping θ_{fc} equal to 0.22 and 0.16 respectively; day 215). When θ_{sm} varied from 0.0735 to 0.0743, the changes in θ_{fc} caused T_{plant} of 0.87 mmd^{-1}, either θ_{fc} was 0.22 or 0.16, respectively (day 216). The reported T_{plant} values for these two days are 2.53 mmd^{-1} and 1.67 mmd^{-1}. In those two days, it could be said that there is a variation in the T_{plant} estimates between 0.43 mmd^{-1} and 0.80 mmd^{-1}.

3.4. Agreement between Methods

The Coniferous site's daily average estimates of E_a and T_{plant} (Tables 2 and 3) are practically the same (1.56 mm/d). For the Deciduous site $\overline{T}_{plant} > \overline{E}_a$ by 0.13 mm/d. Hence, both Equations (9) and (10) draw very similar estimates. Such close similarity could mean that the wet, hot summer conditions of the studied area made the evaporation component negligible. Nevertheless, this should be part of future studies that could observe the agreement between the original Penman–Monteith equation and the stratified model developed by Liu et al. [47].

The comparison between $\overline{E}_a$ and $\overline{T}_{plot}$ is shown in Table 5, while Table 6 shows the comparison between $\overline{T}_{plant}$ and $\overline{T}_{plot}$. For these comparisons, the transpiration values are expressed as the average of the $\overline{J}_{sp}$ [$mm^3_{sap} mm^{-2}_{SA} d^{-1}$] measured in the trees inside of each plot per unit ground area. This unit ground area was estimated as the ratio of SA_{plot} per 1 ha (from now on referred to as "SA_{plot} as unit ground area"). Additionally, E_a and T_{plant} were averaged (i.e., $\overline{E}_a$ and $\overline{T}_{plant}$) on the same days for which $\overline{F}_{plot}$ was computed.

Table 5. $\overline{E}_a$ and $\overline{T}_{plot}$ at the Coniferous (8 days average) and Deciduous (4 days average) sites. SA_{plot} was used as the unit ground area to estimate $\overline{T}_{plot}$. All estimates in mm/d.

Site	$\overline{E}_a$	$\overline{T}_{plot}$	Scale	Agreement
Conifer-4	1.56	1.52	canopy	$\overline{T}_{plot} = 0.97(\overline{E}_a)$
Deciduous-6	4.29	3.14	canopy	$\overline{T}_{plot} = 0.73(\overline{E}_a)$
Deciduous-6 [1]	5.31	3.14	canopy	$\overline{T}_{plot} = 0.59(\overline{E}_a)$

[1] When $K_{VPD} = 0.79$ kPA.

Table 6. $\overline{T}_{plant}$ and $\overline{T}_{plot}$ at the Coniferous (8 days average) and Deciduous (4 days average) sites. SA_{plot} was used as the unit ground area to estimate $\overline{T}_{plot}$. All estimates in mm/d.

Site	$\overline{E}_a$	$\overline{T}_{plot}$	Scale	Agreement
Conifer-4	1.56	1.52	canopy	$\overline{T}_{plot} = 0.97(\overline{T}_{plant})$
Deciduous-6 [1]	4.42	3.14	canopy	$\overline{T}_{plot} = 0.71(\overline{T}_{plant})$

[1] When $K_{VPD} = 0.84$ kPA.

The agreement between the Coniferous $\overline{E}_a$ and $\overline{T}_{plot}$ is acceptable and showed that $\overline{T}_{plot}$ is about 97% of the total forest evapotranspiration. The remaining 3% of $\overline{E}_a$ may be attributed to the other sources of forest evapotranspiration such as surface evaporation and understory transpiration. The contribution of understory evapotranspiration varies and it could be fairly large during the growing season; however, [69] listed different sources that measured understory ET in stands of different pinaceas, and percentages range from 6% to 60% as understory contribution to forest ET. Thus, it is reasonable to attribute the difference between both methods to understory ET. Very similar results were seen in the comparison between $\overline{T}_{plant}$ and $\overline{T}_{plot}$ where $\overline{T}_{plot}$ is 97% of the $\overline{T}_{plant}$ estimates. Although both values are quite similar, the $\overline{T}_{plant}$ is greater than $\overline{T}_{plot}$ by 0.04 mm/d. The agreement is acceptable as well; however, it was expected that both values would be equal (i.e., $\overline{T}_{plant} = \overline{T}_{plot}$).

The Deciduous plot results showed a better agreement with $\overline{E}_a$ when K_{VPD} was set as 0.84 kPa and $VPD_c = 1.0$ kPa. In this case, $\overline{T}_{plot}$ is about 73% of $\overline{E}_a$, and about 71% of $\overline{T}_{plant}$. This value can also be considered acceptable as well, since the days when the $\overline{J}_{sp}$ was measured, the soil moisture was not limiting, and VPD was the driving factor. As it has been shown in other works [52], when this situation happens, the sap flow reaches a plateau and becomes quasi constant along the day. Just as when water is limiting, $\overline{J}_{sp}$ decreases. Thus, the remnant 28% of the $\overline{E}_a$ may be attributed to the understory transpiration and other surfaces of evaporating water.

3.5. Leaf Area Indices as Scaling Factors

Assessing other unit areas that could be helpful in transforming sap flux density values into a canopy transpiration rate, effective leaf area (LA_{eff}) and actual leaf area (LA) were used as unit areas as well:

$$\overline{T}_{plot} = \overline{J}_{plot} \times SAI_{eff} \tag{15}$$

where

$$SAI_{eff} = \frac{SA_{sp}}{LAI_{eff} \times A_{plot}} = \frac{SA_{sp}}{LA_{eff}} \tag{16}$$

or

$$\overline{T}_{plot} = \overline{J}_{plot} \times SAI_{actual} \tag{17}$$

where

$$SAI_{actual} = \frac{SA_{sp}}{LAI \times A_{plot}} = \frac{SA_{sp}}{LA} \tag{18}$$

Leaf area values are the same values as those used to create the regression model with SA_{plot} in [25]. Results are shown in Tables 7 and 8. As it is appreciated, LA_{plot} and LA_{eff} as unit ground areas describe the canopy transpiration of the Coniferous site as 48% and 67% of $\overline{E}_a$, respectively; and the same agreements are shown with $\overline{T}_{plant}$. In the case of the Deciduous site, $\overline{T}_{plot}$ is described as 64% and 83% of $\overline{E}_a$. On the other hand, $\overline{T}_{plot}$ is 62% and 80% of $\overline{T}_{plant}$. The LA_{eff} as a unit area describes the Deciduous $\overline{T}_{plot}$ as a larger proportion of $\overline{E}_a$ than the unit ground area when these values are based on $K_{VPD} = 0.84$ kPa).

Table 7. LA_{plot}, LA_{eff}, and site average canopy transpiration over eight days at the Coniferous site.

Unit Area	Lodgepole Pine	White Spruce	$\overline{T}_{plot}$	Agreement
LA_{plot}	1.56	1.52	0.74	$\overline{T}_{plot} = 0.48(\overline{T}_a)$ $\overline{T}_{plot} = 0.48(\overline{T}_{plant})$
LA_{eff}	0.86	0.18	1.04	$\overline{T}_{plot} = 0.67(\overline{E}_a)$ $\overline{T}_{plot} = 0.67(\overline{T}_{plant})$

Table 8. LA_{plot}, LA_{eff}, and site average canopy transpiration over four days at the Coniferous site.

Unit Area	$\overline{T}_{plot}$	Agreement
LA_{plot}	2.75	$\overline{T}_{plot} = 0.64(\overline{E}_a)$ $\overline{T}_{plot} = 0.62(\overline{T}_{plant})$
LA_{eff}	3.55	$\overline{T}_{plot} = 0.83(\overline{E}_a)$ $\overline{T}_{plot} = 0.80(\overline{T}_{plant})$

4. Discussion

The main objectives of this study involve scaling issues in transpiration: firstly, to identify those parameters influencing transpiration at different scales in order to use them as scaling parameters if adequate models can be developed [25,26,28]; and secondly (but no less important), the improvement of the final transpiration estimates at larger scales. This is a complex task since there often exists large intra- and interspecific variability that, at the same time, is controlled by biophysical characteristics. In this study, these problems were faced and addressed by using more accurate methods to estimate the scaling factors in order to avoid large uncertainty in the final estimates. The effectiveness of using more accurate methods is proven through the validation of $\overline{T}_{plot}$ estimates. That is, $\overline{T}_{plot}$ will be reasonable result if first, $\overline{T}_{plot} = \overline{T}_{plant}$, or at least $\overline{T}_{plot} \approx \overline{T}_{plant}$; and second, $\overline{T}_{plot}$ will be a significant proportion of $\overline{E}_a$.

Each site's $\overline{T}_{plot}$ shows an acceptable agreement with the computed actual forest evapotranspiration and the actual canopy transpiration—Equations (9) and (10). In the Deciduous site case, the obtained $\overline{T}_{plot}$ motivates one to speculate if the agreement is good enough. In this particular case there is an issue worth mentioning here (in case the reader considers the $\overline{T}_{plot}$ fraction to be small). The days in which the $\overline{T}_{plot}$ was calculated, showed large E_a and T_{plant} values because θ_{sm} was not limiting, and VPD was driving E_a and T_{plant} transpiration as well. In this case, the empirical factor K_{VPD} was adjusted as much as possible by respecting previous reports on the influence of VPD on g_s. The actual r_c of the Trembling aspen individuals could go beyond the empirical estimates, but there is no field data that could evince this and allow modification to K_{VPD}. Moreover, BEPS results suggest that at larger scales, a deciduous forest's transpiration is about 67% of the annual actual forest evapotranspiration [47]. Therefore, the Deciduous $\overline{T}_{plot}$ are considered reasonable estimates.

The three different area factors that were used to calculate actual canopy transpiration drew dissimilar results. Still, the three estimated $\overline{T}_{plot}$ values always met the expected agreements with $\overline{E}_a$ and $\overline{T}_{plant}$. The Coniferous site $\overline{T}_{plot}$ estimated by means of SA_{plot} as a unit ground area (i.e., using $SA_{plot}10 \times 10^6$ m^2) implies that there is a significant contribution of canopy transpiration to the total ET of the studied sites. Moreover, the estimated $\overline{T}_{plot} \approx \overline{T}_{plant}$. Thus, the canopy transpiration rates are in good agreement with previous works when using SA_{plot} as the unit ground area.

However, using any leaf area as a unit area factor, it seems that canopy transpiration is underestimated (in the Coniferous site), and overestimated (in the Deciduous site) in comparison with the obtained $\overline{T}_{plot}$ using SA_{plot} as unit area. The LA_{eff} as a unit ground area that defines the Deciduous site's $\overline{T}_{plot}$ as a larger fraction of $\overline{E}_a$ (than that estimated with SA_{plot} as unit ground area). Thus, $\overline{T}_{plot}$ becomes a larger proportion of $\overline{T}_{plant}$ than using SA_{plot} as a unit ground area. Conversely, LA_{eff} as a unit ground area defines the Coniferous site's $\overline{T}_{plot}$ as a smaller fraction. Results suggest that SA_{plot} as unit ground area gives adequate $\overline{T}_{plot}$ estimates for the Coniferous site and the SAI_{eff} gives adequate $\overline{T}_{plot}$ estimates for the Deciduous site. Hence, the chosen unit ground area considerably influences the $\overline{T}_{plot}$ estimates.

With respect to the results from Equations (9) and (10), it was expected that $\overline{T}_{plant}$ will be a significant proportion of $\overline{E}_a$ (i.e., between 70% and 90%). However, the daily values of E_a and T_{plant} slightly differ; and most days showed that indeed $E_a > T_{plant}$ (i.e., Days 212, 215, 216, 234, 225, 226, and 228). As it was expected, T_{plant} is always a significant proportion of E_a for most days (i.e., $T_{plant} > 90\%$ of the E_a. There were also days when $T_{plant} > E_a$ (Days 231, 232, 235, and 227). These results suggest that there were humid days causing some water condensation. Indeed, in those days, the E_a morning estimates (i.e., E_a hourly values) are negative. Moreover, approximately half of each day drew $VPD \leq 0.5$ kPa.

If the volumetric soil moisture approximates its wilting point, there could be significant variations in E_a (as well as T_{plant}). Therefore, the influence of θ_e variations on E_a was studied. During the days that this research was conducted, the soil was either extremely dry (below its θ_{wp}) or very wet ($0.16 \geq \theta_{sm} \geq 0.22$), causing just two days of transition between dryness and wetness to affect E_a values.

Even though the authors' methods help reduce the error carried out by scaling parameters, there is certainly opportunity to improve sap flow measurements. The choice to capture inter-specific variability, that is, changing the thermal probes after 48 h to individuals of different diameters, may have introduced uncertainty into the final data due to the short sampling time. The two summers in which sap flow data were collected were not the most favorable in terms of capturing sap flow patterns (such as those shown in Figures 4 and 5). Overcast days were constantly present, and we were not able to observe expected diurnal sap flow patterns (e.g., Figure 3). Another challenge was that on several occasions, the probes had to be moved to different trees because the probes were not capturing sap flow activity, which was attributed to tree infestation that was not initially visible or obvious. Thus, the sample size and temporal replication of sap flow were reduced to only a few trees and only up to four and eight days of sap flow data for the Coniferous and Deciduous site, respectively. In addition, the calibration method used to account for radial variation was estimated, and many other authors could argue that in situ calibration methods are the most effective methods, but they also indicate that this is a rare practice [15]. The available calibration methods—using potometers—are invasive, and there are concerning factors with these calibration techniques. Among those concerns is the fact that the base of the tree is damaged by the removal of its outermost bark, which means that most likely the tree will suffer from embolism and most likely will die. Injecting pressurized water into a tree could also cause severe damage to the secondary xylem tissue, which is known to be extremely delicate and easily broken by slight changes in pressure. Amid climate change and the global biodiversity crisis, the intention with this research was to propose a scaling approach that will conserve and protect the forest trees and their environment; thus, the authors used those techniques that were the least invasive and did not require sacrifice or damage to the integrity of the trees.

5. Conclusions

The scaling approach proposed here was shown to be an appropriate way to quantify the variation of scaling factors [25,26,28] and to prove their correlation at large scales. The use of these scaling factors and the careful formulation of the scaling approach were fundamental in obtaining canopy transpiration estimates that closely agreed with the estimated actual evapotranspiration using the Penman–Monteith and the modified Penman–Monteith equations.

The canopy transpiration values calculated using the LA_{eff} as a unit ground area factor are meaningful due to the close relationship between the total amount of leaves that fully operate during transpiration. Thus, we suggest that a deeper understanding and testing of this canopy transpiration number would be a significant contribution to the study of the efficiency of trees in water use. It is also recommended that prior characterization of the intraspecific biometrics' variations be made in order to further develop the scaling approach. In addition, future work should focus on observing the behavior of this scaling approach at larger scales.

Many canopy transpiration studies disregard the impact that regression models have in the final estimates of transpiration. Previous research has demonstrated the constant over and underestimations of sapwood depth, sapwood area, leaf area, and leaf area index by using general assumptions, and the concerning level of error that these values carry to larger scales. However, the authors acknowledge that the sap flow data collected for this work would be further improved by increasing the temporal scale and sample size per species. Our suggested approach for capturing sap flow interspecific variability requires further study and thus, for future work, it is advised to measure sap flow simultaneously on a series of trees with different diameters.

Supplementary Materials: The following are available online at http://www.mdpi.com/2073-4433/11/11/1158/s1. Table S1: E_p estimates during the same days that sap flow was measured at each site. Field campaign 2004.

Author Contributions: Conceptualization, M.R.Q.-P. and C.V.; methodology, M.R.Q.-P.; software, M.R.Q.-P.; validation, M.R.Q.-P.; formal analysis, M.R.Q.-P.; investigation, M.R.Q.-P.; data curation, M.R.Q.-P.; writing—original draft preparation, M.R.Q.-P.; writing—review and editing, M.R.Q.-P. and C.V.; supervision, C.V.; funding acquisition, M.R.Q.-P. and C.V. All authors have read and agreed to the published version of the manuscript.

Funding: This research was funded by the Alberta Ingenuity Fund, NSERC, Consejo Nacional de Ciencia y Tecnología, México, and Universidad Autónoma Metropolitana-Iztapalapa, México.

Acknowledgments: Thanks are due to the Biogeoscience Institute of the University of Calgary, Alberta. We would also like to thank the anonymous reviewers for their insightful comments.

Conflicts of Interest: The authors declare no conflict of interest.

Appendix A Computing Actual Evapotranspiration and Canopy Transpiration

Appendix A.1 Actual Evapotranspiration

Since the direct estimation of transpiration is complex, it is more common to estimate evapotranspiration (*ET*) of forested areas as a close estimate of transpiration. For dense, homogenous vegetated areas, transpiration is usually considered the largest portion of total evapotranspiration in forested areas. In Canada, it is estimated that forest transpiration has a large proportion of the total *ET* varying between 45% and 67% of total *ET*), while the rest of the water lost is through soil evaporation or evaporation of water on surfaces (e.g., leaves and trunks) and sublimation [47]. These statements are reinforced with detailed studies of *ET* in the boreal forest that demonstrate the large activity and amounts of energy and mass fluxes [70].

In this study, the Penman–Monteith equation [71] is used to estimate the actual evapotranspiration of the vegetated areas under study. These evapotranspiration estimates will be used to validate the daily transpiration rate estimates at the plot scale. The Penman–Monteith equation estimates the actual evapotranspiration of vegetated surfaces by accounting for all the micrometeorological factors that influence evapotranspiration as well as the influence of the canopy conductance and aerodynamic resistance in the rates of vegetation transpiration:

$$\lambda\ \mathrm{E_a} = \frac{\Delta(R_n - G) + \rho_a c_p (e^o - e_a)/r_a}{\Delta + \gamma\left[1 + \frac{r_c}{r_a}\right]} \tag{A1}$$

where $\lambda\ \mathrm{E_a}$ is the latent heat of evapotranspiration Δ is the slope of the saturation vapor pressure curve (kPa $^\circ$C $^{-1}$), $\lambda\ \mathrm{E_a}$ is the latent heat of actual evapotranspiration (Jkg^{-1}), R_n is the net solar radiation, and G is the soil heat flux (all these terms in units of Jm^{-2}s^{-1}). The air density, ρ_a is in (kgm^{-3}); c_p is the specific heat of air at constant pressure (i.e., 1010 Jkg^{-1} $^\circ$C $^{-1}$). The term $(e^o - e_a)$ is the vapor pressure deficit (*VPD*) calculated by the difference between the saturation vapor pressure (e^o, (kPa)) and the actual vapor pressure (e_a, (kPa)). The psychrometric constant, γ, is in units of (kPa $^\circ$C $^{-1}$). The aerodynamic terms, r_a and r_c are the aerodynamic resistance to vapor and heat transfer, and the bulk canopy resistance (both expressed in sm^{-1}). The following paragraphs explain in detail the calculation of each Penman–Monteith equation's parameter. To convert the latent heat of evapotranspiration to actual evapotranspiration ($\mathrm{E_a}$), use $\mathrm{E_a} = \lambda\ \mathrm{E_a}/\ \lambda$ in units of mms^{-1}.

Appendix A.1.1 Aerodynamic Parameters

To calculate the *VPD* term in the Penman–Monteith equation, the saturation vapor pressure was initially calculated using two different equations:

$$e^o = a_\circ + a_1 T_a + a_2 T_a^2 + a_3 T_a^3 + a_4 T_a^4 + a_5 T_a^5 + a_6 T_a^6 \tag{A2}$$

and

$$e^o = \exp\left(\frac{16.78\ T_a - 116.9}{T_a + 237.3}\right) \tag{A3}$$

In both equations, T_a is the air temperature (°C, field weather station measurements). The first equation is the resultant of a Chebyshev fitting procedure used by [72]. The polynomial coefficients (i.e., $a_\circ$ to a_6) are reported in Lowe's paper [72] and e^o is calculated in mbar units. The latter equation calculates e^o in kPa was derived by [73] and its estimates are considered of high reliability [64]. The average difference between e^o values calculated with both equations was of 0.00017 kPa. Thus, for further estimations, Equation (A3) is applied. The actual vapor pressure is calculated using the estimated e^o and the relative humidity (*RH*, (%)) that was measured in the field [74]:

$$e_a = \frac{RH\ e^o}{100} \tag{A4}$$

The air density, ρ_a, can be derived from [64]:

$$\rho_a = \frac{1000\ P}{T_v\ R} \tag{A5}$$

where P is the daily mean atmospheric pressure calculated with the field measurements (barometer, units of kPa), R is the specific gas constant (287 $Jkg^{-1}K^{-1}$). T_v is the virtual temperature in degrees Kelvin, calculated as [64]:

$$T_v = \frac{\overline{T}_a}{1-(0.378\,\overline{e}_a\,P^{-1})} \tag{A6}$$

where $\overline{e}_a$ and $\overline{T}_a$ are taken as the daily average of e_a and T_a respectively. A sensitivity analysis was performed to observe how $\overline{T}_a$ values affect ρ_a or the evapotranspiration estimates. There were no significant changes in the values. Thus, $\overline{T}_a$ was used in the equation. This analysis was performed since [64] did not specify if an average temperature or temperature at each hourly time-step value should be used.

The psychrometric constant can be expressed as [75]:

$$\gamma = \frac{c_p\,P}{\varepsilon\,\lambda} \tag{A7}$$

where γ is given in units of kPa$^{\circ}$C^{-1}, c_p is entered as 1.010 kJ kg^{-1} $^{\circ}$C^{-1}, P is in kPa. The water vapor ratio molecular weight (ε) is a constant value equal to 0.622, and λ is calculated using the following equation [64]:

$$\lambda = 2.501 - 2.361 \times 10^{-3}\,T_a \tag{A8}$$

where λ is given in units of MJ kg^{-1} (i.e., multiply by 1000 to match units of c_p).

The slope of the saturation vapor pressure curve (Δ) is derived from the following equation:

$$\Delta = \frac{4098\,e^{o}}{(T_a + 237.3)^2} \tag{A9}$$

The aerodynamic resistance to vapor and heat flux, r_a, is estimated with the following equation [64,76]:

$$r_a = \left(\left[\ln\frac{z_u - d}{z_{om}}\right]\left[\ln\frac{z_u - d}{z_{oh}}\right]\right) \div k^2 u_z \tag{A10}$$

where k is von Karman's constant (0.40), z_u is the height (m) at which the wind speed u_z (ms^{-1}) has been recorded (12.19 m in this particular case), d is the zero-plane displacement (m) that is assumed as 67% of the canopy height (i.e., $d = 0.67\ h_c$) for vegetation with $LAI > 2.0$. Here, the average canopy height is 15 m, which is the same height used in previous estimations. The parameters z_{om} and z_{oh} are the roughness lengths for the momentum and heat transfer, respectively. Allen et al. [64] suggested applying $z_{oh} = 0.1\ z_{om}$. In this study, the fact that z_{om} varies with cover has been taken into account; thus, z_{om} is calculated differently for the Deciduous and the Coniferous sites. For the Deciduous sites, whose vegetation is considered dense and homogeneous, the equation suggested by [76] is applied:

$$z_{om} = \frac{1}{e}(h_c - d) = 0.37(h_c - d) \tag{A11}$$

For the Coniferous sites, the equation suggested by [64] is applied:

$$z_{om} = \varsigma(h_c - d) \tag{A12}$$

where ς is an empirical factor that is independent of vegetation height [77]. Based on their calculated values of z_{om} and d for conifers, [77] determined $\varsigma = 0.22$. Table 1 lists the constant terms of the aerodynamic resistance equation. The ratio $z_{om}/h_c = 0.7$ calculated for Coniferous sites concurs with the mean value reported by [64] for this ratio. The Deciduous' sites z_{om} value is between the range of values listed for deciduous trees by [64].

Table A1. Steady parameters in the calculation of the aerodynamic resistance to heat and vapor transfer, r_a. All parameters are reported in meters, with exception of ς, which is unitless.

Parameter	Coniferous Sites	Deciduous Sites
h_c	15	15
ς	0.22	0.37
d	10.05	10.05
z_{om}	1.089	1.82
z_{oh}	0.1089	0.1821

The canopy resistance is more complicated to estimate since it varies along the day and it is a function of several atmospheric parameters [78]:

$$g_c = g_{c_{max}}[minimum(g(LAI), g(R_s), g(VPD), g(T_a), g(\theta_{sm}))] \tag{A13}$$

This equation implies that the canopy conductance (g_c) is a function of the environmental parameters: LAI, R_s (Wm^{-2}), VPD (kPa), T_a (°C), and volumetric soil moisture (θ_{sm}, in m^3m^{-3}). The parameter that reaches its minimum at a specific time (g_{env}), drives the canopy conductance. The lower the value of the environmental parameter reduction function, the lower the value of g_c, therefore the higher the r_c. Each parameter is represented by a reduction function that computes the value of the function between zero and one (i.e., $0 \leq g_c \leq 1$). Different authors have developed and calibrated reduction functions for calculating each one of the parameters in Equation (A13). Allen et al. [64] suggested that these equations can be replaced in the function above. Here, a set of equations was chosen and presented below. Most of the equations and empirical factors are taken from [79]; otherwise, the author is cited. Stewart [79] developed and calibrated these functions for Scots pine. This is the closest species to the species studied in this work with reported functions. In the case of the Deciduous site, the empirical factors were adjusted according to the response of r_c or g_c to the environmental parameters. This task was performed based on previous results and results that were obtained in this study.

The $g_{c_{max}}$ is the reciprocal of the minimum canopy or surface resistance ($r_{c_{min}}$). Typical values reported for coniferous forests $r_{c_{min}}$ range from 30 sm^{-1} to 60 sm^{-1} [64]. Here, an average value of the reported ranges was taken for the Coniferous site (i.e., 51 sm^{-1}). Ref. [80] reported maximum values of canopy conductance for Trembling aspen (31 ms^{-1}) and it is the one applied here for the Deciduous site. To compute $g(LAI)$:

$$g(LAI) = \frac{LAI}{LAI_{max}} \tag{A14}$$

where LAI_{max} is the maximum LAI Along the year. Since data collection occurred during the peak of the summer (July and August), it is assumed that $g(LAI) \approx 1.0$ for both the Coniferous and the Deciduous sites. The $g(R_s)$ is calculated with

$$g(R_s) = \frac{R_s(1000 + K_R)}{1000(R_s + K_R)} \tag{A15}$$

where R_s is in Wm^{-2} and K_R is an empirical factor that was set up as 104.4 Wm^{-2}.

The VPD function is established based on the two following equations:

$$g(VPD) = 1 - K_{VPD}VPD \text{ for } 0 < VPD < VPD_c \tag{A16}$$

and

$$g(VPD) = 1 - K_{VPD_c}VPD \text{ for } VPD \geq VPD_c \tag{A17}$$

with $K_{VPD} = 0.5$ kPa. The VPD_c is called the "threshold vapor pressure deficit" and is set up as 1.5kPa for the Coniferous site. For the Deciduous site, [52] reported the sap flow trend of four hardwood species in relation to *VPD*. One of the species studied is from the genus *Populus*. For that result, it was reported that the *Populus* sap flow did not significantly vary when *VPD* was greater than 1 kPa, unless the soil moisture content was limiting. The results presented by [52] perfectly concur with our study results. Thus, the threshold for the Deciduous site was assumed as 1 kPa. Since a K_{VPD} factor was not found in the literature, its value was determined by using previously reported trends of g_c versus *VPD*. Thus, the value was assumed as $K_{VPD} = 0.79$ kPa initially. This decision was somehow conservative and based on the fact that deciduous r_c reported values have reached 160 sm^{-1} [64]. Therefore, K_{VPD} was set up to make the reciprocal of $g_{max}g_{env}$ to quasi match r_c to 160 sm^{-1} when VPD_c is greater than 1 kPa and becomes the driving environmental parameter of r_c. Using graphs by [80] of half-hourly changes in g_c and *VPD*, it was observed that r_c can change from 81 sm^{-1} to 200 sm^{-1} as *VPD* reaches values greater than 1 kPa. In this case, a second run for E_a was performed assuming $K_{VPD} = 0.84$ kPa, to make $r_c \approx 200$ sm^{-1} when $VPD > 1$ kPa. Values of E_a obtained with both parameters are presented here.

For calculating $g(T_a)$, a maximum and a minimum temperature (T_M and T_N, in °C) is required that constrain the stomas process, plus another empirical factor, K_T (called the "optimum conductance temperature"):

$$g(T_a) = \frac{(T - T_N)(T_M - T)^\rho}{(K_T - T_N)(T_M - K_T)^\rho} \tag{A18}$$

where

$$\rho = \frac{T_M - K_T}{K_T - T_N} \tag{A19}$$

and K_T is 18.35 °C for the Coniferous site. In the case of the Deciduous site, reported half-hour Trembling aspen g_c and temperature values [80] were used to estimate the optimum conductance temperature for Trembling aspen g_c. An average optimum temperature of 18.29 °C was obtained.

Finally, to estimate the $g(\theta_{sm})$, a function reported by [64], which is a slightly modified version of the one suggested by [79], was used:

$$g(\theta_{sm}) = 1 - e^{-K_\theta \theta_e} \tag{A20}$$

where K_θ ($K_\theta = 6.7$) is the empirical factor used to calculate $g(\theta_{sm})$; and θ_e is the fraction available for transpiration, also called the "effective fraction of available soil moisture" [64]:

$$\theta_e = \frac{\theta_{sm} - \theta_{wp}}{\theta_{fc} - \theta_{wp}} \tag{A21}$$

where θ_{sm} is the volumetric soil moisture (field measurements, m^3m^{-3}), θ_{wp} is the soil wilting point and θ_{fc} is the soil field capacity. The values of θ_{fc} and θ_{wp} are obtained based on the soil texture. Direct studies of the soil type and texture in the area of Kananaskis [81–84] were used to define the soil texture in the Coniferous and Deciduous sites. The soil texture, generally defined as fine sandy loam (for both areas), drew a soil field capacity ranging between 0.16 and 0.22, while the soil wilting point was estimated as 0.07 (all values in volumetric fraction).

Appendix A.1.2 Energy Parameters

The soil heat flux is calculated using a "universal relationship" developed by [85]:

$$G = 0.4\left(e^{-0.5LAI}\right)R_n \tag{A22}$$

G has the units of R_n. The net solar radiation is derived from the following equation [64]:

$$R_n = (1 - \alpha)R_s + R_{nl} \tag{A23}$$

where R_s is the shortwave solar radiation (measured in the field with a pyranometer), R_{nl} is the net outgoing longwave solar radiation, and α is the surface's albedo value. The term $(1-\alpha)$ helps to calculate the fraction of incident net shortwave solar radiation that is absorbed by a specific surface. For coniferous forests, mean α values are in the range of 0.09–0.15 [66,76], and deciduous forests are in the range of 0.15–0.25 [76]. Monthly albedo values for mid-latitude forests are of 0.14 during the months of July and August [86–88]. The net longwave solar radiation is calculated based on the emissivity values of four different surfaces and the air temperature, T_a [47]:

$$R_{nl} = \left\{\varepsilon_o\left[\varepsilon_a\,\sigma_{sb}\,T_a^4 + \varepsilon_u\,\sigma_{sb}\,T_a^4\left(1-e^{-0.5LAI_u\,\Omega_u/\cos\overline{\theta}_u}\right) + \varepsilon_g\,\sigma_{sb}\,T_a^4\left(e^{-0.5LAI_u\,\Omega_u/\cos\overline{\theta}_u}\right)\right] - 2\varepsilon_o\,\sigma_{sb}\,T_a^4\right\}\left(1-e^{-0.5LAI_o\Omega_E/\cos\overline{\theta}_o}\right) \quad \text{(A24)}$$

where σ_{sb} is the Stefan–Boltzmann constant (5.675×10^{-8} Jm^{-2}K^{-4}s^{-1}), T_a is the air temperature (units of K). LAI_o and LAI_u are the Leaf Area indices of the overstory and understory respectively; Ω_E and Ω_u are the clumping indices of the overstory and understory; $\cos\overline{\theta}_u$ and $\cos\overline{\theta}_o$ are estimations of the transmission of diffuse radiant energy through the understory and overstory. The emissivity of the overstory, the ground, the understory, and the atmosphere are respectively represented by ε_o, ε_g, ε_u, and ε_a. Emissivity values for the first three surfaces are assigned from [47,89] as 0.98, 0.95 and 0.98, respectively. These emissivity values concur with values reported by [64]. Emissivity from the atmosphere is calculated with the following equation [76]:

$$\varepsilon_a = 1.24\left(\frac{e_a}{T_a}\right)^{1/7} \quad \text{(A25)}$$

where e_a is in [mba] and T_a is in degrees Kelvin. The transmission of diffuse radiant energy through the understory and overstory is given by the following two equations that were derived by [47]:

$$\cos\overline{\theta}_u = 0.537 + 0.025LAI_u \quad \text{(A26)}$$

$$\cos\overline{\theta}_o = 0.537 + 0.025LAI_o \quad \text{(A27)}$$

LAI_o was measured for every coniferous and deciduous site (i.e., $LAI_{plot} = LAI_o$); LAI_u is more complex to measure directly and it was derived from previous reports of understory NDVI and LAI values. Buerman et al. [90] used the reflectance values to estimate the understory NDVI and calculate LAI indices based on understory NDVI-LAI scatterplots developed by [91]. The LAI_u values reported by [90] range between 0.6 and 1.0 (being the largest values for Black spruce and the smallest for Jack Pine). Conifers understory NDVI ($NDVI_u$) values reported by [90] were compared with the studied Coniferous sites $NDVI_u$ calculated from the understory spectral reflectance that was recorded in the 2003 field campaign at two Coniferous and two Deciduous sites [92]. For both Coniferous and Deciduous sites, the average $NDVI_u$ is 0.8, which is 0.3 larger than the values reported by [90] in 2002 (their $NDVI_u$ range is 0.35–0.50). Using information reported by [91], ref. [90] established that an $NDVI_u$ of 0.5 corresponded to an LAI_u of 1.0. On the other hand, ref. [92] established a standard LAI_u value of 0.5 for broadleaf and needle-leaf forests.

Therefore, based on these previous results, LAI_u for the Coniferous sites in Kananaskis is assumed 1.0, and for Deciduous sites, 0.6. The latter value is also in the LAI_u range reported by [69] for deciduous stands in a boreal forest. Figure A1 is the typical understory spectral response at a Coniferous and a Deciduous site in Kananaskis Field Station. It is convenient to stress the fact that these LAI_u values are approximate; however, the main objective is to acknowledge the importance of understory in the overall evapotranspiration estimates. Thus, as [47] thought, it is convenient to somehow include the understory evapotranspiration based on assumptions about its LAI_u.

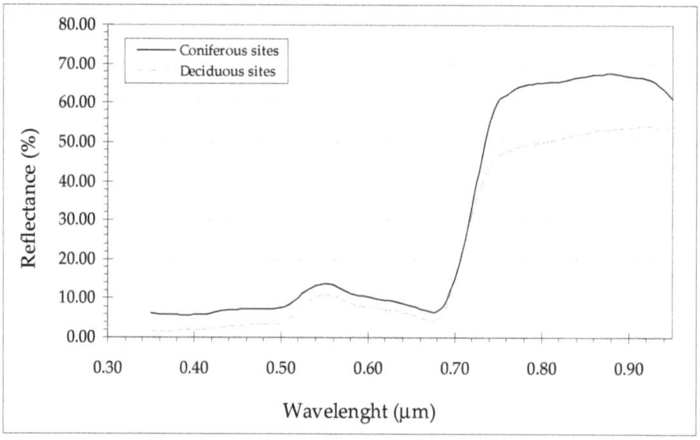

Figure A1. Typical understory spectral reflectance in Kananaskis Field Station study sites during the summer of 2003.

The understory clumping index Ω_u, was derived by modifying the former Chen's equation:

$$LAI = (1 - \alpha_l)LAI_{eff}\ \gamma_E \Omega_E \tag{A28}$$

where $\gamma_E \Omega_E = 1\Omega_E$ in vascular vegetation [93]. Thus, for understory vegetation Ω_E does not have to be partitioned into fractions that account for the shoot effect. At the same time, the α_l value is zero since there is no fraction of wood to account for in the understory vegetation present at the study sites. Thus,

$$LAI_u = LAI_{eff}\ \Omega_u \tag{A29}$$

As LAI_u is known, LAI_{eff} can be approximated as 50% of LAI_u as suggested by [94] for grasses (the closest that can be found to a forest understory). Hence,

$$\Omega_u = LAI_{eff}/LAI_u = 0.5LAI_u/LAI_u = 0.5 \tag{A30}$$

Appendix A.2 Potential Evapotranspiration

The potential evapotranspiration results are provided in the Supplementary Material section for the reader to compare the great disparity between potential evapotranspiration and actual evapotranspiration estimated with the Penman–Monteith equation. The Penman combination equation estimates the potential evapotranspiration, or also, the free water evaporation. Potential rates of evapotranspiration assume that the water is never a limiting factor, the plant completely shades the ground (thus, there is no soil evaporation) and it has the optimal environmental conditions to transpire at its maximum rate (there is no canopy resistance). Two versions of the Penman–Monteith equation are used here to estimate the Potential Evapotranspiration (E_p), the combination equation for free water evaporation [95,96], and the Penman–Monteith equation that includes the aerodynamic parameter but sets $r_c = 0$ [97]. The former equation is computed in the following form:

$$E_p = \frac{\Delta(R_n - G) + \rho_a c_p \left(e^{\circ} - e_a\right) u_2}{\lambda(\Delta + \gamma)\rho_w} \tag{A31}$$

where ρ_w is the water density in units of kg m^{-3}, and u_2 is the wind speed at 2 m height. Wind speed measured at 3 m height was scaled down to 2 m using the aerodynamic function [98]:

$$\frac{u_2}{u_\circ} = \frac{z_2}{z_\circ} \tag{A32}$$

where u_2 is the wind speed to be estimated at height $z_2 = 2$ m and $u_\circ$ is the wind speed at the reference height $z_\circ$ (in this case, 3 m). Wind differences of ±6 cm were registered between the two heights. The rest of the parameters were already defined. The G parameter is not included in the original equation; however, it was decided to slightly modify the method and include G. Equation (A31) gives E_p in units of ms^{-1}. The second one is Equation (A1), making $r_c = 0$, and E_p is given in mms^{-1}:

$$E_\mathrm{p} = \frac{\Delta(R_n - G) + \rho_a c_p (e^o - e_a)/r_a}{\lambda[\Delta + \gamma]} \tag{A33}$$

The obtained E_a and E_p daily values were averaged along the eight days (for the Coniferous site) and the four days (for the Deciduous site) and compared with the average $\overline{T}_{plot}$ value obtained for their respective period of time.

Appendix A.3 Canopy Transpiration Using Modified Penman–Monteith Equation

Liu et al. [47] used a slightly modified version of the Penman–Monteith equation in order to estimate actual canopy transpiration at large scales. According to [47], a model such as Penman–Monteith should be adjusted by separately estimating the transpiration of shaded and sunlit leaves as follows (stratified model):

$$T_{plant} = T_{sun} LAI_{sun} + T_{shade} LAI_{shade} \tag{A34}$$

where T_{sun} and T_{shade} are the actual transpiration of sunlit and shaded leaves respectively; LAI_{sun} and LAI_{shade} are the leaf area indexes for sunlit and shaded leaves as well. The Penman–Monteith equation is then used by [47] to estimate T_{sun} and T_{shade}:

$$\lambda T_{sun} = \frac{\Delta(R_{n,\ sun}) + \rho_a c_p (e^o - e_a)/r_a}{\Delta + \gamma[1 + r_s/r_a]} \tag{A35}$$

and

$$\lambda T_{shade} = \frac{\Delta\left(R_{n,\ shade}\right) + \rho_a c_p (e^o - e_a)/r_a}{\Delta + \gamma[1 + r_s/r_a]} \tag{A36}$$

where $R_{n,\ sun}$ and $R_{n,\ shade}$ are the net solar radiation available for sunlit and shaded leaves in Jm^{-2}s^{-1}, and r_s is the stomatal resistance in units of sm^{-1}. The rest of the parameters and units remain the same as in Equation (A1). The boreal ecosystem productivity simulator (BEPS) sets up a set of equations to calculate $R_{n,\ sun}$ and $R_{n,\ shade}$ [47,63]. The equations compute the shortwave solar radiation for sunlit and shaded leaves as well. The net longwave solar radiation is assumed to behave equally for sunlit and shaded leaves; therefore, a single equation is used to calculate net longwave solar radiation. Thus, $R_{n,\ sun}$ and $R_{n,\ shade}$ are respectively given by

$$R_{n,\ sun} = R_{s,\ sun} + R_{nl,\ sun} \tag{A37}$$

and

$$R_{n,\ shade} = R_{s,\ shade} + R_{nl,\ shade} \tag{A38}$$

where $R_{s,\ sun}$ and $R_{s,\ shade}$ are the shortwave solar radiation for sunlit and shaded leaves, respectively, and $R_{nl,\ sun}$ and $R_{nl,\ shade}$ are the net longwave solar radiation for sunlit and shaded leaves, respectively. The shortwave solar radiation terms are calculated by the following equations:

$$R_{s,\ sun} = (1 - \alpha_L)\left(R_{s,\ dir} \cos \alpha_{sa} / \cos \theta\right) + R_{s,\ shade} \quad \text{(A39)}$$

where α_L is the leaf scattering coefficient (constant that equals 0.25); α_{sa} is the mean leaf–sun angle, which is taken as 60° [47]; θ is the solar zenith angle; and $R_{s,\ dir}$ is the direct shortwave solar radiation. $R_{s,\ shade}$ is calculated with

$$R_{s,\ shade} = \left(R_{s,\ dif} - R_{s,\ dif-under}\right) / LAI_o + C \quad \text{(A40)}$$

where $R_{s,\ dif}$ is the diffuse shortwave solar radiation; $R_{s,\ dif-under}$ is the diffuse shortwave solar radiation under the overstory; and C accounts for the multiple scattering of direct radiation, which is calculated by

$$C = \alpha_L\ \Omega_E\ R_{s,\ dir}(1.1 - 0.1LAI_o)\ e^{-\cos\theta} \quad \text{(A41)}$$

$R_{s,\ dir}$ is a function of R_s and $R_{s,\ dif}$:

$$R_{s,\ dir} = R_s - R_{s,\ dif} \quad \text{(A42)}$$

and $R_{s,\ dif}$ can be estimated using the following cases:

$$\frac{R_{s,\ dif}}{R_s} = \begin{cases} 0.13 & \text{if r} \sim\ \geq 0.8 \\ 0.943 + 0.734\tilde{r} - 4.9\tilde{r}^2 + 1.796\tilde{r}^3 + 2.058\tilde{r}^4 & \text{if r} \sim\ < 0.8 \end{cases} \quad \text{(A43)}$$

where $\tilde{r}$ is calculated as a function of the solar constant ($SC = 1367$ Wm^{-2}), R_s and θ:

$$\tilde{r} = \frac{R_s}{SC \cos\theta} \quad \text{(A44)}$$

and finally, $R_{s,\ dif-under}$ can be calculated as a function of $R_{s,\ dif}$, Ω_E, LAI_o, and the angle for diffuse radiation ($\overline{\theta}_o$):

$$R_{s,\ dif-under} = R_{s,\ dif}\left(e^{-0.5\Omega_E LAI_o / \cos\overline{\theta}_o}\right) \quad \text{(A45)}$$

where $\cos\overline{\theta}_o$ is calculated using Equation (A27). The Ω_E is of course the clumping index of the overstory, which is taken as 0.83 and 0.64 for the Coniferous and Deciduous site respectively (values obtained in situ with the TRAC optical device). As mentioned, the net longwave radiation terms are considered to behave the same for sunlit and shaded leaves. Thus, $R_{nl,\ sun} = R_{nl,\ shade}$, and their value is calculated by

$$R_{nl,\ sun} = R_{nl,\ shade} = \frac{R_{nl}}{LAI_o} \quad \text{(A46)}$$

and Equation (A24) calculates R_{nl}. As it is noticed, Equations (A35) and (A36) include the term r_s instead of r_c. The stomatal resistance is calculated based on the r_c values obtained with the set of reduction functions that resolve g_c (Equations (13)–(21)) and with the LAI_o:

$$r_s = LAI_o\ r_c \quad \text{(A47)}$$

Allen et al. [94] reported the previous equation using a *LAI* value which is standardized for crops and relatively tall grasses (i.e., 0.5 *LAI*). Here, the equation is modified to make it applicable to overstory. In addition, it is considered that shaded and sunlit leaves have similar stomatal resistances responses.

Appendix A.4 Calibration of Tree Sap Flow Measurements with TDPs

In theory, when the sap flow is constant, [52] assumed that the sap's velocity is

$$J_i = \frac{1}{\alpha}\left[\frac{\Delta T_m - \Delta T}{\Delta T}\right] \quad \text{(A48)}$$

where ΔT_m is the maximum temperature difference given when the sap flow is null (i.e., $J_i = 0$), ΔT is the difference in temperature between the two probes at a specific time. The ratio between the temperature differences becomes the calibrated constant K (flux index) in Granier's 1985 technique [52]. With a sample size of 53 trees of three different species and diameters, [52] determined that the flux index has an exponential relationship with the velocity of sap flow:

$$K = 0.0206 J_i^{\ 0.8124} \quad \text{(A49)}$$

with a $R^2 = 0.96$, and units of J_i are 10^{-6} ms^{-1}. J_i is expressed in the same way as sap flux density; that is, flow rate of sap volume per unit of sapwood area (i.e., $10^{-6} m_{sap}^3\ m_{SA}^2\ s^{-1}$). Substituting K into Equation (A48) by Equation (A49), and considering the α term independent of the experimentation the sap flow velocity is estimated by

$$J_i = 0.0119\left[\frac{\Delta T_m - \Delta T}{\Delta T}\right]^{1.231} \quad \text{(A50)}$$

where J_i is in units of cms^{-1}. Granier [52] validated his results with the Penman equation outcomes, finding a good agreement between both set of results (of course, Penman potential evapotranspiration estimates were greater than the ones obtained with the Granier method). The studied species were Douglas fir (*Pseudotsuga menziesii*), European black pine (*Pinus nigra*), and Oak tree (*Quercus pedunculata*). In this study, J_i radial flow is correct and azimuthal sap flow variation was corrected by measuring sapwood depth in four sides of the tree to obtain an average sapwood depth and capture sapwood depth variation around the tree trunk [25,26,28].

References

1. Jia, G.E.; Shevliakova, P.; Artaxo, N.; De Noblet-Ducoudré, R.; Houghton, J.; House, K.; Kitajima, C.; Lennard, A.; Popp, A.; Sirin, R.; et al. Land–climate interactions. In *Climate Change and Land: An IPCC Special Report on Climate Change, Desertification, Land Degradation, Sustainable Land Management, Food Security, and Greenhouse Gas Fluxes in Terrestrial Ecosystems*; IPCC—Intergovernmental Panel on Climate Change: Geneva, Switzerland, 2019.
2. Bates, B.C.; Kundzewicz, Z.W.; Wu, S.; Palutikof, J.P. Climate Change and Water. In *Technical Paper of the Intergovernmental Panel on Climate Change*; IPCC Secretariat: Geneva, Switzerland, 2008; 210p, Available online: https://www.ipcc.ch/publication/climate-change-and-water-2/ (accessed on 2 September 2020).
3. Qu, Y.; Zhuang, Q. Evapotranspiration in North America: Implications for water resources in a changing climate. *Mitig. Adapt Strag. Glob.* **2019**, *2019*, 1–16. [CrossRef]
4. Zhang, K.; Kimball, J.S.; Ramakrishna, R.N.; Running, S.W.; Hong, Y.; Gourley, J.J.; Yu, Z. Vegetation Greening and Climate Change Promote Multidecal Rises of Global Land Evapotranspiration. *Sci. Rep.* **2015**, *10*, 2–9. [CrossRef]
5. Boulanger, Y.; Taylor, A.R.; Price, D.T.; Cyr, D.; McGarrigle, E.; Rammer, W.; Sainte-Marie, G.; Beaudoin, A.; Guindon, L.; Mansuy, N. Climate change impacts on forest landscapes along the Canadian southern boreal forest transition zone. *Land. Ecol.* **2017**, *32*, 1415–1431. [CrossRef]
6. Drobyshev, I.; Gewehr, S.; Berninger, F.; Bergeron, Y. Species specific growth responses of black spruce and trembling aspen may enhance resilience of boreal forest to climate change. *J. Ecol.* **2013**, *101*, 231–242. [CrossRef]
7. Allen, C.D.; Breshears, D.D.; Mcdowell, N.G. On underestimation of global vulnerability to tree mortality and forest die-off from hotter drought in the Anthropocene. *Ecosphere* **2015**, *6*, 1–55. [CrossRef]

8. Kingston, D.G.; Todd, M.C.; Taylor, R.G.; Thompson, J.R.; Arnell, N.W. Uncertainty in the estimation of potential evapotranspiration under climate change. *Geophys. Res. Lett.* **2009**, *36*, L20403. [CrossRef]
9. Lijie, S.; Puyu, F.; Wang, B.; Liub, D.L.; Cleverly, J.; Fang, Q.; Yud, Q. Projecting potential evapotranspiration change and quantifying its uncertainty under future climate scenarios: A case study in southeastern Australia. *J. Hydrol.* **2020**, *584*, 1–14. [CrossRef]
10. Mobilia, M.; Schmidt, M.; Longobardi, A. Modelling Actual Evapotranspiration Seasonal Variability by Meteorological Data-Based Models. *Hydrology* **2020**, *3*, 50. [CrossRef]
11. Shwetha, H.R.; Kumar, D.N. Estimation of Daily Actual Evapotranspiration Using Vegetation Coefficient Method for Clear and Cloudy Sky Conditions. *IEEE J. Sel. Top. Appl. Earth Obs. Remote Sens.* **2020**, *3*, 2385–2395. [CrossRef]
12. Lu, P.; Urban, L.; Zhao, P. Granier's Thermal Dissipation Probe Method for Measuring Sap Flow in Trees: Theory and Practice. *Acta Botanica Sinica* **2004**, *466*, 631–646.
13. Flo, V.; Martinez-Vilalta, J.; Sttepe, K.; Schuldt, B.; Poyatos, R. A Synthesis of Bias and Uncertainty in sap flow methods. *Agric. For. Meteorol.* **2019**, *271*, 362–374. [CrossRef]
14. Peters, R.L.; Fonti, P.; Frank, D.C.; Poyatos, R.; Pappas, C.; Kahmen, A.; Carraro, V.; Prendin, A.L.; Schneider, L.; Baltzer, J.L.; et al. Quantification of Uncertainties in conifer sap flow measured with the thermal dissipation method. *New Phytol.* **2018**, *219*, 1283–1299. [CrossRef] [PubMed]
15. Pasqualotto, G.; Carraro, V.; Menardi, R.; Anfodillo, T. Calibration of Granier Type (TDP) Sap Flow Probes by a High Precision Electronic Potometer. *Sensors* **2019**, *19*, 2419. [CrossRef] [PubMed]
16. Nhean, S.; Ayutthaya, S.I.N.; Rocheteau, A.; Do, F.C. Multi-species test and calibration of an improved transient thermal dissipation system of sap flow measurement with a single probe. *Tree Physiol.* **2019**, *39*, 106161070. [CrossRef] [PubMed]
17. Granier, A.; Anfondillo, T.; Sabatti, M.; Cochard, H.; Dreyer, E.; Tomasi, M.; Valentini, R.; Bréda, N. Axial and radial water flow in the trunks of oak trees: A quantitative and qualitative analysis. *Tree Physiol.* **1994**, *14*, 1383–1396. [CrossRef]
18. Phillips, N.; Oren, R.; Zimmerman, R. Radial patterns of xylem sap flow in non-diffuse and ring-porous tree species. *Plant Cell Environ.* **1996**, *19*, 983–990. [CrossRef]
19. Nadezhdina, N.; Čermák, J.; Ceulemans, R. Radial patterns of sap flow in woody stems of dominant and understory species: Scaling errors associated with positioning of sensors. *Tree Physiol.* **2002**, *22*, 907–918. [CrossRef]
20. Ford, C.R.; Hubbard, R.M.; Kloeppel, B.D.; Vose, J.M. A comparison of sap-flux based evapotranspiration estimates with basin-scale water balance. *Agric. For. Meteorol.* **2007**, *145*, 176–185. [CrossRef]
21. Zhao, H.; Yang, S.; Guo, X.; Peng, C.; Gu, X.; Deng, C.; Chen, L. Anatomical explanations for fcute depressions in radial pattern of axial sap flow in two diffuse-porous mangrove species: Implications for water use. *Tree Physiol.* **2020**, *38*, 277–287.
22. Baiamonte, G.; Motisi, A. Analytical Approach Extending the Granier method to radial sap flow patterns. *Agric. Water Manag.* **2020**, *231*, 105988. [CrossRef]
23. Fan, J.; Guyot, A.; Ostergaard, K.T.; Lokington, D.A. Effects of early wood and latewood on sap flux density-based transpiration estimates in conifers. *Agric. For. Meteorol.* **2018**, *15*, 264–274. [CrossRef]
24. Cermak, J.; Nadezhdina, N. Sapwood as the scaling parameter defining according to xylem water content or radial pattern of sap flow? *Ann. Sci. For.* **1998**, *55*, 509–521. [CrossRef]
25. Quiñonez-Piñón, M.R.; Valeo, C. Assessing the Translucence and Color-Change Methods for Estimating Sapwood Depth in Three Boreal Species. *Forests* **2018**, *9*, 686. [CrossRef]
26. Quiñonez-Piñón, M.R.; Valeo, C. Allometry of Sapwood Depth in Five Boreal Species. *Forests* **2017**, *8*, 457. [CrossRef]
27. Wang, H.; Gaun, H.; Simmons, C.T.; Lockington, D.A. Quantifying sapwood width for three Australian native species using electrical resistivity tomography. *Ecohydrology* **2016**, *9*, 83–92. [CrossRef]
28. Quiñonez-Piñón, M.R.; Valeo, C. Scaling Approach for Estimating Stand Sapwood Area from Leaf Area Index in Five Boreal species. *Forests* **2019**, *10*, 829. [CrossRef]
29. Aparecido, L.M.T.; dos Santos, J.; Higuchi, N.; Kunert, N. Relevance of wood anatomy and size of Amazonian trees in the determination and allometry of sapwood area. *Acta Amazonica* **2019**, *49*, 1–10. [CrossRef]
30. Lubczynski, M.W.; Chavarro-Rincon, D.C.; Rossiter, D.G. Conductive sapwood area prediction from stem and canopy areas—Allometric equations of Kalahari trees, Botswana. *Ecohydrology* **2017**, *10*, e1856. [CrossRef]

31. Forrester, D.I.; Benneter, A.; Bouriaud, O.; Bauhus, J. Diversity and competition influence tree allometric relationships—Developing functions for mixed-species forests. *J. Ecol.* **2017**, *105*, 761–774. [CrossRef]
32. Mitra, B.; Papuga, S.A.; Alexander, M.R.; Swetnam, T.L.; Abramson, N. Allometric relationships between primary size measures and sapwood area for six common tree species in snow-dependent ecosystems in the Southwest United States. *J. For. Res.* **2019**. [CrossRef]
33. Renner, M.; Hassler, S.K.; Blume, T.; Weiler, M.; Hildenbrandt, A.; Guderle, M.; Schymanski, S.J.; Kleidon, A. Dominant controls of Transpiration along a Hillslope transect inferred from ecohydrological measurements and thermodynamic limits. *Hydrol. Earth Syst. Sci.* **2016**, *20*. [CrossRef]
34. Dean, T.J.; Long, J.N. Variation in sapwood area-leaf area relations within two stands of Lodgepole pine. *For. Sci.* **1986**, *32*, 749–758.
35. Dean, T.J.; Long, J.N.; Smith, F.W. Bias in leaf area-sapwood area ratios and its impact on growth analysis in Pinus contorta. *Trees* **1988**, *2*, 104–109. [CrossRef]
36. Granier, A.; Huc, R.; Barigah, S.T. Transpiration of natural rain forest and its dependence on climatic factors. *Agric. For. Meteorol.* **1996**, *78*, 19–29. [CrossRef]
37. Ewers, B.E.; Mackay, D.S.; Gower, S.T.; Ahl, D.E.; Burrows, S.N.; Samanta, S.S. Tree species effects on stand transpiration in northern Wisconsin. *Water Resour. Res.* **2002**, *38*, 1103. [CrossRef]
38. Kumagai, T.; Nagasawaa, H.; Mabuchia, T.; Ohsakia, S.; Kubotaa, K.; Kogia, K.; Utsumia, Y.; Kogaa, S.; Otsuki, K. Sources of error in estimating stand transpiration using allometric relationships between stem diameter and sapwood area for Criptomeria japonica and Chamaecyparis obtuse. *For. Ecol. Manag.* **2005**, *206*, 191–195. [CrossRef]
39. Denny, C.; Nielsen, S. Spatial Heterogeneity of the Forest Canopy Scales with the Heterogeneity of an Understory Shrub Based on Fractal Analysis. *Forests* **2017**, *8*, 146. [CrossRef]
40. Montesano, P.M.; Niegh, C.S.R.; Wagner, W.; Wooten, M.; Cook, B.D. Boreal Canopy Surfaces from spaceborne stereogrammetry. *Remote Sens. Environ.* **2019**, *225*, 148–159. [CrossRef]
41. Saito, K.; Iwahana, G.; Ikawa, H.; Nagano, H.; Busey, R.C. Links between annual surface temperature variation and land cover heterogeneity for a boreal forest as characterized by continuous, fibre-optic DTS monitoring. *Geosci. Instrum. Methods Data Syst.* **2018**, *7*, 223–234. [CrossRef]
42. Molina, E.; Valeria, O.; De Grandpre, L. Twenty-Eight Years of Changes in Landscape Heterogeneity of Mixedwood Boreal Forest Under Management in Quebec, Canada. *Can. J. Remote Sens.* **2018**, *44*, 26–39. [CrossRef]
43. Legendre, P. Spatial Autocorrelation: Trouble or New Paradigm? *Ecology* **1993**, *74*, 1659–1673. [CrossRef]
44. Loranty, M.M.; Ewers, B.E.; Adelman, J.D.; Kruger, E.L.; Mackay, D.S. Environmental drivers of spatial variation in whole-tree transpiration in an aspen-dominated upland-to-wetland forest gradient. *Water Resour. Res.* **2008**, *44*. [CrossRef]
45. Tyree, M.T. Water Relations of Plants. In *Eco-Hydrology, Plants and Water in Terrestrial Aquatic Environments*; Bard, S.J., Wilby, R.L., Eds.; Taylor & Francis Group, Routledge Publishers: Abingdon, UK, 1999.
46. Elliot-Frisk, D.L. The boreal forest. In *North American Terrestrial Vegetation*, 1st ed.; Barbour, M.G., Billings, W.D., Eds.; Cambridge University Press: New York, NY, USA, 1988; Chapter 2; pp. 33–62.
47. Liu, J.; Chen, J.M.; Cihlar, J. Mapping evapotranspiration based on remote sensing: An application to Canada's landmass. *Water Resour. Res.* **2003**, *39*, 1189–1203. [CrossRef]
48. Peet, R.K. Forests of the Rocky Mountains. In *North American Terrestrial Vegetation*, 1st ed.; Barbour, M.G., Billings, W.D., Eds.; Cambridge University Press: New York, NY, USA, 1988; pp. 33–62. ISBN 978-0521261982.
49. Rowe, J.S. *Forest Regions of Canada*; Publication No. 1300; Department of the Environment, Canadian Forestry Service: Ottawa, ON, Canada, 1972; p. 177. Available online: http://cfs.nrcan.gc.ca/pubwarehouse/pdfs/24040.pdf (accessed on 2 September 2020).
50. Strong, W.L.; Leggat, K.R. *Ecoregions of Alberta*, 1st ed.; Report No. T/245; Alberta Forestry, Lands and Wildlife: Edmonton, AB, Canada, 1992; p. 59.
51. Granier, A.; Bobay, V.; Gash, J.; Gelpe, J.; Saugier, B.; Shuttleworth, W. Vapour flux density and transpiration rate comparisons in a stand of Maritime pine (*Pinus pinaster* Ait.) in Les Landes forest. *Agric. For. Meteorol.* **1990**, *51*, 309–319. [CrossRef]
52. Granier, A. Une nouvelle methode pour la mesure du flux de sève brute dans letronc des arbres. *Ann. Sci. For.* **1985**, *42*, 193–200. [CrossRef]

53. Bovard, B.D.; Curtis, P.S.; Vogel, C.S.; Su, H.-B.; Schmid, H.P. Environmental controls on sap flow in a northern hardwood forest. *Tree Physiol.* **2005**, *25*, 31–38. [CrossRef] [PubMed]
54. Booker, R.E. Dye-flow apparatus to measure the variation in axial xylem permeability over a stem cross section. *Plant Cell Environ.* **1984**, *7*, 623–628.
55. James, S.A.; Clearwater, M.J.; Meinzer, F.C.; Goldstein, G. Heat dissipation sensors of variable length for the measurement of sap flow in trees with deep sapwood. *Tree Physiol.* **2002**, *22*, 277–283. [CrossRef]
56. Čermák, J.; Kučera, J.; Nadezhdina, N. Sap flow measurements with some thermodynamic methods, flow integration within trees and scaling up from sample trees to entire forest stands. *Trees* **2004**, *18*, 529–546. [CrossRef]
57. Ford, C.R.; McGuire, M.A.; Mitchell, R.J.; Teskey, R.O. Assessing variation in the radial profile of sap flux density in Pinus species and its effect on daily water use. *Tree Physiol.* **2004**, *24*, 241–249. [CrossRef]
58. Mark, W.R.; Crews, D.L. Heat-Pulse velocity and bordered pit condition in living Engelman spruce and Lodgepole pine trees. *For. Sci.* **1973**, *19*, 291–296.
59. Delzon, S.; Sartore, M.; Burlett, R.; Dewar, R.; Loustau, D. Hydraulic responses to height growth in maritime pine trees. *Plant Cell Environ.* **2004**, *27*, 1077–1087. [CrossRef]
60. Loustau, D.; Berbigier, P.; Roumagnac, P.; Arruda-Pacheco, C.; David, J.S.; Ferreira, M.I.; Pereira, J.S.; Tavares, R. Transpiration of a 64-year-old maritime pine stand in Portugal. *Oecologia* **1996**, *107*, 33–42. [CrossRef] [PubMed]
61. Hogg, E.; Hartog, G.D.; Neumann, H.H.; Zimmermann, R.; Hurdle, P.A.; Blanken, P.D.; Nesic, Z.; Staebler, R.M.; McDonald, K.C.; Black, T.A.; et al. A comparison of sap flow and eddy fluxes of water vapor from a boreal deciduous forest. *J. Geophys. Res. Space Phys.* **1997**, *102*, 28929–28937. [CrossRef]
62. Goldstein, G.; Andrade, J.L.; Meinzer, F.C.; Holbrook, N.M.; Cavelier, J.; Jackson, P.; Celis, A. Stem water storage and diurnal patterns of water use in tropical forest canopy trees. *Plant Cell Environ.* **1998**, *21*, 397–406. [CrossRef]
63. Liu, J. A process-based boreal ecosystem productivity simulator using remote sensing inputs. *Remote Sens. Environ.* **1997**, *62*, 158–175. [CrossRef]
64. Allen, R.G.; Pruitt, W.O.; Businger, J.A.; Fritschen, L.J.; Jensen, M.E.; Quinn, F.H. Evaporation and transpiration. In *Hydrology Handbook, ASCE Manuals and Reports on Engineering Practice No. 28*, 2nd ed.; Heggen, R.J., Ed.; ASCE: New York, NY, USA, 1996; Chapter 4.
65. Perrier, A. Land surfaces processes: Vegetation. In *Land Surface Processes in Atmospheric General Circulation Model*; Eagleson, P.S., Ed.; Cambridge University Press: Greenbelt, MA, USA, 1982; pp. 395–448.
66. Jarvis, P.G.; James, G.B.; Landsberg, J.J. Coniferous forest. In *Vegetation and the Atmosphere*, 1st ed.; Monteith, J.L., Ed.; Academic Press Inc.: New York, NY, USA, 1976; Chapter 7; pp. 171–240.
67. EC. *Canadian Climate Normals 1971–2000, Report*; Environment Canada: Ottawa, ON, Canada, 2006. Available online: https://climate.weather.gc.ca/climate_normals/index_e.html (accessed on 2 September 2020).
68. Dunne, T.; Leopold, L.B. *Water in Environmental Planning*, 15th ed.; Freeman and Company Publishers: New York, NY, USA, 1998.
69. Black, T.A.; Kelliher, F.M.; Wallace, J.S.; Stewart, J.B.; Monteith, J.L.; Jarvis, P.G. Processes controlling understorey evapotranspiration. *Philos. Trans. R. Soc. Lond. Ser. B Biol. Sci.* **1989**, *324*, 207–231.
70. Baldocchi, D.D.; Vogel, C.A. Energy and CO_2 flux densities above and below the temperature broad-leaved forest and a boreal pine forest. *Tree Physiol.* **1996**, *16*, 5–16. [CrossRef]
71. Monteith, J.L. Evaporation and environment. In *Proceedings during Symposia of the Society for Experimental Biology. The State and Movement of Water Living Organisms, Number XIX*; Fogg, G.E., Ed.; Cambridge University Press: London, UK, 1965; pp. 205–234.
72. Lowe, P.R. An approximating polynomial for the computation of saturation vapour pressure. *J. Appl. Meteorol.* **1977**, *16*, 100–103. [CrossRef]
73. Murray, F.W. On the computation of saturation vapour pressure. *J. Appl. Meteorol.* **1967**, *6*, 203–204. [CrossRef]
74. Dingman, S.L. *Physical Hydrology*, 2nd ed.; Prentice-Hall, Inc.: Upper Saddle River, NJ, USA, 2002.
75. Smith, M. *Report on the Expert Consultation on Revision of FAO Methodologies for Crop Water Requirements Report*; Food and Agriculture Organization of the United Nations (FAO): Rome, Italy, 1990.
76. *Evaporation into the Atmosphere: Theory, History, and Applications*, 1st ed.; Reidel: Dordrecht, The Netherlands, 1982.

77. De Bruin, H.A.R.; Moore, C.J. Zero-plane displacement and droughness length fir tall vegetation, derived from a simple mass conservation hypothesis. *Bound. Layer Meteorol.* **1985**, *31*, 39–49. [CrossRef]
78. Price, D.T.; Black, T.A. Estimation of forest transpiration and CO_2 uptake using the Penman-Monteith equation and a physiological photosynthesis model. In *Estimation of Aereal Evapotranspiration, Number 177 in IAHS, Red Books, Proceedings of an International Workshop Held during the XIXth General Assembly of the International Union of Geodesy and Geophysics, Vancouver, BC, Canada, 9–22 August 1987*, 1st ed.; Black, T.A., Spittlehouse, D.A., Novak, M.D., Eds.; International Association of Hydrological Sciences: Wallingford, UK, 1987; pp. 213–227.
79. Stewart, J.B. Modelling surface conductance of pine forest. *Agric. For. Meteorol.* **1988**, *43*, 19–35. [CrossRef]
80. Blanken, P.; Black, T.A.; Yang, P.C.; Newmann, H.H.; Nesic, Z.; Staebler, R.; de Hartog, G.; Novak, M.D.; Lee, X. Energy balance and canopy conductance of a boreal aspen forest: Partitioning overstory and understory components. *J. Geophys. Res.* **1997**, *102*, 28915–28927. [CrossRef]
81. Greenlee, G.M. Soils map of Bow Valley Provincial Park and adjacent Kananaskis area. In *Soils Map;* Alberta Research Council, Soils Division, Alberta Institute of Pedology: Edmonton, AB, Canada, 1973.
82. Greenlee, G.M. Soil survey of Kananaskis Lakes area and intrepretation for recreational use. In *Final Report M-76-1;* Albeta Research Council, Soils Division, Albeta Institute of Pedology: Edmonton, AB, Canada, 1976.
83. McGregor, C.A. Ecological land classification and evaluation. Kananaskis country. In *Natural Resources Summary, Alberta Energy and Natural Resources;* Resource Appraisal Section, Resource Evaluation Branch: Edmonton, AB, Canada, 1984.
84. Archibald, J.; Klappstein, G.D.; Corns, I.G.W. Field guide to ecosites of Southwestern Alberta. In *Special Report 8;* Canadian Forest Service, Northwest Region, Northern Forestry Centre: Edmonton, AB, Canada, 1996.
85. Choudhury, B.J. Estimating evaporation and carbon assimilation using infrared temperature data: Vistas in modeling. In *Theory and Applications of Optical Remote Sensing, Remote Sensing Series*, 1st ed.; Asrar, G., Ed.; John Wiley & Sons: Hoboken, NJ, USA, 1989; Chapter 17; p. 628.
86. Dooge, J.C.I. Modelling the behaviour of water. In *Forests, Climate, and Hydrology*, 1st ed.; Reynolds, E.R.C., Thompson, F.B., Eds.; The United Nations University, Kefford Press: Singapore, 1988; Chapter 7; p. 227.
87. Kondratyev, K.Y.; Korzov, V.I.; Mukhenberg, V.V.; Dyachenko, L.N. The shortwave albedo and the surface emissivity. In *Land Surface Processes in Atmospheric General Circulation Models;* Eagleson, P.S., Ed.; WMO/ICSO Joint Scientific Committee, Cambridge University Press: Greenbelt, MD, USA, 1982; pp. 463–514.
88. Henderson-Sellers, A.; Wilson, M.F. Surface albedo data for climatic modeling. *Rev. Geophys. Space Phys.* **1983**, *21*, 1743–1778. [CrossRef]
89. Chen, J.M.; Zhang, R.-H. Studies on the measurements of crop emissivity and sky temperature. *Agric. For. Meteorol.* **1989**, *49*, 23–34. [CrossRef]
90. Buermann, W.; Wang, Y.; Dong, J.; Zhou, L.; Zeng, X.; Dickinson, R.E.; Potter, C.; Myneni, R.B. Analysis of a multiyear global vegetation leaf area index data set. *J. Geophys. Res. Space Phys.* **2002**, *107*. [CrossRef]
91. Myneni, R.B.; Ramakrishna, R.; Nemani, R.; Running, S.W. Estimation of global leaf area index and absorbed par using radiative transfer models. *IEEE Trans. Geosci. Remote Sens.* **1997**, *35*, 1380–1393. [CrossRef]
92. McAllister, D. Remote Estimation of Leaf Area Index in Forested Ecosystems. Master's Thesis, University of Calgary, Calgary, AB, Canada, May 2005.
93. Leblanc, S.G.; Chen, J.M.; Kwong, M. Tracing Radiation and Architecture of Canopies. In *TRAC Manual Version 2.1.3*, 1st ed.; Natural Resources Canada, Canada Centre for Remote Sensing: Ottawa, ON, Canada, 2002.
94. Allen, R.G.; Jensen, M.E.; Wright, J.L.; Burman, R.D. Operation estimates of reference evapotranspiration. *Agron. J.* **1989**, *81*, 650–662. [CrossRef]
95. Bladon, K.D.; Silins, U.; Landhäusser, S.M.; Lieffers, V.J. Differential transpiration by three boreal tree species in response to increased evaporative demand after variable retention harvesting. *Agric. For. Meteorol.* **2006**, *138*, 104–119. [CrossRef]
96. Van Bavel, C.H.M. Potential evaporation: The combination concept and its experimental verification. *Water Resour. Res.* **1966**, *2*, 455–467. [CrossRef]

97. Chang, M. Forests and vaporization. In *Forest Hydrology, an Introduction to Water and Forests*, 1st ed.; Chang, M., Ed.; CRC Press LLC: Boca Raton, FL, USA, 2002; Chapter 9; pp. 151–174.
98. McCuen, R.H. *Hydrological Analysis and Design*, 3rd ed.; Prentice Hall: Upper Saddle River, NJ, USA, 1989.

Publisher's Note: MDPI stays neutral with regard to jurisdictional claims in published maps and institutional affiliations.

atmosphere

MDPI

Article

Assessing Suitable Areas of Common Grapevine (*Vitis vinifera* L.) for Current and Future Climate Situations: The CDS Toolbox SDM

Guillermo Hinojos Mendoza [1,2], Cesar Arturo Gutierrez Ramos [2], Dulce María Heredia Corral [2], Ricardo Soto Cruz [2] and Emmanuel Garbolino [3,*]

1 ASES Ecological and Sustainable Services, Pépinière d'Entreprises l'Espélidou, Parc d'Activités du Vinobre, 555 Chemin des Traverses, Lachapelle-sous-Aubenas, 07200 Aubenas, France; ghinojos@asessc.net

2 ASES Ecological and Sustainable Services, Avenida Francisco Villa #7701 Plaza Bambú, Local 10 Chihuahua, Chih 31210, Mexico; cesar.gutierrez@asessc.net (C.A.G.R.); dulcehc@asessc.net (D.M.H.C.); ricardo.soto@asessc.net (R.S.C.)

3 Climpact Data Science (CDS), Nova Sophia-Regus Nova, 291 rue Albert Caquot, CS 40095, 06902 Sophia Antipolis CEDEX, France

* Correspondence: emmanuel.garbolino@asessc.net

Received: 5 October 2020; Accepted: 30 October 2020; Published: 6 November 2020

check for updates

Abstract: Climate Data Science (CDS) Toolbox Species Distribution Model (SDM) aims identifying the suitable areas for species, community of species and landscape units. This model is based on the use of 23 variables available over the Internet, for which any assumptions are formulated about their relationships with the spatial distribution of species. The application of CDS Toolbox SDM on the assessment of the potential impact of two scenarios of climate change (Representative Concentration Pathways RCP4.5 and RCP6.0) on the suitability of grapevine crops in France shows a general decrease of the most suitable areas for grapevine crops between 41% and 83% towards 2070 according to the current location of the vineyard parcels. The results underline a potential shift of the suitable areas in northern part of the French territory. They also show a potential shift of the most suitable areas in altitude (60 m in average) for RCP6.0 scenario. Finally, the model shows that RCP4.5 scenario should be more drastic than RCP6.0 scenario by 2050 and 2070. In effect, the model underlines a significant potential decrease of cultivated crops in the areas of high probably of suitable areas, according to the baseline scenario. This decrease would be of 630,000 ha for 2070 RCP4.5 scenario and 330,000 ha for 2070 RCP6.0 scenario.

Keywords: species distribution model; climate change; scenarios; GIS; ecological niche; grapevine

1. Introduction

Species Distribution Models (SDM) have shown a significant development in the last decades, especially due to the needs of scientists to provide methods and tools in order to assess the potential impacts of climate change on the distribution of species or communities of species [1].

Also, public and private sectors, and the public in general interested on the potential impacts of climate change on ecosystems services, expressed the need to have more access to studies, tools and results from the experts.

Currently different methodologies are in use to estimate the potential impact of climate change on the distribution and assemble of species at different spatio-temporal scales. Among these methods are the regression trees [2], Artificial Neural Networks [3–8], and Bayesian approaches [9–11]. ANN (Artificial Neural Network) are able to learn complex non-linear relations and can help estimate parameters like suitable areas of a territory for species. However, they require huge datasets in order to be efficient, which is not always possible when scientists have to assess the spatial distribution of few observed species. Bayesian approaches are also developed when the model uses random variables or observed data, or when the assumption of fixed variables is not verified, which is often the case for data records from long term and/or huge areas [9–11]. Most of these approaches can be aggregated in order to improve the models. A key step of the models is calibration of the relationships of the species and environmental variables using ad hoc of wildlife with environmental data and taking into account the quantitative and intermittent nature of the relationships of the data. As mentioned by [12] some approaches are based on geometrical statistics that do not really respect the intermittent nature of the relationships between species or communities of species and the parameters of their environment, like climate and soil parameters [13].

Other approaches are based on probabilistic methods that take into account the intermittent nature of the data better [14,15]. We propose a model integrated into a set of other models and tools named CDS toolbox SDM (CDS for Climate Data Science) in order to assess the potential suitable areas for species, community of species, or landscape units according to current and future scenarios of climate change. We started to develop this model in 2009, in the frame of an exploratory project called *"Climpact"* in order to assess the potential consequences of climate change scenarios on the risk of wildland fires in Corsica [14,15]. This first prototype, computerized in C++, was initially based on three climatic variables (minimum temperature—Tmin; maximum temperatures—Tmax, precipitations—P). This project led us to improve our model by integrating more bioclimatic and environmental variables in order to characterize, in a more accurate way, the ecological niche of species and landscape units [16–19]. The current version is an ArcGis© tool (ArcGis, ESRI, Redlands, CA, USA) developed with the model builder of GIS (Geographic Information System) application. This tool is only available through a collaboration agreement and an online version is under development.

Since the beginning of its design, this SDM provided three main benefits:

(i) It respects the intermittent nature of species occurrences into environmental variables;
(ii) It is a GIS (Geographic Information System) based application that does not require a high level of expertise in computer systems in order to implement it;
(iii) It shows gradients of probabilities to find suitable areas for each species, communities of species, or landscape units.

The next sections introduce the model structure, its functioning, and an example of a species distribution modelling (grapevine, *Vitis vinifera* L.) according to a baseline climatic situation and two scenarios of climate change provided by the IPCC (Intergovernmental Panel on Climate Change) and downscaled thanks to the WorldClim 2 contributors [20]. The discussion is based on the comments of the models results and a comparison with other studies on the potential impact of climate change on grapevine crops.

2. Model Description and Functioning

The aim of the CDS toolbox SDM is to identify, on a territory, the potential suitable areas where a species or a community of species could grow.

CDS toolbox SDM is based on ecological niche theory where an ecological niche can be considered as *"the position of a species within an ecosystem, describing both the range of conditions necessary for persistence of the species, and its ecological role in the ecosystem"* [21].

It is required to calibrate the relationships between the spatial distribution of a species (or group of species) with the spatial distribution of environmental variables that seems relevant for its development, like climate, soil types, slope, etc. This calibration represents the first step of the CDS toolbox SDM (Figure 1) which consists with the overlap of the spatial distribution of 23 environmental variables with the observations of a species. 19 are related to climatic and bioclimatic variables that are considered as relevant for the development and survival of species.

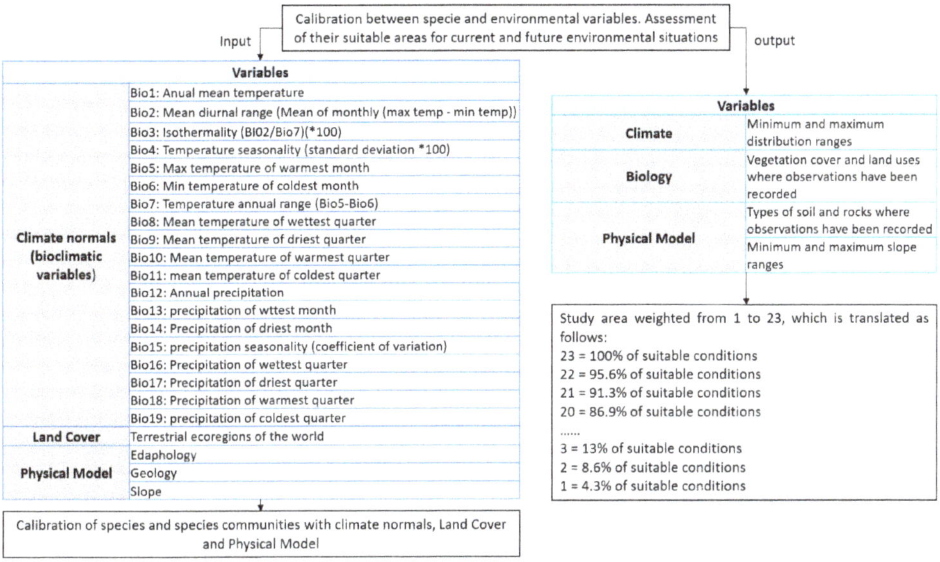

Figure 1. Data and steps related of CDS Toolbox SDM.

The other 4 environmental variables are related to the land use and vegetation type in which the species are observed, the type of soil, and rocks that are relevant for their ecology and the range of slope where species can be observed.

The CDS toolbox SDM does not formulate any assumption on the relationship between a species with its environment: it just considers the occurrence of the species on the values or category of each variable. In this process, each variable has the same weight in order to avoid conjectural assumptions. Like other SDM [1], the accuracy of the calibration belongs to the spatial resolution and the amount of observations and measures. The result of this first step is the identification of the range of variable values that are considered significant in order to ensure species development and survival. In another terms, this step allows establishing the ecological niche of a species. The 19 selected climatic variables are related to temperature and precipitation statistics that give a synthetic description of climatic envelop of species and can also be considered as limiting factors [22]. Temperature plays a role on plant lethality: temperatures that are too low slow down or stop the growth of plants. For example, the frost causes a mechanical action on plant cells, resulting in the formation of ice crystals which destroy cell walls. The frost also causes water loss, which leads to desiccation of certain organs. In contrast, high temperatures have an effect on the evaporation of water reserves contained in the soil and generate excessive leaf transpiration causing water stress (evapotranspiration phenomenon). Depending on its duration and intensity, it can be lethal for non-adapted or poorly adapted plants. Scorching episodes, such as the one that occurred in France and Europe in 2003, resulted in increased mortality of plants [23,24].

In this frame, we give more importance to climatic variables because they influence largely the survival of plants especially for areas where climatic gradients are significant, according to the spatial distribution of their observations and the spatial distribution of the potential suitable areas for their development. Appendix A provides a statistical description (min, max, mean, standard deviation) of the quantitative variables and the description of the classes of nominal variables.

The second step is the identification of potential suitable areas for the species survival and development according to current or future environmental situations. In our approach, we perform the two assessments (baseline and future environmental contexts) in order to identify the potential trends (increase, decrease, stability) of the spatial distribution of suitable areas for species, community of species, or landscape units.

The calculation is based on the finding of favorable conditions on a territory. In this step, the algorithm looks for the pixels where the reference conditions are the same as the one observed for species. The algorithm selects the pixels that have the same categories of nominal variables (land use, vegetation cover, edaphology, geology) and the pixels that fall in the range of quantitative variables (temperature, precipitation, slope). However, in order to take into account the uncertainty of finding similar environmental conditions for the species and their capacity to adapt to the environmental changes, the model considers 3 ecological situations:

1. The species or community of species adapt slightly to the new environmental conditions and they select the areas where the conditions are closest to the optimum of reference with a contraction of the populations or the community;
2. The species or community of species adapts drastically to the new environmental conditions and they can remain in the same areas;
3. The species or community of species are not able to adapt to changes and disappears locally.

Thus, the algorithm uses a linear equation in which each of the 23 variables is summed in order to calculate the level of suitability of each area for the ecological niche of species or community of species. This parameter, named "*Potential Ecological Distribution*" (PED), is given by the following expression:

$$\text{Potential Ecological Distribution} = \text{Variable 1} + \text{Variable 2} \ldots + \text{Variable n} \ldots + \text{Variable 23} \quad (1)$$

Figure 2 shows a hypothetical example of the application of the algorithm on four pixels of a territory and with only three variables (pattern 1 = maximum temperature, pattern 2 = minimum temperature, and pattern 3 = precipitations). For pattern 1, there is only one pixel with a value corresponding to the ecological niche of a species. For pattern 2, there are two pixels with a favorable value for the development of a species, and for pattern 3, all four pixels have a suitable value. Then, the algorithm calculates the sum of each value of variables for each pixel. The result shows that 1 pixel has the best suitability (amount = 300), another has an average level of suitability (amount = 200) and two pixels have a low level of suitability (amount = 100).

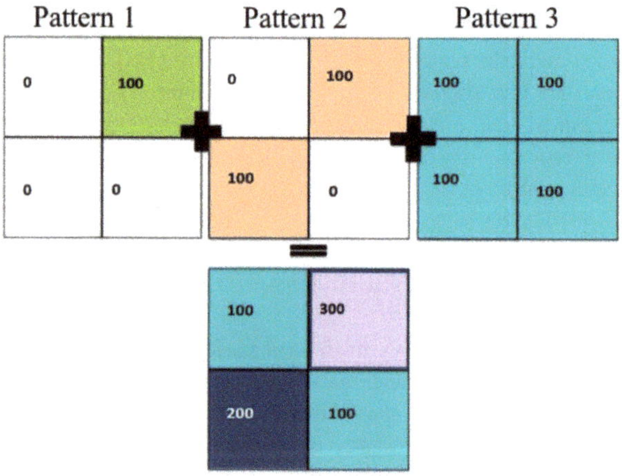

Figure 2. Potential ecological distribution of a species in 4 pixels according to 3 variables (patterns).

This step allows identifying the level of environmental similarity of each part of a territory that will support the decision process for ecosystems and biological resource management. This activity is related to different categories of stakeholders and decision-makers at local, regional, and national levels like Ministers of Ecology and Agriculture, Mayors and public authorities, farmers, forest managers, policy makers, protected areas administrators, supply chain supervisors, etc.

In the CDS toolbox SDM, the algorithm carries out this calculation for each pixel of a particular territory according to the 23 variables taken into account for defining the ecological niche of a species or a community of species. This process is applied for baseline and future environmental conditions especially the climatic ones based on the IPCC scenarios. The result allows identifying trends of species spatial dynamics and can help support decisions in order to manage potential changes in ecosystem services.

According to [25], we present, in Figure 3, the three modalities that contribute to the decision process. These decision rules are based on arithmetic and statistical procedures allowing the integration of stabilized criteria into a unique index. This index aims to help decision makers for making comparisons of alternatives of spatial distribution of species, community of species and landscape units.

Decision criteria										
1/23	2/23	3/23	4/23	5/23	6/23	7/23	8/23	9/23	10/23	11/23
Extremely low		Very low			Low			Rather low		
Weighting										
100	200	300	400	500	600	700	800	900	1000	1100
Global similarity in %										
4.34	8.69	13.04	17.39	21.73	26.08	30.43	34.78	39.13	43.47	47.82

Decision criteria											
12/23	13/23	14/23	15/23	16/23	17/23	18/23	19/23	20/23	21/23	22/23	23/23
Moderate			Rather high		High		Very high		Extremely high		Equal
Weighting											
1200	1300	1400	1500	1600	1700	1800	1900	2000	2100	2200	2300
Global similarity in %											
52.17	56.52	60.86	65.21	69.56	73.91	78.26	82.6	86.95	91.3	95.65	100.00

Figure 3. Decision criteria, weighting, and global similarities of pixels according to the ecological niche of species, a community of species, or landscape units.

In order to respect the three ecological situations mentioned previously, the interpretation of the table follows the coming logic:

- When a pixel has the 100% of global similarity (weighting = 23) it means that the pixel has 100% of suitability for species or community of species. In this area the environmental parameters correspond to the ecological niche of species or community of species (decision criteria = equal, that means equality of environmental parameters). In the case of the assessment of the potential impact of climate change on species distribution, 100% of global similarity means that species would not find problems for their life and their development.
- When a pixel has global similarity values between 82.6 and 100% (weighting = 1900, 2000, 2100 or 2200), this indicates that the environmental conditions for the species are slightly similar to their ecological niche. The potential impact of climate change should not be significant on their life and development, and the adaptation of species to the future environmental conditions should be appropriate.
- When a pixel has global similarity values between 52.17 and 82.6% (weighting between 1200 and 1800), the pixels represent an area where the species should adapt to the new environmental conditions but showing some slight periods of stress. In this case, the uncertainty for the adaption of species to the new ecological situation is more important than in the other part of the range of the global similarity values.

- Finally, when a pixel has global similarity values between 1 and 52.17% (weighting between 100 and 1200), the area can be considered as poorly suitable for the development of the species or community of species. The possibility of adaptation of species to the future environmental conditions decreases significantly.

The final result of the application of the CDS Toolbox SDM is a map showing the probability to find suitable areas for species, community of species, and landscape units.

We present hereinafter an example of the application of CDS Toolbox SDM in order to assess the spatial distribution of suitable areas for *Vitis vinifera* L., the common grapevine, in France. Viticulture is a key socio-economic sector in Europe. Due to the strong sensitivity of grapevines to atmospheric factors, climate change may represent an important challenge for this sector [26].

According to the CnIV (National Committee of Interprofessions of Wines), France is the leading wine and wine brandy exporter, and is the second economic sector in the trade This economic sector employs 500,000 people and it can be considered as a key sector for the economy. With 750,000 ha of grapevine crops, France represents 11% of the world surface area for wine production.

For this case study, the spatial resolution of the data is around 1 km. They come from:

- WorldClim for the 19 climatic/bioclimatic data and for the slopes (elevation layer);
- The Ecoregion Layer for the data on land cover provided by WWF (World Wildlife Fund) [27];
- GeoTypes.net for geology layer [28];
- FAO GeoNetwork (Food and Agriculture Organization of the United Nations) for the edaphology layer.

The proposed assessment aims to identify the potential problems or opportunities on such crop and, if necessary, to aware stakeholders for adapting their practices on crops, on supply chain management, and on the selection of the best areas for the cultivation of grapevine.

3. Results

The field observations of *Vitis vinifera* L. in France (Figure 4) come from the iNaturalist Internet platform and they represent an amount of 35 observations. We also present the official map of grapevine crops in France (Figure 4) provided by the RGP (Parcells Geographic Register provided by the French National Geographic Institute–IGN) in 2018. All of this data is free of charge and open source.

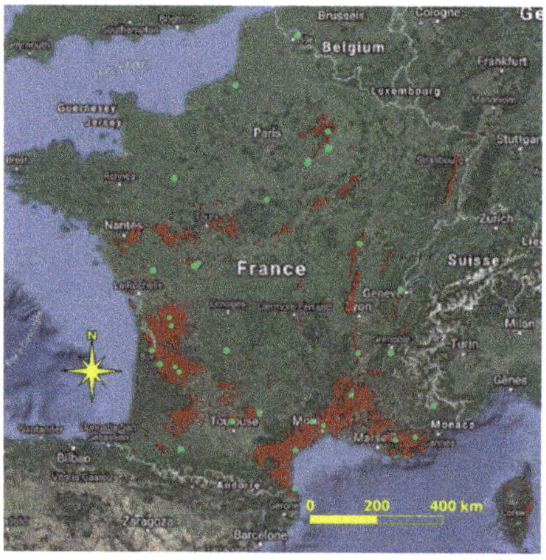

Figure 4. Locations of the observations of *Vitis vinifera* L. in France (in green, source: iNaturalist) and the grapevine crops declared in 2018 (in red).

The assessment of the suitable areas for *Vitis vinifera* L. has been performed for the closest climatic situation of the current period (called "baseline") provided by the WorldClim platform and for 2050 and 2070 by the use of IPCC scenarios corresponding to two average Representative Concentration Pathway: RCP4.5 and RCP6.0. We decided to use those two scenarios because they represent moderately optimistic (RCP4.5) and moderately pessimistic (RCP6.0) scenarios of climate change according to the current and future adoption of policies and application of means dedicated to GHG (Greenhouse Gas) emissions reduction by countries, industries, and organizations.

The contributors of WorldClim dataset provided the baseline scenario that we use in our study [20] by using climatic data from 60,000 weather stations for a temporal range of 1970 to 2000. They interpolated these measures with thin-plate splines and covariates (elevation, distance to the coast, maximum and minimum temperature, cloud cover from MODIS satellite). For the scenarios of climate change, the WorldClim contributors used IPCC data for which they apply a statistical downscaling method based on interpolations [29].

These maps show the different level of probabilities that have been classified into three classes according to the Jenks Natural Break classification method [30]. This method is often used in GIS project because it allows underlying differences of different objects of a same data set in a map. This classification is based on an iterative process in order to define classes with significant differences in the values of the data. Table 1 presents the probability ranges of the different classes.

Table 1. Ranges of the different classes of probabilities.

Classes	2050 RCP4.5		2050 RCP6.0		2070 RCP4.5		2070 RCP6.0		Baseline	
1	4.34	60.86	4.34	56.52	4.34	56.52	4.34	52.17	4.34	60.86
2	60.86	82.6	56.52	82.6	56.52	78.26	52.17	78.26	60.86	86.95
3	82.6	100	82.6	100	78.26	95.65	78.26	100	86.95	100

Class 1 corresponds to the low level of probabilities to find environmental conditions required for the development of a species, class 2 corresponds to the average level of probabilities, and class 3 corresponds to the high level of probabilities. The results of the model are analyzed by taking into account the different levels of probabilities to find suitable environmental conditions for the selected specie. These classes of probabilities are expressed by a colored code described below each figure.

The aim of these maps is to compare the potential spatial distribution of the suitable areas for cultivating the grapevine according to the current climatic situation and the climate change scenarios. The maps of probabilities to find suitable areas are presented on the left side of the figure while the right side of the figure represents the current probabilities to find suitable areas and the current locations of grapevine crops.

Figure 5 shows the potential suitable areas for *Vitis vinifera* L. in France according to the baseline climate. It also presents the location of the map of grapevine crops in order to compare its spatial distribution with the probabilities to find suitable areas.

A simple visual analysis of the maps shows that the current declared grapevine crops are mainly localized in the high probability level to find suitable areas for the development of *Vitis vinifera* L. The quantitative analysis of the overlap of these two information layers (Table 2) shows that 93% of the surface of grapevine crops are located into the high level of probability to find suitable areas for this cultivation. This first result demonstrates the relevance of the model in order to assess the spatial distribution of suitable territories of this specie.

The other areas of grapevine crops (7%) are located into to average probability of occurrence of suitable areas and there are no declared crops in the low probability class. Only 0.1% of the crops' surfaces are in areas that do not present any probability of occurrence. For the other results relating to the assessment for the future (2050 and 2070), they present a value of 1% of grapevine crops in areas without any probability to find suitable areas.

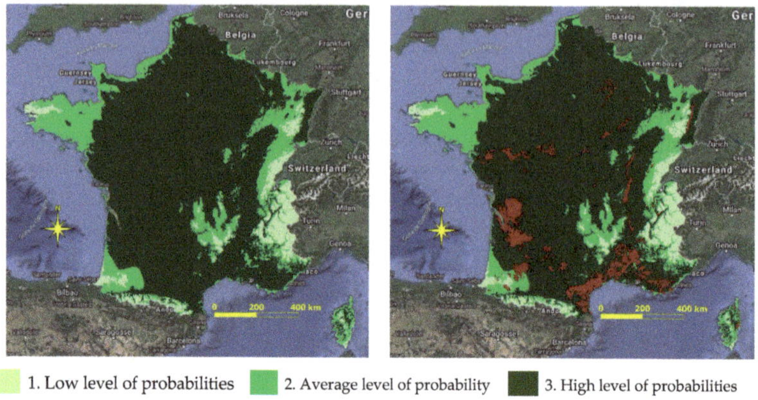

Figure 5. Potential suitable areas for grapevine crop according to baseline climate. Grapevine crops declared in 2018 are represented in red.

Table 2. Percentage and ha of grapevine crops in the probability classes to find suitable area.

	% of Polygons Outside of Suitable Areas	ha Outside of Suitable Areas	% of Polygons in Low Proba. Class	ha in Low Proba. Class	% of Polygons in Average Proba. Class	ha in Average Proba. Class	% of Polygons in High Proba. Class	ha in High Proba. Class
Baseline	<1	0	0	0	7	52,500	93	697,500
2050 RCP 4.5	1	7500	16	120,000	64	480,000	19	142,500
2050 RCP 6.0	1	7500	1	7500	46	345,000	52	390,000
2070 RCP 4.5	1	7500	5	37,500	85	637,500	9	67,500
2070 RCP 6.0	1	7500	1	7500	49	367,500	49	367,500

The analysis of Table 2 underlines a significant potential decrease of cultivated crops in the areas of high probably of suitable areas, according to the baseline scenario: this decrease would be of 555,000 ha for 2050 RCP4.5 scenario, 307,500 ha for 2050 RCP6.0 scenario, 630,000 ha for 2070 RCP4.5 and 330,000 ha for 2070 RCP6.0.

According to these first results, it is possible to compare the baseline situation with the potential future situations. The simulation of the potential impact of RCP4.5 climate scenario for 2050 (Figure 6) on the spatial distribution of suitable areas for *Vitis vinifera* L. shows a significant decrease of high level of probabilities on the French territory.

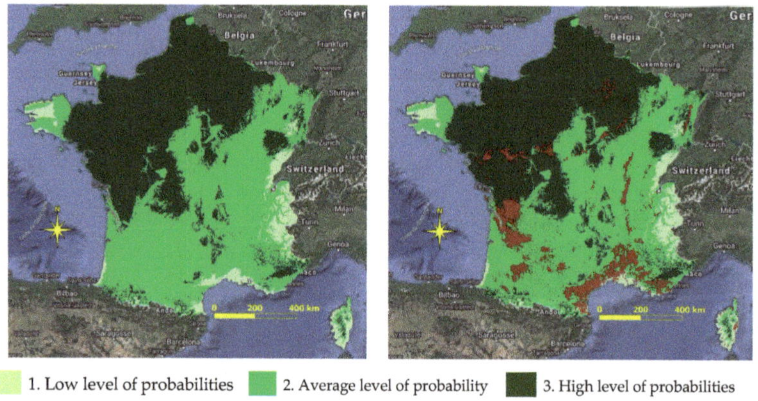

Figure 6. Potential suitable areas for grapevine crop according to RCP 4.5 climate scenario for 2050. Grapevine crops declared in 2018 are represented in red.

The comparison with the current grapevine crops (Table 2) shows that only 19% of the crops should be in the high level of probability to find suitable areas, 64% in the average level, and 16% in the low level.

At the opposite of the previous result, the simulation of the potential impact of RCP6.0 climate scenario for 2050 on the spatial distribution of suitable areas for *Vitis vinifera* L. (Figure 7) presents a less contrasted situation. The comparison with the current grapevine crops (Table 2) shows that 52% of the crops should be in the high level of probability to find suitable areas, 46% in the average level, and only 1% in the low level.

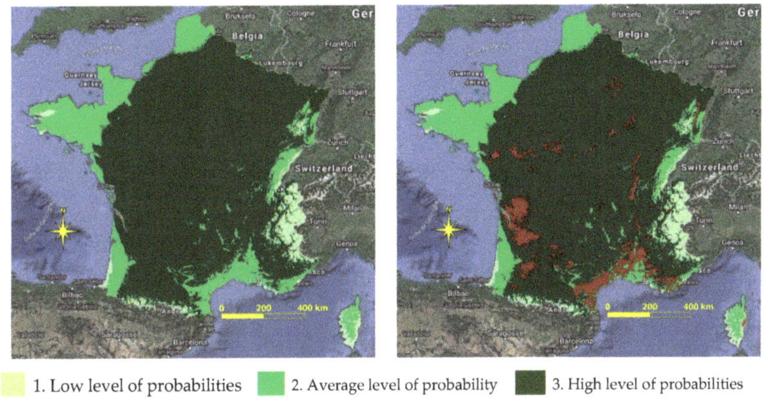

Figure 7. Potential suitable areas for grapevine crop according to RCP 6.0 climate scenario for 2050. Grapevine crops declared in 2018 are represented in red.

The modeling of the potential spatial distribution of suitable areas for *Vitis vinifera* L. in 2070 with RCP4.5 scenario (Figure 8) confirms the decrease of high probability class on the French territory. The comparison with the current crops areas shows that they should be located mainly in average probability class (85%) for this scenario (Table 2). The rest of the spatial distribution should correspond to 9% in high probability class and 5% in the low probability class.

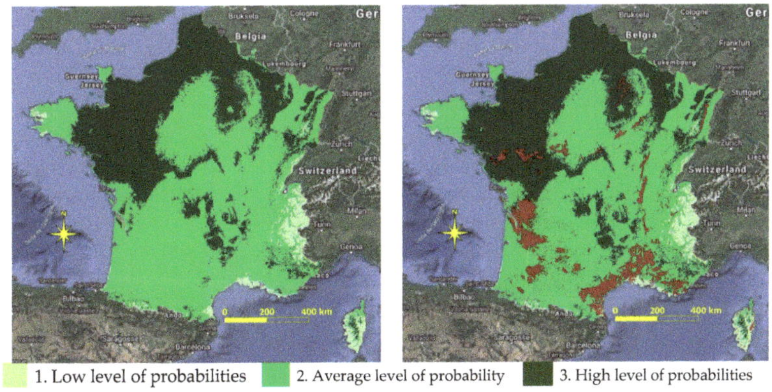

Figure 8. Potential suitable areas for grapevine crop according to RCP 4.5 climate scenario for 2070. Grapevine crops declared in 2018 are represented in red.

The assessment of potential suitable areas for 2070 according to RCP6.0 scenario (Figure 9) presents a potential equivalent spatial distribution of grapevine crops of 49% for high and average classes of probability. Only 1% of current cultivations should be located in the lowest probability class.

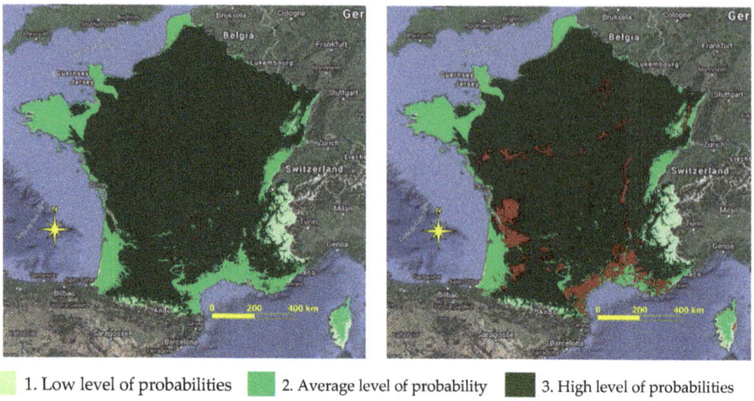

Figure 9. Potential suitable areas for grapevine crop according to RCP 6.0 climate scenario for 2070. Grapevine crops declared in 2018 are represented in red.

The results presented in Table 3 are related to the statistical distribution of the average altitude in each class of probability and according to the different climatic situations (baseline, 2050 and 2070).

Table 3. Average altitude of polygons into the different probability classes according to baseline and future climate scenarios.

Title Climate Scenarios	Avg. Alti. (m) Outside of Suitable Areas	Avg. Alti. (m) in Low Proba. Class	Avg. Alti. (m) in Average Proba. Class	Avg. Alti. (m) in High Proba. Class
Baseline	390	1306	421	233
2050 RCP 4.5	387	934	386	204
2050 RCP 6.0	384	1408	269	297
2070 RCP 4.5	387	1207	342	231
2070 RCP 6.0	384	1601	302	297

This table highlights that the areas with a high level of probability to find suitable areas for grapevine crops should be located in higher altitudes than the current ones for scenario RCP6.0 in 2050 and 2070. For scenario RCP4.5, there could be a global decrease of average altitudes for 2050 and 2070 according to the baseline situation. This decrease should be significant with the low and average classes of probability.

The analysis of the whole results underlines that the current location of grapevine crops are mainly situated in high level probability class (93%) but, according to the potential impact of climate change, these areas should become less favorable to its cultivation towards 2050 and 2070, even if the RCP4.5 and RCP6.0 scenarios show contrasted future situations. The variation of the surface located in the high probability class would decrease from 41% to 84% of the amount of the current areas situated in this class which would represent an amount between 307,500 ha and 630,000 ha where grapevines would face some perturbations on its growth and its mortality rate.

The results also show that RCP4.5 scenario would have a more drastic impact on the spatial distribution of suitable areas for grapevine than RCP6.0 scenario for both 2050 and 2070 periods in France because most of the crops would be situated in average and low probability classes with RCP4.5 scenario.

4. Discussion

The application of CDS Toolbox SDM on the potential suitability of grapevine crops in France shows two main ecological and biogeographical mechanisms.

The first one is the selection pressure that leads to the contraction of the spatial distribution of species. This appears when the suitable areas are very few (global similarity values between 1 and 52.17%) like it is for class 1. In this case, species cannot adapt or may face very difficult problems to adapt to the future environmental conditions. The result is a decrease of the surface they previously colonized.

The second one is the environmental pressure on the phenotypic plasticity that can lead or not to the expansion of the areas colonized by the species or community of species. This process is complex because there are different ways of expression of the phenotypic plasticity. One of these possibilities is the contraction of the distribution area of species that correspond to the resistance of new environmental conditions. This process can also generate a migration of species to other areas that are more suitable but the areas colonized remain lower than the previous ecological situation (classes 2 and 3). Another type of expression of phenotypic plasticity is the expansion of the specie in more areas than before because the environmental changes provides areas that are more suitable.

With the use of RCP4.5 scenarios for the 2050 and 2070 periods, CDS Toolbox SDM shows that climate change would have a significant negative role on the spatial distribution of suitable areas for grapevine crops. RCP4.5 scenario seems to have a more drastic impact on the spatial distribution of grapevine than RCP6.0 scenario. Nevertheless, those two scenarios also show a significant decrease of suitable areas for grapevine in 2050 and 2070 according to its current distribution. This result is coherent with the conclusions of [31], which identified that new territories should be suitable for grapevine cultivation in the northern part of France by using A1B from penultimate IPCC climate scenarios version (scenario similar to RCP6.0, [32]) with three downscaling methods (weather type—WT; Quantile-Quantile—QQ; and Anomalies—ANO) in 2050 and 2100, but without proposing a mapping method. The model developed by [32] is based on the use of annual means and standard deviation in order to calculate the climate change impact on phenology, transpiration ratio, and climatic water balance [33], using a GIS approach, also shows a potential shift of suitable areas for viniculture towards 2100 in the northern part of France. They use the penultimate IPCC scenarios B1 (scenario close to RCP2.6 scenario, the most optimistic one, [31]) and A1B with a spatial resolution of 18 km. The estimation of the potential distribution of suitable areas is based on the spatial distribution of bioclimatic variables, but without calibrating the relationships between vineyards areas and these variables.

Fraga et al. [34] present similar trends with RCP4.5 and RCP8.5 scenarios in order to show the potential impact of climate change on climatic suitability of 44 varieties of grapevine in Portugal. Their results show a potential shift of suitable areas in the northern part of Portugal and other European countries and in higher altitudes than currently. Their model is based on a spatial resolution of 1 km using the WorldClim dataset and they focus their analysis on the spatial distribution of bioclimatic indexes and their correlation with the viticultural regions of Portugal.

Moriondo et al. [35] argue that climate change would provoke a shift in the north and north-west of their current location in France, Germany, Italy, Portugal and Spain using A2 and B2 scenarios towards 2050 at 1 km of spatial resolution. They also underline the potential expansion or contraction of some suitable areas for grapevine crops according to the potential impact of climate change.

In the frame of the European CORDEX project (Coordinated Downscaling Experiment—European Domain) Cardell et al. [36] studied the evolution of 11 bioclimatic indices for 3 periods (2021–2045, 2046–2070, 2071–2095) by using RCP4.5 and RCP8.5 scenarios at 12 km of spatial resolution. Their results show that climate change would induce a shift of suitable areas for grape wine crops in Central and Northern parts of Europe like Germany, North of France, Belgium, Poland, Southern England and Czech Republic due to better temperatures around 2050. Our study presents similar conclusions but with a more accurate spatial resolution (1 km) at the scale of the French territory.

In Italy, Caffarra and Eccel [37], by using A2 (scenario similar to RCP6.0, IPCC 2013) and B2 (scenario similar to RCP8.5, [31]) IPCC scenarios from the penultimate version of climatic assessment, mention that mountain areas at an elevation of around 1000 m in the region of Trentino (Italian Alps) would be suitable for the cultivation of grapevine due climate change towards 2100. In our model, the potential development of the grapevine in mountains is more nuanced, especially for the territories situated in high levels of probability to find suitable conditions. According to RCP6.0 in 2050 and 2070, there could be a potential increase of the average altitude but only around 300m. The areas situated around 1000 m mainly match with low probability class to find suitable environmental conditions for the cultivation of grapevine.

However, CDS Toolbox SDM also underlines that the potential impact of climate change may be less significant than the other studies suppose. This is particularly the case by the use of RCP6.0 scenario for 2050 and 2070. In this case, it seems that the future climate conditions related to RCP6.0 scenario would be more favorable for the grapevine crops than the one related to RCP4.5 scenario. These results show the ability of CDS Toolbox SDM to render the bioclimatic dimension of the relationship between the species and the climatic variables, which is a relevant aspect in order to help decision-makers establish their strategy to make their activities resilient to climate change.

5. Conclusions

CDS Toolbox SDM has been developed in order to help decision-makers adapt their strategy and activities concerning the biological resources (biodiversity, agriculture, landscape, etc.) to the potential impacts of climate change. The aim of this model is to support a prospective approach that can be considered as a key process to ensure territorial resilience.

The application of CDS Toolbox SDM on the viniculture in France has shown the capability of the model to take into account the different potential impacts of climate change on the spatial distribution of suitable areas for grapevine crops according to the scenarios RCP4.5 and RCP6.0 for 2050 and 2070 periods. These results can contribute to define practical actions in order to adapt viniculture to climate change. For example, some of these actions can be related to developing irrigation infrastructures and techniques as suggested by [31]. Another strategy could be based on the diversification of cultivars. In that frame, [38] demonstrated that the use of 11 grapevine cultivars may help to reduce the potential losses of suitable areas by a half (for a 2 °C scenario) or by a third (for a 4 °C) for 2100. Finally, CDS Toolbox SDM can be used in order to identify the territories that are and would still be suitable for grapevine crops towards 2050 and 2070 in order to plan the development of wine supply chain in regions where this cultivation has not been developed or only slightly so.

As an improvement of CDS Toolbox SDM, we plan to add a climatic downscaling module allowing the use of climate change scenarios at very high spatial resolution (75 m) in order to provide the results at parcels scale. Because Morales-Castilla et al. [38] underlines that climate change since the 2000s affected the production of wine, especially when drought occurs during the growing season in summer, we also plan to combine our approach with the use of other climate based indexes like the Net Primary Productivity (NPP). This index is able to assess the potential productivity of vineyard through the next decades and its potential impact on the expected grapevine harvest. As suggested by [39,40], the implementation of the Huglin index, used at high spatial resolution, could also be helpful in order to map the potential evolution of the thermal requirements of grape varieties and the potential sugar content of grapes, this last parameter being relevant for wine producers.

Author Contributions: Conceptualization, G.H.M.; methodology, G.H.M., E.G., C.A.G.R. and D.M.H.C., software, C.A.G.R. and D.M.H.C.; validation, G.H.M., E.G., R.S.C.; formal analysis, G.H.M., E.G., C.A.G.R. and D.M.H.C.; writing—original draft preparation, G.H.M. and E.G.; writing—review and editing, G.H.M., E.G., C.A.G.R., D.M.H.C. and R.S.C.; supervision, G.H.M. and D.M.H.C. All authors have read and agreed to the published version of the manuscript.

Funding: This research received no external funding.

Conflicts of Interest: The authors declare no conflict of interest.

Appendix A

We present here the statistical description of the quantitative variables used by CDS Toolbox SDM in order to assess the potential distribution of suitable areas for the development of species like Vitis vinifera L. on the French territory. We also give the description of the nominal variables classes.

Quantitative variables

Table A1. 19 Bioclimatic variables of baseline scenario (see Figure 1 for variables names).

Variable	Min	Max	Mean	Std
Bio1	−10.26	17.21	10.73	2.1
Bio2	3.11	12.1	8.59	0.99
Bio3	17.12	45.05	35.97	2.49
Bio4	312.1	714.12	567.33	57.3
Bio5	−0.69	30.2	22.34	2.43
Bio6	−19.2	8	−1.55	2.33
Bio7	12.7	31.4	23.89	2.31
Bio8	−15.33	19.39	9.14	4.25
Bio9	−6	−24.88	12.7	6.58
Bio10	−3.15	24.88	17.9	2.09
Bio11	−15.98	11.18	4.1	2.23
Bio12	468	2104	848.73	182.72
Bio13	55	274	92.88	22.43
Bio14	7	127	49.12	13.58
Bio15	6.74	53.78	18.61	7.83
Bio16	157	692	256.64	61.23
Bio17	38	401	165.9	41.17
Bio18	39	401	182.91	43.94
Bio19	90	592	224.35	58.41

Table A2. 19 Bioclimatic variables of 2050 RCP4.5 scenario (see Figure 1 for variables names).

Variable	Min	Max	Mean	Std
Bio1	−4.7	19.2	13.577	1.985
Bio2	4.3	12.1	9.897	1.274
Bio3	19	40	33.92	2.39
Bio4	409.4	783	655.102	60.41
Bio5	8.8	36.4	30.662	2.684
Bio6	−13.9	10	1.948	2.159
Bio7	17.8	34.6	28.714	2.865
Bio8	−9.8	20	9.331	4.308
Bio9	−3.9	27.8	20.89	5.262
Bio10	3.6	27.8	22.466	2.086
Bio11	−10.9	13	5.726	2.148
Bio12	415	2442	745.58	153.37
Bio13	60	262	87.16	18.03
Bio14	4	141	35.52	12.9
Bio15	10	55	24.37	7.67
Bio16	167	727	239.85	50.56
Bio17	28	460	129.75	37.61
Bio18	48	460	137.81	39.73
Bio19	84	697	205.64	56.1

Table A3. 19 Bioclimatic variables of 2050 RCP6.0 scenario (see Figure 1 for variables names).

Variable	Min	Max	Mean	Std
Bio1	−4.9	18.7	13.088	1.97
Bio2	3.9	10.7	8.251	1.139
Bio3	19	42	34.56	2.82
Bio4	351.3	651.7	553.378	45.459
Bio5	4.3	32.5	26.755	2.507
Bio6	−11.9	9.9	3.246	2.021
Bio7	14.5	30	23.508	2.426
Bio8	−8.8	21.5	9.8513	4.722
Bio9	−1.9	25.7	13.371	5.822
Bio10	1	25.7	20.202	2.033
Bio11	−9.9	12.8	6.246	2.001
Bio12	490	2702	826.5	173.43
Bio13	62	332	96.33	23.56
Bio14	4	182	45.74	13.34
Bio15	11	62	20.92	7.388
Bio16	168	856	258.09	61.47
Bio17	26	562	161.57	41.03
Bio18	52	563	187.95	45.47
Bio19	107	759	231.01	62.53

Table A4. 19 Bioclimatic variables of 2070 RCP4.5 scenario (see Figure 1 for variables names).

Variable	Min	Max	Mean	Std
Bio1	−4.2	19.7	14.103	1.991
Bio2	4.3	12.4	10.063	1.296
Bio3	18	38	32.18	2.19
Bio4	443.6	822.5	698.475	59.803
Bio5	10.2	37.5	32.146	2.709
Bio6	−14.8	9.5	1.424	2.122
Bio7	19.4	36.3	30.722	2.804
Bio8	−11.4	20.4	9.888	4.174
Bio9	2.8	28.7	23.053	3.45
Bio10	4.9	28.8	23.677	2.036
Bio11	−11.4	13	5.72	2.145
Bio12	471	2435	737.69	151.05
Bio13	58	278	88.8	18.77
Bio14	3	114	32.79	11.08
Bio15	13	60	26.2	7.65
Bio16	161	755	238.46	54.45
Bio17	23	399	119.27	31.6
Bio18	38	399	127.56	26.81
Bio19	97	755	211.43	57

Table A5. 19 Bioclimatic variables of 2070 RCP6.0 scenario (see Figure 1 for variables names).

Variable	Min	Max	Mean	Std
Bio1	−4.3	19.2	13.509	1.956
Bio2	3.9	10.9	8.267	1.23
Bio3	19	42	34.4	3.29
Bio4	334.2	653.6	552.964	50.36
Bio5	5.8	33.7	27.634	2.616
Bio6	−10.8	11.2	3.982	1.809
Bio7	14.1	29.6	23.652	2.344
Bio8	−8.4	22	9.78	4.207
Bio9	1.7	26.4	16.722	6.071
Bio10	1.8	26.5	20.814	2.096
Bio11	−9.1	13.6	6.861	1.963
Bio12	479	2702	791.68	171.24
Bio13	60	359	95.16	25.6
Bio14	3	165	43.45	13.56
Bio15	10	66	23.62	9.62
Bio16	163	928	253.14	66.23
Bio17	18	524	148.1	41.92
Bio18	37	524	166.2	47.11
Bio19	97	834	233.58	66.75

Table A6. Slopes.

	Min	Max	Mean	Std
Slopes	0	45.66	3.59	4.95

Nominal variables

Table A7. Ecoregions.

Ecoregions		
Mediterranean Forests, Woodlands and Scrub	Temperate Broadleaf and Mixed Forests	Temperate Conifer Forests

Table A8. Edaphology.

Acronym	Soil type	Acronym	Soil type
Bc	Chromic Cambisols	Lg	Gleyic Luvisols
Bd	Dystric Cambisols	Lo	Orthic Luvisols
Be	Eutric Cambisols	Oe	Eutric Histosols
Bh	Humic Cambisols	Ph	Humic Podzols
Bk	Calcic Cambisols	Pl	Leptic Podzols
Dd	Dystric Podzoluvisols	Po	Orthic Podzols
E	Rendzinas	Ql	Luvic Arenosols
I	Lithosols	Rc	Calcaric Regosols
Jc	Calcaric Fluvisols	WR	Planosols
Je	Eutric Fluvisols	Zg	Gleyic Solonchaks
Lc	Chromic Luvisols		

Table A9. Geology.

Rock Type
Cenozoic
Lower paleozoic (Cam, Ord, Sil)
Mesozoic - Jurassic and Cretaceous
Mesozoic - Triassic
Metamorphic formations
Paleozoic or older volcanic formations
Plutonic rocks
Quaternary
Recent volcanic formations
Upper paleozoic (Dev, Car, Per)

References

1. Norberg, A.; Abrego, N.; Blanchet, F.G.; Adler, F.R.; Anderson, B.J.; Anttila, J.; Araújo, M.B.; Dallas, T.; Dunson, D.; Elith, J.; et al. A comprehensive evaluation of predictive performance of 33 species distribution models at species and community levels. *Ecol. Monogr.* **2019**, *89*, 1–24. [CrossRef]
2. Iverson, L.R.; Prasad, A.M.; Hale, B.J.; Sutherland, E.K. *Atlas of Current and Potential Future distributions of Common Trees of the Eastern United States*; General Technical Report NE-265; United States Department of Agriculture, Forest Service, Northeastern Research Station: Radnor PA, USA, 1999; 125p.
3. Thuiller, W.; Lavorel, S.; Sykes, M.T.; Araújo, M.B. Using niche based modelling to assess the impact of climate change on tree functional diversity in Europe. *Divers. Distrib.* **2006**, *12*, 49–60. [CrossRef]
4. Pearson, R.G.; Thuiller, W.; Araujo, M.B.; Martinez-Meyer, E.; Brotons, L.; McClean, C.; Miles, L.; Segurado, P.; Dawson, T.P.; Lees, D.C. Model based uncertainty in speciesrange prediction. *J. Biogeogr.* **2006**, *33*, 1704–1711. [CrossRef]
5. Wenger, S.J.; Olden, J.D. Assessing transferability of ecological models: An underappreciated aspect of statistical validation. *Methods Ecol. Evol.* **2012**, *3*, 260–267. [CrossRef]
6. Maguire, K.C.; Nieto-Lugilde, D.; Blois, J.L.; Fitzpatrick, M.C.; Williams, J.W.; Ferrier, S.; Lorenz, D.J. Controlled comparison of species- and community-level models across novel climates and communities. *Proc. R. B* **2016**, *283*, 1–10. [CrossRef]
7. Sor, R.; Park, Y.S.; Boets, P.; Goethals, P.L.; Lek, S. Effects of species prevalence on the performance of predictive models. *Ecol. Model.* **2017**, *354*, 11–19. [CrossRef]
8. Peterson, A.T.; Soberón, J.; Pearson, R.G.; Anderson, R.P.; Martínez-Meyer, E.; Nakamura, M.; Araújo, M.B. Ecological niches and geographical distributions: A modeling perspective. In *Monographs in Population Biology*; Princeton University Press: Princeton, NJ, USA, 2011.
9. Thuiller, W.; Araujo, M.B.; Lavorel, S. Generalized models vs. classification tree analysis: Predicting spatial distributions of plant species at different scales. *J. Veg. Sci.* **2003**, *14*, 669–680. [CrossRef]
10. Austin, M.; Belbin, L.; Meyers, J.A.; Doherty, M.D.; Luoto, M. Evaluation of statistical models used for predicting plant species distributions: Role of artificial data and theory. *Ecol. Model.* **2006**, *199*, 197–216. [CrossRef]
11. D'Amen, M.; Pradervand, J.N.; Guisan, A. Predicting richness and composition in mountain insect communities at high resolution: A new test of the SESAM framework. *Glob. Ecol. Biogeogr.* **2015**, *24*, 1443–1453. [CrossRef]
12. Garbolino, E.; de Ruffray, P.; Brisse, H.; Grandjouan, G. Relationships between plants and climate in France: Calibration of 1874 bio-indicators. *Comptes Rendus Biol.* **2007**, *330*, 159–170. [CrossRef]
13. Elith, J.; Phillips, S.J.; Hastie, T.; Dudik, M.; Chee, Y.E.; Yates, C.J. A statistical explanation of MaxEnt for ecologists. *Divers. Distrib.* **2011**, *17*, 43–57. [CrossRef]
14. Garbolino, E.; Hinojos, G.; De Ruffray, P.; Brisse, H. Forest Fire Risk in Corsica at the End of the XXIst Century: Impact of Global Warming on the Spatial Distribution of Botanical Taxa Mainly Involved in Forest Fires. In Proceedings of the 21st Workshop European Vegetation Survey, Vienna, Austria, 24–27 May 2012.

15. Garbolino, E.; Sanseverino-Godfrin, V.; Hinojos-Mendoza, G. Describing and predicting of the vegetation development of Corsica due to expected climate change and its impact on forest fire risk evolution. *Saf. Sci.* **2016**, *88*, 180–186. [CrossRef]
16. Hinojos-Mendoza, G.; Garbolino, E.; Soto-Cruz, R.; Borderon-Carrez, S.; Mariscal-Guerra, J.; Morales-Nieto, C. Response of the grassland community to climatic change: Simulation of scenarios towards the temporary horizon 2050 and 2100 using the Climpact model in central valleys, Chihuahua, Mexico. In Proceedings of the Southwestern Association of Naturalists, 66th Annual Meeting, Chihuahua, Mexico, 11–14 April 2019.
17. Hinojos-Mendoza, G.; Garbolino, E.; Soto-Cruz, R.; Borderon-Carrez, S.; Mariscal-Guerra, J.; Fernández, J.A. Response of pine-oak communities to climatic change: Simulation of scenarios towards the temporary horizon 2050 and 2100 using the model Climpact in the Sierra del Nido, Chihuahua, Mexico. In Proceedings of the Southwestern Association of Naturalists, 66th Annual Meeting, Chihuahua, Mexico, 11–14 April 2019.
18. Hinojos-Mendoza, G.; Garbolino, E.; Mariscal-Guerra, J.; Soto-Cruz, R.; Borderon-Carrez, S.; De La Maza-Beningnos, M. From the conservation of the isolation to the ecology of the connectivity: Example of application in the first ecological network of conservation in the north of Mexico. In Proceedings of the Southwestern Association of Naturalists, 66th Annual Meeting, Chihuahua, Mexico, 11–14 April 2019.
19. Hinojos-Mendoza, G.; Garbolino, E.; Sanseverino-Godfrin, V.; Carrega, P.; Martin, N. Impacts synergiques du changement climatique et du développement urbain sur la biodiversité des Alpes-Maritimes. In *Provence-Alpes-Côte d'Azur, une région face au changement climatique*; Groupe régional d'experts sur le climat en Provence-Alpes-Côte d'Azur (GREC-PACA): Marseille, France, 2015; Volume 1, p. 15.
20. Fick, S.E.; Hijmans, R.J. WorldClim 2: New 1km spatial resolution climate surfaces for global land areas. *Int. J. Climatol.* **2017**, *37*, 4302–4315. [CrossRef]
21. Polechová, J.; Storch, D. *Ecological Niche, Encyclopedia of Ecology*, 2nd ed.; Elsevier: Oxford, UK, 2019.
22. Woodward, F.I. *Climate and Plant Distribution*; Cambridge Studies in Ecology; Cambridge University Press: Cambridge, UK, 1987; 174p.
23. Fink, A.H.; Brücher, T.; Krüger, A.; Leckebusch, G.C.; Pinto, J.G.; Ulbrich, U. The 2003 European summer heatwaves and drought? Synoptic diagnosis and impacts. *Weather* **2004**, *59*, 209–216. [CrossRef]
24. Reichstein, M. Severe Impact of the 2003 European Heat Wave on Ecosystems. Available online: https://www.pik-potsdam.de/news/press-releases/archive/2005/severe-impact-of-the-2003-european-heat-wave-on-ecosystems (accessed on 20 April 2018).
25. Eastman, J.R. *IDRISI 15.0, The Andes Edition*; Clark University: Worcester, MA, USA, 2006.
26. Fraga, H.; de Cortázar-Atauri, I.G.; Malheiro, A.C.; Santos, J.A. Modelling climate change impacts on viticultural yield, phenology and stress conditions in Europe. *Glob. Chang. Biol.* **2016**, *22*, 3774–3788. [CrossRef]
27. Olson, D.M.; Dinerstein, E.; Wikramanayake, E.D.; Burgess, N.D.; Powell, G.V.N.; Underwood, E.C.; D'Amico, J.A.; Itoua, I.; Strand, H.E.; Morrison, J.C.; et al. Terrestrial ecoregions of the world: A new map of life on Earth. *Bioscience* **2001**, *51*, 933–938. [CrossRef]
28. Dürr, H.H.; Meybeck, M.; Dürr, S.H. Lithologic composition of the Earth's continental surfaces derived from a new digital map emphasizing riverine material transfer. *Glob. Biogeochem. Cycles* **2005**, *19*, GB4S10. [CrossRef]
29. Flint, L.E.; Flint, A.L. Downscaling future climate scenarios to fine scales for hydrologic and ecological modeling and analysis. *Ecol. Process.* **2012**, *1*, 2. [CrossRef]
30. Jenks, G.F.; Caspall, F.C. Error on Chloroplethic Maps: Definition, Measurement, Reduction. *Ann. Assoc. Am. Geogr.* **1971**, *61*, 217–244. [CrossRef]
31. Pieri, P.; Lebon, E.; Brisson, N. Climate change impact on French vineyards as predicted by models. *Acta Hortic.* **2012**, *931*, 29–38. [CrossRef]
32. Stocker, T.F.; Qin, D.; Plattner, G.-K.; Tignor, M.; Allen, S.K.; Boschung, J.; Nauels, A.; Xia, Y.; Bex, V.; Midgley, P.M. (Eds.) IPCC, Annex II: Climate System Scenario Tables. [Prather, M.; Flato, G.; Friedlingstein, P.; Jones, C.; Lamarque, J.-F.; Liao, H.; Rasch, P. (Eds.)]. In *Climate Change 2013: The Physical Science Basis. Contribution of Working Group I to the Fifth Assessment Report of the Intergovernmental Panel on Climate Change*; Cambridge University Press: Cambridge, UK; New York, NY, USA, 2013.
33. Malheiro, A.C.; Santos, J.A.; Fraga, H.; Pinto, J.G. Climate change scenarios applied to viticultural zoning in Europe. *Clim. Res.* **2010**, *43*, 163–177. [CrossRef]

34. Fraga, H.; Santos, J.A.; Malheiro, A.C.; Oliveira, A.A.; Moutinho-Pereira, J.; Jones, G.V. Climatic suitability of Portuguese grapevine varieties and climate change adaptation. *Int. J. Climatol.* **2016**, *36*, 1–12. [CrossRef]
35. Moriondo, M.; Jones, G.V.; Bois, B.; Dibari, C.; Ferrise, R.; Trombi, G.; Bindi, M. Projected shifts of wine regions in response to climate change. *Clim. Chang.* **2013**, *119*, 825–839. [CrossRef]
36. Cardell, M.F.; Amengual, A.; Romero, R. Future effects of climate change on the suitability of wine grape production across Europe. *Reg. Environ. Chang.* **2019**, *19*, 2299–2310. [CrossRef]
37. Caffarra, A.; Eccel, E. Projecting the impacts of climate change on the phenology of grapevine in a mountain area. *Aust. J. Grape Wine Res.* **2011**, *17*, 52–61. [CrossRef]
38. Morales-Castilla, I.; de Cortázar-Atauri, I.G.; Cook, B.I.; Lacombe, T.; Parker, A.; van Leeuwen, C.; Nicholas, K.A.; Wolkovich, E.M. Diversity buffers winegrowing regions from climate change losses. *PNAS* **2020**, *117*, 2864–2869. [CrossRef]
39. Lereboullet, A.-L.; Beltrando, G.; Bardsley, D.K.; Rouvellac, E. The viticultural system and climate change: Coping with long-term trends in temperature and rainfall in Roussillon, France. *Reg. Environ. Chang.* **2014**, *14*, 1951–1966. [CrossRef]
40. Quénol, H.; de Cortazar Atauri, I.G.; Bois, B.; Sturman, A.; Bonnardot, V.; Le Roux, R. Which climatic modeling to assess climate change impacts on vineyards? *OENO One, Institut des Sciences de la Vigne et du Vin (Université de Bordeaux)* **2017**, *51*, 91–97.

atmosphere

MDPI

Article

Predicting Suitable Habitats of the African Cherry (*Prunus africana*) under Climate Change in Tanzania

Richard A. Giliba [1,*] and Genesis Tambang Yengoh [2]

1 School of Life Sciences and Bio-Engineering, The Nelson Mandela African Institution of Science and Technology, P.O. Box 447, Arusha, Tanzania
2 Lund University Centre for Sustainability Studies–LUCSUS, Lund University, Biskopsgatan 5, SE 223 62 Lund, Sweden; yengoh.genesis@lucsus.lu.se
* Correspondence: richard.giliba@nm-aist.ac.tz; Tel.: +255-76771-0622

Received: 5 August 2020; Accepted: 25 August 2020; Published: 15 September 2020

Abstract: *Prunus africana* is a fast-growing, evergreen canopy tree with several medicinal, household, and agroforestry uses, as well as ecological value for over 22 countries in sub-Saharan Africa. This species is under immense pressure from human activity, compounding its vulnerability to the effects of climate change. Predicting suitable habitats for *P. africana* under changing climate is essential for conservation monitoring and planning. This study intends to predict the impact of climate change on the suitable habitats for the vulnerable *P. africana* in Tanzania. We used maximum entropy modeling to predict future habitat distribution based on the representative concentration pathways scenario 4.5 and 8.5 for the mid-century 2050 and late-century 2070. Species occurrence records and environmental variables were used as a dependent variable and predictor variables respectively. The model performance was excellent with the area under curve (AUC) and true skill statistics (TSS) values of 0.96 and 0.85 respectively. The mean annual temperature (51.7%) and terrain ruggedness. index (31.6%) are the most important variables in predicting the current and future habitat distribution for *P. africana*. Our results show a decrease in suitable habitats for *P. africana* under all future representative concentration pathways scenario when compared with current distributions. These results have policy implications for over 22 countries of sub-Saharan Africa that are facing problems associated with the sustainability of this species. Institutional, policy, and conservation management approaches are proposed to support sustainable practices in favor of *P. africana*.

Keywords: habitat suitability; species distribution; climate change; conservation; *P. africana*

---

1. Introduction

Climate change is a serious threat to floral biodiversity conservation [1–3]. Over the next century, the global temperature is projected to rise by 0.3–4.8 °C [4]. The projected increase in temperature is anticipated to influence both species and habitat distribution in several different ways. Some of the climate change impacts on flora include changes in species distribution, the increased extinction rate of species, changes in the length of growing season, and reproduction timings for plants [5]. With a rise in temperature, plant species are likely to shift their distribution patterns depending on resource availability [6]. This may lead to species range expansion or range contraction, or range shift when respond to changing climate [7]. Predicting suitable habitat of species under climate change using species distribution models (SDM) is one of the key important steps to undertake conservation planning and management [6,8]. Greenhouse gases (GHGs) from various human activities are the primary agents answerable for climate change and the current emission rates are at the highest level in the recorded history [9].

SDM have been widely used to monitor the impacts of climate change on the floral distribution [10] and identification of suitable habitats of species [11]. SDM uses species occurrence and environmental variables [12] to predict the distribution of a species in a geographical or an environmental space [13]. SDM engage a variety of methods for predicting species distribution and mapping habitat suitability [14]. These include: MAXENT, BIOCLIM, DOMAIN artificial neural networks, generalized linear models, and generalized additive models. Maxent, Bioclim, and Domain use presence-only data [15] while others require both presence absent data [14]. Presence-only methods include bioclimatic envelope algorithm BIOCLIM, DOMAIN.

Maxent has been widely used and gives better results when compared to other different modeling methods that use presence-only data are used [14,16,17]. Several studies have demonstrated Maxent's ability to accurately predict species distribution in a wide range of ecological and geographical regions [10,18,19]. Subsequently, conservation practitioners have been increasingly using habitat suitability models from Maxent to make conservation management decisions [20,21]. We used Maxent: first to identify the most important variables that govern the current distribution of *P. africana*; second to predict current suitable habitats; third to predict future suitable habitats based on two representative concentration pathways (RCP 4.5 and RCP 8.5) for the mid-century 2050 and late-century 2070 in Tanzania. We selected *P. africana* due to local and international economic importance for medicinal purposes, contributing to its overexploitation. Due to the changing climate and overexploitation and, there is a need to use species occurrence data for identification of key conservation sites so as to develop a countrywide conservation strategy.

1.1. *P. africana*–Its Value, Demand and Conservation Pressures

P. africana is a fast-growing, evergreen canopy tree about 30–40 m in height. It has a wide range, spanning several countries in central, western, southern, and eastern Africa (including Madagascar). It occupies habitats in upland rain-forest, montane and riverine forests, moist evergreen forest, and the edges of dry gallery forests [22]. The uses of *P. africana* are many, and explain the huge demand for this tree and its products throughout the world:

Medicinal uses: The tree is widely used in both traditional and modern medicine to treat a variety of ailments [23–25]. *P. africana* contains several medically active compounds including the cyanogenic glycoside amygdalin, which is found in the bark, leaf, and fruit; phytosterols such as. β-sitosterol 15–18%, and its 3-O-glycoside, β-sitostenone, campesterol, and aucosterol; pentacyclic triterpenoids [26,27]. The bark is highly valued for its medicinal properties, particularly as a treatment for benign prostatic hyperplasia and prostate gland hypertrophy, diseases that commonly affect older men in Europe and North America [28]. In traditional medicinal practices throughout sub-Saharan Africa, the bark is used in traditional medicine as a purgative and as a remedy for stomachache, while the leaves are used as an inhalant for fever or are drunk as an infusion to improve appetite. Demand for *P. africana* bark for medicinal uses has been high and growing, putting immense pressure on the species throughout sub-Saharan Africa. In Tanzania, this has raised issues of sustainability for the species [29]. While data on current demand and supply of *P. africana* barks in the international and local markets is scarce, it was expected in 2000 that this demand would triple or quadruple to 7000 to 11,000 tons/year in export and about 500 tons/year for use in Africa the years after the report [30].

Household uses: Domesticated trees serve as shade in compounds, as windbreaks, and as ornamental trees. The tree yields a high-quality fuel, and so is a favorite for the production of charcoal or for use as fuel wood in many communities. Regarding household socioeconomics, *P. africana* supports revenue streams of communities in a wide variety of ways. As an input to furniture production, the seasoned wood saws easily and cleanly; works well with hand and machine tools; and polishes and finishes well. It also serves as highly desirable timber for flooring and heavy construction where durability is not required.

Agroforestry uses: Where the tree has been domesticated for integration into agricultural systems, the tree is used for erosion control, while the leaves are incorporated into the system of organic manure complex of crop production. There have been efforts to more tightly incorporate *P. africana* into the agroforestry mix of agricultural systems in some parts of the continent. For example, in Cameroon where the harvesting and commercialization have come under more targeted scrutiny given the large volumes involved, the government established a "National Management Plan for *P. africana*" that supported greater efforts towards reducing the pressures on wild tree species with locally cultivated trees [31].

Ecological value: One of the most cited ecological value of *P. africana* is its preference by the black and white colobus monkeys (*Colobus guereza*) as their top food species [32]. In forests, agricultural fields, as in homes, *P. africana* also serves as shelter for a variety of bird species. It is a valuable species for beekeeping, and therefore an important contributor in supporting pollination services that are relied on by forest species, and for the success of agriculture.

The heavy pressures to which *P. africana* has recently come under in most African countries because of wild harvesting for the medicinal plant trade have not gone unnoticed in the international and biodiversity community. In 1995, *P. africana* was added to Appendix A of the Convention on International Trade in the Endangered Species of Wild Fauna and Flora's (CITES) list of endangered species, for the regulation of its trade from wild harvesting [33]. Currently, all exports of *P. africana* should therefore subject to a CITES export permit to protect the tree from depletion in Africa. In response, a European Union (EU) ban on imports of *P. africana* bark came into force in 2007 to help stocks to recover [34]. The International Union for Conservation of Nature (IUCN) recognize *P. africana* as a vulnerable species on its red list (https://www.iucnredlist.org) and it was categorized as a species "of urgent concerns" by CITES [22].

1.2. Forests and the Circular Economy

The circular economy is defined as "the concept can, in principle, be applied to all kinds of natural resources, including biotic and abiotic materials, water and land. Eco-design, repair, reuse, refurbishment, remanufacture, product sharing, waste prevention and waste recycling are all important in a circular economy" [35]. The concept of the circular economy seeks to achieve a shift from the linear economy, which is characterized by less than optimal recycling and reuse of materials and resources in human societies. The overall goal of the circular economic model is to reduce the undesirable impacts of the linear economy by achieving a systemic transition into a more sustainable approach to natural resource exploitation and use built on long-term sustainability. One of the main objectives of the circular economy is to reduce the impact of human activities on the planet's ecosystems by reducing the excessive exploitation of natural resources and minimizing the pressure of human actions on the functioning of these ecosystems.

The role of forests in human societies and development, as well as the nature of forestry and forestry-based industries makes them a prime candidate for contributing to the global drive towards achieving the goals of a circular economy [36,37]. By striving towards achieving the objectives of the circular economy, positive contributions can be made towards achieving several sustainable development goals [38], many of which are relevant for Tanzania's development.

2. Material and Methods

2.1. Study Area

The United Republic of Tanzania is located in East Africa between longitude 29° and 41° East and latitude 1° and 12° South (Figure 1). Tanzania is endowed with a wide range of natural resources as well as ecological and cultural diversity including extensive areas of arable land, wildlife reserves and parks, mountains, forest reserves, rivers, and lakes. The mean annual rainfall varies from below 500 mm to over 2000 mm per annum while the mean temperature ranges from −4.9 °C to 27.9 °C per annum. The central and western plateau is relatively dry while, the northern and southern highland are cool. Rainfall for large parts of the country is bimodal with short rains from October-December and long rains from March to May [39]. The country has 7 hotspots including forest reserves, nature reserves game reserves, and national parks that are recognized by The United Nations Educational, Scientific and Cultural Organization as World Heritage sites (https://whc.unesco.org). A significant number of world endemic and threatened species are reported from Tanzania (https://www.cbd.int). However, the country has lost its forest cover from land use change and it is threatened by changing climate [40].

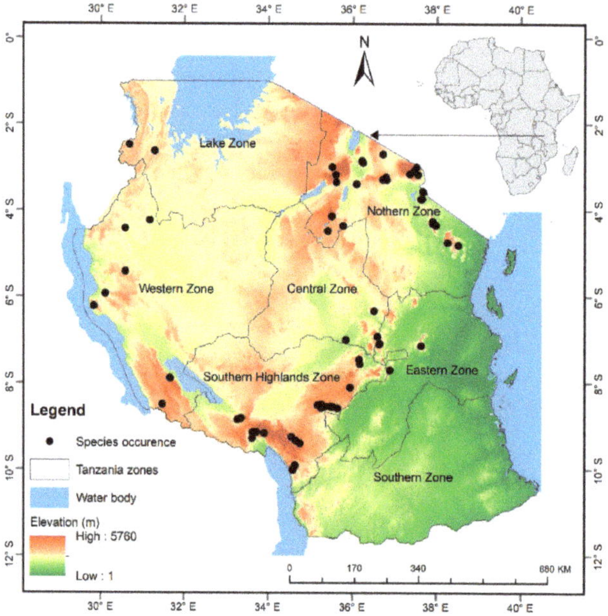

Figure 1. Location map of the study area Tanzania. Black points show species occurrence. Note that the data for *P. africana* is superimposed on top of elevation layer.

2.2. Species Presence Records

We obtained the present locations of *P. africana* from a 5-year field survey done across the country by National Forest Resources Monitoring and Assessment Project, and different online sources, TROPICOS an online botanical database containing taxonomic information on plants (http://www.tropicos.org) and Global Biodiversity Information Facility database (http://www.gbif.org). A total of 187 records were collected, and after screening, 57 duplicate records were removed, and finally 120 records were used to run the model (Figure 1). To model potential attribution of *P. africana* across the country.

2.3. Environmental Variables

We collected 19 bioclimatic variables from WorldClim dataset (https://www.worldclim.org). To derive elevation and terrain ruggedness index, we downloaded a digital elevation model from Shuttle Radar Topography Mission dataset (http://srtm.csi.cgiar.org). Soil type (Table 1) were obtained from the International Soil Reference and Information Centre database (https://www.isric.org). We resampled both soils and topographic layers to the resolution of bioclimatic variables (~1 km) using ArcGIS 10.6. To reduce multi-collinearity of climate variables, the two variables that found to have a high correlation coefficient ($|r| > 0.7$), as suggested by [41], we selected "one variable for modeling due to its ecological importance for the survival of *P. africana*" [42]. This resulted in the inclusion of eight variables for modeling. Table 2 lists the general statistics of the major environmental variables used in this study. We used Climate Community Climate System version four (CCSM4) bioclimatic variables to predict the future distribution of *P. africana* under future climate scenarios, namely representative concetration pathway (RCP) 4.5 a moderate greenhouse gas emission scenario and RCP 8.5 a extreme greenhouse gas emission scenario for mid-century and late-century. The CCSM4 is among the most commonly used bioclimatic variables to predict the impact of climate change on plant distribution [1,3].

Table 1. Summary of dominant tropical soil groups for Southern Africa (Batjes, 2004).

Code	Major Soil Group	Descriptions
1	Acrisols	Strongly weathered acid soils, with low base saturation
2	Andosols	Black soils of volcanic landscapes, rich in organic matters
3	Arenosols	Sandy soils with limited soil development, under scattered (mostly grassy) vegetation to very old plateaus of light forest
4	Cambisols	Weakly to moderately developed soil soils occurring from sea level to the highlands and under all kind of vegetation (savanna woodland and forests)
5	Chernozems	Black soil rich in organic matter, occurring in flat to undulating plains with forest and tall grass vegetation
6	Ferralsols	Deep, strongly weathered, physically stable but chemically depleted
7	Fluvisols	Associated with important river plains, periodically flooded areas
8	Gleysols	Temporary or permanent wetness near soil surface, support swamp forests or permanent grass cover
9	Histosols	Peat and muck soils with incompletely decomposed plant remains
10	Leptosols	Shallow soils over hard rock/gravel, at medium to high altitude landscapes, suitable for forestry and nature conservation
11	Lixisols	Strongly weathered and leached, finely textured materials support natural savanna or open woodland vegetation
12	Luvisols	Common in flat or gently sloping land with unconsolidated alluvial, colluvial, aeolian deposits in cooler environments and young surface
13	Nitisols	Deep, red, well-drained tropical soils with a clayey, well defined nut-shaped peds with shiny surface. Found in level to highland under tropical rain forest or savanna vegetation
14	Phaeozems	Dark soils, rich in organic matter. Occur on flat to undulating land in a warm to cool (tropical highland). Support natural vegetation with tall grass steppe and or/forest
15	Planosols	Clayey alluvial and colluvial deposits and support light forest or grass vegetation
16	Regosols	Contain gravelly lateritic materials (murrum) with low suitability for plant growth
17	Solonchanks	Occur in seasonally or permanently water logged areas with grasses and/or halophytic herbs
18	Solonetz	Associated with flat lands in a hot climate, dry summers, coastal deposit. Contain a high proportional of sodium ions
19	Vertisols	Contain sediments with a high proportion of smectite clay, high swelling and shrinking of results in deep cracks during dry season. Climax vegetation is savanna, natural grass and/or woodland
20	Water	Areas covered by water bodies

Table 2. Environmental variables used to model distribution of *P. africana*.

Variable	Code	Mean	Standard Error	Minimum	Maximum
Annual mean temperature (°C)	bio1	17.10	3.46	3.70	24.00
Isothermality (dimensionless)	bio3	6.64	0.48	6.10	8.40
Annual precipitation (mm)	bio12	1237	38	503	2287
Precipitation of warmest quarter (mm)	bio18	364	12	140	576
Precipitation of driest month (mm)	bio14	7	0.9	0	57
Terrain ruggedness index (m)	tri	104.43	9.47	0.13	418.75
Elevation (m)	eleva	1903	56	698	4249

2.4. Species Distribution Modeling

We used Maxent version 3.3.3; [17] to model the distribution of *P. africana* in this study due to the unavailability of absence records. Maxent uses presence records in combination with environmental conditions the species is present to model the spatial distribution based on the theory of maximum entropy [14]. During modeling, we selected 75% of presence records to training the model and 25% for testing the model [1,3,15], while changing Maxent setting. We tried to set various values for the regularization multiplier and the number of iterations and changed feature types. We obtained the good results with the following settings; cross-validate with iterations set to 5000, regularization multiplier set to 1, and feature type set to quadratic, hinge, and linear. Further, the maximum number of background points was set to 10,000, and replicates were set to 30. Afterward, we imported current and future predicted maps for *P. africana* from Maxent models into ArcGIS 10.6 and reclassified into five classes of potential habitats according to [43]: unsuitable habitat (0–0.2); barely suitable habitat (0.2–0.4); suitable habitat (0.4–0.6); highly suitable habitat (0.6–0.8); very highly suitable habitat (0.8–1). Finally, current distribution maps were subtracted from future maps to compute the relative changes in species range (decreasing or increasing) [2].

2.5. Model Evaluation and Validation

We used the area under receiver operating characteristic (AUC) and true skill statistic (TSS) to assess the performance of model. The AUC values range between 0–1; higher AUC values suggest the better and higher performance of a model [14,17]. "TSS values range between +1 to −1; a values > 0.8 suggest excellent, 0.4–0.8 useful, and <0.4 poor model performance" [2]. Finally, we selected the model with highest AUC and TSS. Besides, we used jackknife test to identify important variable governing the distribution of *P. africana*. Further, we use response curves to show how the predicted probability of presence changes as each environmental variable is varied.

3. Results

3.1. Model Validation and Influencing Bioclimatic Variables

Model for *P. africana* provided satisfactory results, with AUC and TSS values of 0.957 and 0.845 respectively. These suggest that the model for *P. africana* produced good results. Annual mean temperature (bio1) contributed most to the model, followed by terrain ruggedness index (tri) (Table 3). The cumulative contribution of these two variables is 83.30%. On the other hand, the variable with the highest gain when used in isolation is bio1, this implies this variable has the most useful information by itself. The variable that decreases the gain most when it is omitted is tri, this implies that the tri has the most information that is not present in the other variables (Figure 2). These results signify that bio1 and tri a proxy measure of topographic heterogeneity are the master variables governing the current and future distribution of *P. africana* in Tanzania.

Table 3. Environmental variables used in the study and their percentage contributions, and the maps that show the spatial distribution of the important variables are presented in Figure A1.

Variable	Code	Percent Contribution (%)
Annual mean temperature	bio1	51.7
Terrain ruggedness index	tri	31.6
Elevation	eleva	5.7
Soil type	soils	5.5
Annual precipitation	bio12	3.4
Precipitation of warmest quarter	bio18	0.9
Precipitation of driest month	bio14	0.8
Isothermality	bio3	0.5

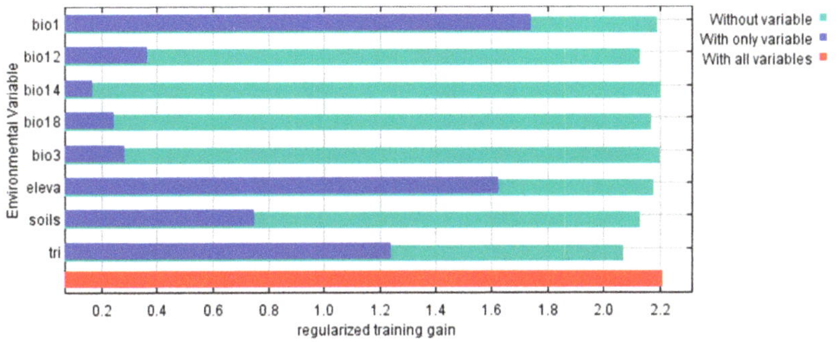

Figure 2. Jackknife test results indicating variable with highest gain when used in isolation and variable that decreases the gain the most when it is omitted. The test results indicate annual mean temperature (bio1) has the most useful information by itself while terrain ruggedness index (tri) has the most information that is not present in the other variables. Jackknife of regularized training gain for *P. africana*.

Figure 3, below, shows how the predictions depend on the variables, as mean annual temperature (bio1), increases habitat suitability for *P. africana* decreases while its habitat suitability increases with annual precipitation (bio12). On the other hand, as elevation (eleva) and terrain ruggedness index (tri), increases habitat suitability for *P. africana* increases indicating that it prefers undulating upland areas. Further, *P. africana* prefers to reside on nitisols, histosols, leptosols, and acrisols soils, which is widely distributed in undulating upland areas.

3.2. Current and Future Distribution of *P. africana*

Predicted distributions under current conditions revealed that highly and moderately suitable areas for *P. africana* covers only 6.69% (62,388.75 sq. km), while low and very low suitable areas cover the large portion, 93.31% (869,937.04 sq. km), of the study area (Table 4; Figure 4). The highly suitable areas to a large extent are identified in the southern highlands, western, and northern zones of the study area (Tanzania, Figure 4). Predicted distributions under future conditions indicates decline in suitable areas and increase in suitable areas under RCP 4.5 and RCP 8.5 scenarios for mid-century 2050 and late-century 2070 (Table 4; Figure 4). Climatically highly and moderately suitable areas will decline by 2.29% and 3.07% under RCP 8.5 for 2050 and 2070, respectively while under RCP 4.5 for 2050 and 2070 highly and moderately suitable areas will decline by 2.10% and 2.20% respectively. Climatically very low areas will increase by 6.85% and 8.59% under RCP 8.5 for 2050 and 2070, respectively while under RCP 4.5 for 2050 and 2070 suitable conditions will decline by 6.25% and 6.59% respectively (Table 4; Figure 4). Southern highlands, western, and northern zones are anticipated to lose large portion of suitable areas in the future for all climate scenarios under mid-century and late-century (Table 5; Figure 4).

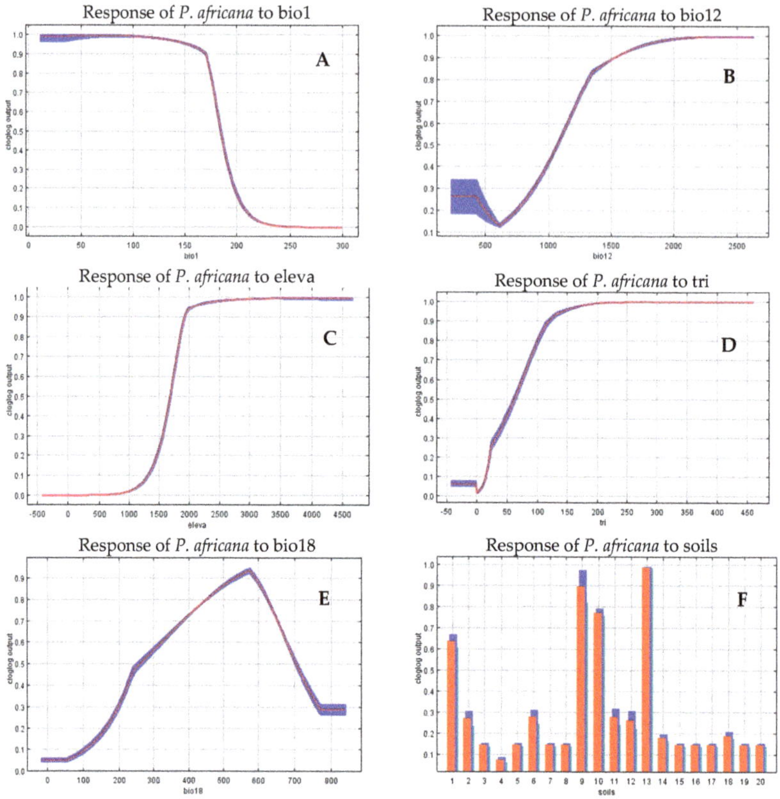

Figure 3. Relationships between selected environmental variables and probability of species suitability of *P. africana* (**A**) mean annual temperature (bio1) (**B**) annual precipitation (bio12), (**C**) elevation (eleva), (**D**) terrain ruggedness index (tri), (**E**) precipitation of warmest quarter (bio18), (**F**) soil types (soils). The y-axis represents the probability of presence (cloglog output). Red curves show the average response and blue margins are ± SD calculated by 30 replicate runs. For the interpretation of soil type legend, refer Table 1.

Table 4. Change in suitable areas of *P. africana* country-wide for mid-century and late-century under representative concertation pathway (RCP) 4.5 and RCP 8.5 scenarios.

	Suitability Class	Species Distribution Area (km^2)				
		Current	RCP 4.5	Area Change	RCP 8.5	Area Change
2050	Very low	767,755.74	826,010.1	58,254.37	831,659.80	63,904.06
	Low	102,181.30	64,349.85	−37,831.46	59,628.51	−42,552.80
	Moderate	32,044.01	19,254.13	−12,789.88	19,514.24	−12,529.77
	High	15,758.05	11,585.16	−4172.88	11,224.09	−4533.96
	Very high	14,586.70	11,126.55	−3460.15	10,299.16	−4287.54
2070	Very low	767,755.74	829,251.22	61,495.48	847,873.04	80,117.30
	Low	102,181.30	61,217.40	−40,963.90	50,670.11	−51,511.19
	Moderate	32,044.01	19,591.25	−12,452.77	15,309.70	−16,734.31
	High	15,758.05	11,456.82	−4301.23	9021.71	−6736.34
	Very high	14,586.70	10,809.11−	3777.58	9451.23	−5135.46

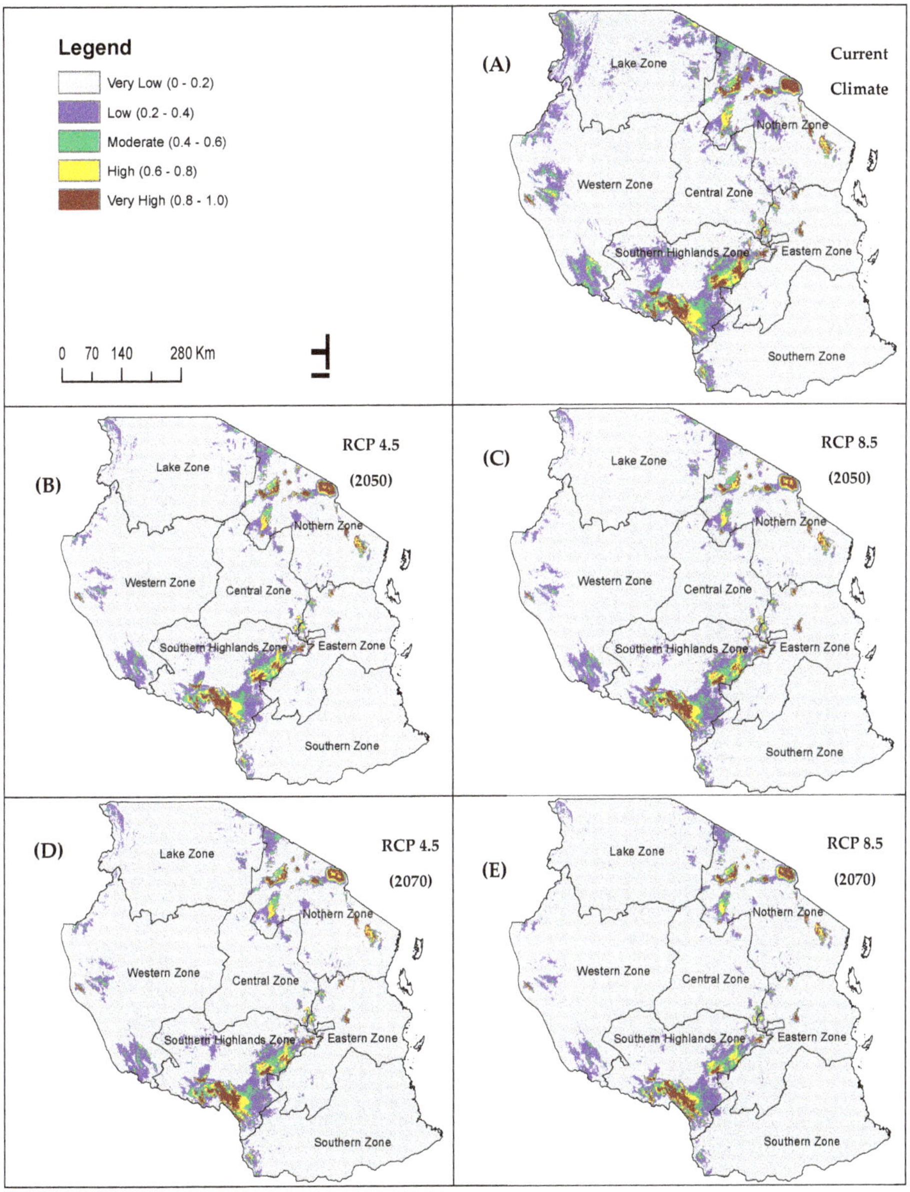

Figure 4. Climate Community Climate System version four (CCSM4) climate model based predicted future suitability of *P. africana* species: (**A**) current potential distribution; (**B**,**C**) RCP 4.5, and 8.5 emission scenario for 2050; (**E**,**D**) RCP 4.5, and 8.5 scenario for 2070.

Table 5. Change in suitable areas of *P. africana* zone wide for mid-century and late-century under RCP 4.5 and RCP 8.5 scenarios.

Scenario	Suitability Class	Species Distribution Area (km^2)						
		Eastern Zone	Southern Zone	Northern Zone	Central Zone	Southern Highlands Zone	Western Zone	Lake Zone
RCP 4.5 (2050)	Very low	1089.21	1216.69	12,694.85	2972.43	11,410.56	12,119.02	16,740.23
	Low	−416.69	−556.15	−8740.17	−2262.26	−5417.79	−6645.61	−13,783.20
	Moderate	−219.89	−412.41	−2170.71	−404.71	−2750.82	−4373.08	−2457.34
	High	−176.26	−175.40	−885.57	−145.46	−1374.98	-941.18	−473.16
	Very high	−276.37	−72.73	−898.40	−160.00	−1866.97	−159.15	−26.52
RCP 4.5 (2070)	Very low	1156.80	1082.36	14,000.53	2985.26	10,945.11	12,734.21	18,578.96
	Low	−479.15	−416.69	−9586.38	−2332.42	−5392.98	−7310.43	−15,435.41
	Moderate	−229.31	−403.85	−2176.70	−396.15	−2286.22	−4326.88	−2632.75
	High	−169.41	−183.10	−860.75	−114.65	−1546.96	−942.04	−483.43
	Very high	−278.93	−78.72	−1376.69	−142.03	−1718.94	−154.87	−27.38
RCP 8.5 (2050)	Very low	1333.06	1315.09	13,487.16	3082.80	12,644.37	13,412.72	18,614.89
	Low	−582.68	−624.60	−8825.73	−2387.18	−6618.23	−7879.42	−15,622.79
	Moderate	−250.70	−390.16	−2244.29	−399.58	−2331.57	−4409.02	−2503.55
	High	−175.40	−213.91	−765.78	−133.48	−1827.61	−954.02	−462.89
	Very high	−324.28	−86.42	−1651.35	−162.57	−1866.97	−170.27	−25.67
RCP 8.5 (2070)	Very low	1911.46	1935.42	17,698.52	4095.86	16,881.41	16,414.24	21,166.36
	Low	−748.67	−975.41	−11,529.49	−3080.24	−7644.12	−9855.04	−17,666.87
	Moderate	−421.82	−577.54	−3021.20	−521.07	−3902.49	−5315.97	−2973.28
	High	−274.65	−280.64	−1523.86	−278.93	−2831.25	−1048.14	−497.12
	Very high	−466.31	−101.82	−1623.97	−215.62	−2503.55	−195.08	−29.09

4. Discussion

The findings of the study show that the suitability distribution of the *P. africana* is largely controlled by annual temperature, terrain ruggedness index, elevation, and soil type (contributed more than 90%). *P. africana* showed sharp decline in response to an increase in annual temperature beyond signifying that the probability of occurrence of the species may be affected with higher temperature. This is in line with the results from a study in eastern arc mountains Tanzania, where the studied tree species showed a declining trend in response to an increase in temperature [44]. With warming trends, plant species are expected to track the changing climate and shift their distributions to the extent that resource availability allows [6]. *P. africana* showed an increasing trend in response to an increase of the terrain ruggedness index suggesting that the species prefers rugged or undulating areas [45]. "Terrain ruggedness index as a measure of terrain heterogeneity is an important variable for predicting which habitats are used by a species and the density at which species occur in a variety of environments" [46]. *P. africana* showed an increasing trend in response to the increase of elevation indicating that the species prefers high elevation areas. Higher temperature is stated to cause shifts in plant distribution along the elevation gradients [47]. *P. africana* distribution appears to associate with Nitisols, Histosols, and Leptosols soils that are found in undulating upland areas. Soil type plays a major role in the heterogeneity of habitats, thus determining the distribution of plant species [48].

4.1. Management Implications

Our results indicate that climate change will pose a severe impact on the future distribution of *P. africana* in Tanzania. Research institutions and public universities can take an interest in both in situ and ex situ long-term monitoring trends of *P. africana* distribution in a country. In-situ interventions should focus on the "recruitment and regeneration of the species while ex-situ interventions should target to promote tree retention on farms, or advocate further planting, collect specimens, and to establish gene banks and botanical gardens to ensure the survival of the species" [42].

4.2. Institutional and Policy Context for Addressing Challenges Associated with P. africana

Addressing challenges associated with the vulnerability of *P. africana* in Tanzania can benefit from the institutional context already in place in the country. The government has put in place institutional frameworks to manage natural resources and environment-related initiatives and challenges countrywide. The President's Office-Regional Administration and Local Government (PORALG) works closely with Local Government Authorities through their various departments in collaboration with the respective sector ministries to implement the strategic interventions at the local level (municipalities, districts, wards, villages, and sub-villages). This is important for addressing the immediate management decisions that directly affect the health and survival of *P. africana*, as the trees are directly impacted by community demand for livelihoods at the local level. Successful implementation of policies, laws, and plans also requires enhanced engagement with Civil Society Organizations, development partners, the private sector, and academic and research institutions.

Addressing the more systemic and long-term environmental challenges facing *P. africana* would be best addressed within the institutional arrangements for environmental management functions in Tanzania. These are two basic types of such functions: (i) Sectoral Environmental Management Functions (also known as Type A functions) that are concerned with the management of specific natural resources or environmental services, such as forestry, agriculture, fisheries, wildlife, mining, and waste management. These functions are to a large extent operational and guided by sector-specific policies and acts such as the Forest Act (2002), which should be directly relevant for addressing challenges of *P. africana*. (ii) Coordinating and Supporting Environmental Management Functions (commonly referred to as Type B functions) involve the task of providing central support functions by coordinating and supporting the different and sometimes conflicting Type A activities and integrating them into an overall sustainable system. Specific tools within this Type B functions relevant for addressing challenges of *P. africana* are the National Environmental Policy (1997) and the Environmental Management Act (2004), which provide policy and legislative framework for the coordination of the implementation of policies and laws on environmental and natural resources management.

4.3. *Conservation and Management Approaches to Support Sustainable Practices in Favor of P. africana*

Three main conservation and management approaches can offer possibilities to address the vulnerability of *P. africana* in Tanzania. These draw from experiences in other parts of the developing world facing conservation challenges of their own and the lessons learned from their conservation and management approaches.

4.3.1. Supporting Inclusive Conservation Approaches

While climate changing is increasingly representing a challenge to the distribution and health of *P. africana*, this challenge comes to compound existing pressures imposed by anthropogenic pressures of the species. Given the high degree of anthropic intervention contributing to the vulnerability of the species, conservation and management strategies need to be closely linked to the needs of indigenous peoples and local communities. There is therefore a serious need to consider the genuine and effective participation of indigenous peoples and local communities in the definition and application of resource management options when addressing challenges to the sustainable management of *P. africana*. Given the importance of this specie to the health of local populations, socio-economic and environmental welfare, emphasis on supporting its sustainable and adaptive use, supported by initiatives that reduce local reliance on the harvesting of wild resources (such as agroforestry) may prove to be more locally acceptable. This principle of "conservation through use" is an example of the application of community-based natural resources management models to address issues of resource degradation. These types of co-management models have been applied to the successful management of protected resources in other parts of the world [41]. Such an approach will be in alignment with current governmental efforts towards a more inclusive management of forests and natural resources. A program of Participatory Forest Management has been introduced and operationalized through the Joint Forest Management (JFM) as well as a Community Based Forest Management (CBFM) processes across the country. Under JFM, agreements between community groups and the Government have been developed with a view to promoting the participation of communities in the management and utilization of forest resources. The Community Based Forest Management program encourages communities to set up forest reserves from the general lands for economic and conservation activities.

4.3.2. Collaboration to Streamline and Align Regional and International Efforts

Given the regional and international character of challenges of *P. africana* vulnerability, there is a need for collaboration to improving forest management, share best practices, and support effective conservation as well as the production and trade of forest products. The Collaborative Partnership on Forests (CPF) is an informal, voluntary arrangement among 14 international organizations to share experiences and build on them to produce new benefits for their respective national stakeholders in the forest resources sector [41]. Addressing the challenges of *P. africana* has the potential of benefiting from the CPF whose mission is to promote sustainable management of all types of forests and to strengthen long-term political commitment to this end.

4.3.3. Leveraging the Potential of Payments for Ecosystem Services (PES)

Payments for ecosystem services (PES) refer to voluntary transactions between users and suppliers of environmental services, such that suppliers are subject to natural resource management and handling rules within and outside of service provision areas [49]. Under a PES scheme, users of land upstream may agree to voluntary limitation or diversification of their activities in return for an economic benefit. In many parts of the world, the positive impact on forest cover and species diversity of the implementation of PES) schemes have been documented [50,51]. Developing PES schemes that specifically target compounding factors contributing to the vulnerability of *P. africana* can contribute to reducing these vulnerabilities. Such schemes would also concomitantly contribute to combatting the degradation or loss of essential ecosystems and ecosystem services without sacrificing the well-being of people—an essential element in the portfolio of Tanzania's sustainable development goals. It must be noted however that the design and implementation of PES programs must be carefully done, adopting best practices as well as the best and most recent scientific guidance on the subject matter. This is because PES per se is not a panacea for addressing underlying deficiencies in natural resources governance policies and practices where they exist. For example, Tuanmu Viña [50] observed that the effectiveness of a PES program depends on who receives the payment and on whether the payment provides sufficient incentives.

4.3.4. Incorporating Forest Management into the Circular Economy

In striving to incorporate forests and forestry into the circular economy, it is important to understand some of the strategies required to achieve the transition from a linear to a circular economy. In a study aimed at understanding strategies for an effective transition into sustainable forest-based bioeconomy in Italy, Falcon, Tani [52] identified that four strategies are most effective. These include defining viable methods of circular management to improve environmental and forest planning; investing in forest infrastructure; supporting entrepreneurship programs for professionals in the forestry sector; and enhancing the development and application of innovative forest-based value chains.

One of the key challenges to transitioning the exploitation and use of *P. africana* in Tanzania from a linear to a circular economic model is that of adding value to the main product, as well as to bio-residuals. Much of the chain for its value-addition for its many uses (medicinal, furniture, fuelwood, etc.) is minimal to non-existent. This is in line with the recognition of that weak market pull, needs for big investments, and the adoption of risk-averse approaches among the few incumbent firms in the sector are reducing the potential for the forestry industry to invest in technological and market capabilities for valorizing residuals [53].

5. Conclusions

Bioclimatic predictors mainly mean annual temperature presented high contribution and important information in predicting distribution and mapping habitat suitability for *P. africana* in Tanzania. Suitable habitats for *P. africana* will decline in mid- and late-century for both RCP 4.5 and RCP 8.5 scenarios when compared with baseline conditions. For instance, southern highlands and northern zones will constantly lose much more suitable habitats for *P. africana* in the future. The areas mapped in this study as suitable habitats for the tree species could be advantageous for conservation planning and afforestation interventions.

Author Contributions: R.A.G. and G.T.Y. contributed equally in the following tasks: conceptualization of the study; methodology development; investigation, data analysis; writing—original draft preparation; as well as the review and editing of the final draft. All authors have read and agreed to the published version of the manuscript.

Funding: This research was not funded by any external party.

Acknowledgments: The authors acknowledge the occurrence data providers Tanzania Forest Service Agency, Global Biodiversity Information Facility and TROPICOS. Further, the authors are grateful to Mathew Mpanda for his valuable suggestions and comments on the manuscripts. Finally, the authors are grateful to Elikana John for his guidance to access the soils data.

Conflicts of Interest: The authors declare no conflict of interest.

Appendix A

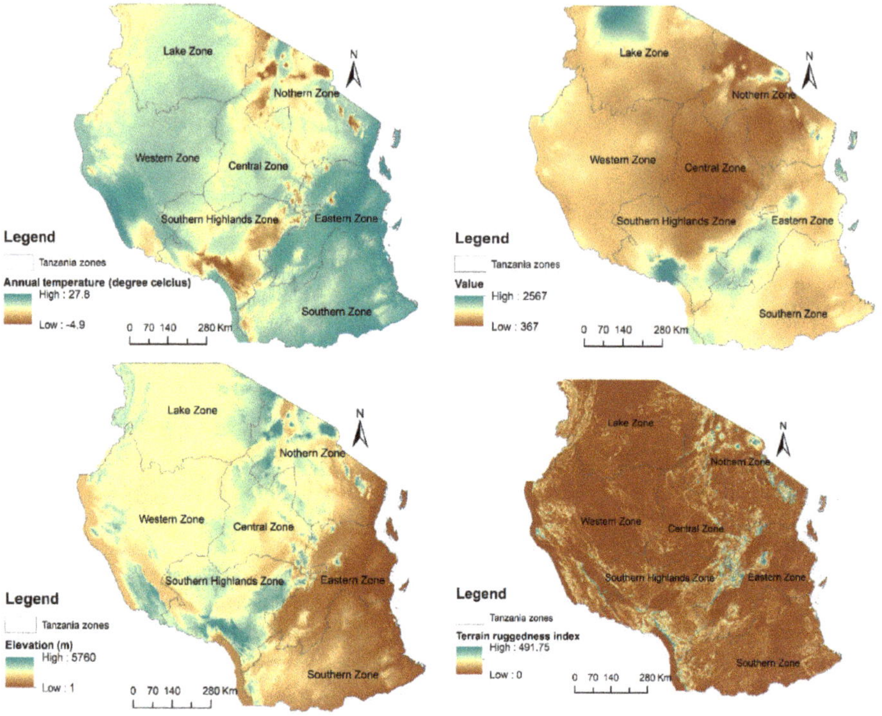

Figure A1. Distribution of the selected environmental variable in the study area.

References

1. Heshmati, I.; Khorasani, N.; Shams-Esfandabad, B.; Riazi, B. Forthcoming risk of Prosopis juliflora global invasion triggered by climate change: Implications for environmental monitoring and risk assessment. *Environ. Monit. Assess.* **2019**, *191*, 72. [CrossRef] [PubMed]
2. Pramanik, M.; Paudel, U.; Mondal, B.; Chakraborti, S.; Deb, P. Predicting climate change impacts on the distribution of the threatened Garcinia indica in the Western Ghats, India. *Clim. Risk Manag.* **2018**, *19*, 94–105. [CrossRef]
3. Abrha, H.; Birhane, E.; Hagos, H.; Manaye, A. Predicting suitable habitats of endangered Juniperus procera tree under climate change in Northern Ethiopia. *J. Sustain. For.* **2018**, *37*, 842–853. [CrossRef]
4. Priti, H.; Aravind, N.A.; Uma Shaanker, R.; Ravikanth, G. Modeling impacts of future climate on the distribution of Myristicaceae species in the Western Ghats, India. *Ecol. Eng.* **2016**, *89*, 14–23. [CrossRef]
5. Thompson, I.; Mackey, B.; McNulty, S.; Mosseler, A. Forest Resilience, Biodiversity, and Climate Change. *Secr. Conv. Biol. Divers. Montr. Tech. Ser.* **2009**, *43*, 1–67.
6. Berry, P.M.; Dawson, T.P.; Harrison, P.A.; Pearson, R.G. Modelling potential impacts of climate change on the bioclimatic envelope of species in Britain and Ireland. *Glob. Ecol. Biogeogr.* **2002**, *11*, 453–462. [CrossRef]
7. Sommer, J.H.; Kreft, H.; Kier, G.; Jetz, W.; Mutke, J.; Barthlott, W. Projected impacts of climate change on regional capacities for global plant species richness. *Proc. R. Soc. B Biol. Sci.* **2010**, *277*, 2271–2280. [CrossRef]
8. Lawler, J.J.; White, D.; Neilson, R.P.; Blaustein, A.R. Predicting climate-induced range shifts: Model differences and model reliability. *Glob. Chang. Biol.* **2006**, *12*, 1568–1584. [CrossRef]
9. Lindner, M.; Maroschek, M.; Netherer, S.; Kremer, A.; Barbati, A.; Garcia-Gonzalo, J.; Seidl, R.; Delzon, S.; Corona, P.; Kolström, M.; et al. Climate change impacts, adaptive capacity, and vulnerability of European forest ecosystems. *For. Ecol. Manag.* **2010**, *259*, 698–709. [CrossRef]

10. Thuiller, W.; Lavorel, S.; Araújo, M.B.; Sykes, M.T.; Prentice, I.C. Climate change threats to plant diversity in Europe. *Proc. Natl. Acad. Sci. USA* **2005**, *102*, 8245–8250. [CrossRef]
11. Porfirio, L.L.; Harris, R.M.B.; Lefroy, E.C.; Hugh, S.; Gould, S.F.; Lee, G.; Bindoff, N.L.; Mackey, B. Improving the Use of Species Distribution Models in Conservation Planning and Management under Climate Change. *PLoS ONE* **2014**, *9*, e113749. [CrossRef] [PubMed]
12. Kumar, S.; Stohlgren, T.J. Maxent modeling for predicting suitable habitat for threatened and endangered tree Canacomyrica monticola in New Caledonia. *J. Ecol. Nat. Environ.* **2009**, *1*, 94–98.
13. Botkin, D.B.; Saxe, H.; Araujo, M.B.; Betts, R.; Bradshaw, R.H.; Cedhagen, T.; Chesson, P.; Dawson, T.P.; Etterson, J.R.; Faith, D.P.; et al. Forecasting the effects of global warming on biodiversity. *Bioscience* **2007**, *57*, 227–236. [CrossRef]
14. Elith, J.H.; Graham, C.P.; Anderson, R.; Dudík, M.; Ferrier, S.; Guisan, A.J.; Hijmans, R.; Huettmann, F.R.; Leathwick, J.; Lehmann, A.; et al. Novel methods improve prediction of species' distributions from occurrence data. *Ecography* **2006**, *29*, 129–151. [CrossRef]
15. Phillips, S.J.; Dudík, M. Modeling of species distributions with Maxent: New extensions and a comprehensive evaluation. *Ecography* **2008**, *31*, 161–175. [CrossRef]
16. Ortega-Huerta, M.A.; Peterson, A.T. Modeling ecological niches and predicting geographic distributions: A test of six presence-only methods. *Rev. Mex. Biodivers.* **2008**, *79*, 205–216.
17. Phillips, S.J.; Anderson, R.P.; Schapire, R.E. Maximum entropy modeling of species geographic distributions. *Ecol. Model.* **2006**, *190*, 231–259. [CrossRef]
18. Mwakapeje, E.R.; Ndimuligo, S.A.; Mosomtai, G.; Ayebare, S.; Nyakarahuka, L.; Nonga, H.E.; Mdegela, R.H.; Skjerve, E. Ecological niche modeling as a tool for prediction of the potential geographic distribution of Bacillus anthracis spores in Tanzania. *Int. J. Infect. Dis.* **2019**, *79*, 142–151. [CrossRef]
19. Yi, Y.J.; Zhou, Y.; Cai, Y.P.; Yang, W.; Li, Z.W.; Zhao, X. The influence of climate change on an endangered riparian plant species: The root of riparian Homonoia. *Ecol. Indic.* **2018**, *92*, 40–50. [CrossRef]
20. Loiselle, B.A.; Howell, C.A.; Graham, C.H.; Goerck, J.M.; Brooks, T.; Smith, K.G.; Williams, P.H. Avoiding Pitfalls of Using Species Distribution Models in Conservation Planning. *Conserv. Biol.* **2003**, *17*, 1591–1600. [CrossRef]
21. Saatchi, S.; Buermann, W.; Ter Steege, H.; Mori, S.; Smith, T.B. Modeling distribution of Amazonian tree species and diversity using remote sensing measurements. *Remote Sens. Environ.* **2008**, *112*, 2000–2017. [CrossRef]
22. Cheboiwo, J.K.; Mugabe, R.; Langat, D. Review of conservation of *Prunus africana* and international trade opportunities for its bark in Kenya. *J. Emerg. Trends Eng. Appl. Sci.* **2014**, *5*, 372–377.
23. Bii, C.; Korir, K.R.; Rugutt, J.; Mutai, C. The potential use of *Prunus africana* for the control, treatment and management of common fungal and bacterial infections. *J. Med. Plants Res.* **2010**, *4*, 995–998. [CrossRef]
24. Jena, A.K.; Vasisht, K.; Sharma, N.; Kaur, R.; Dhingra, M.S.; Karan, M. Amelioration of testosterone induced benign prostatic hyperplasia by Prunus species. *J. Ethnopharmacol.* **2016**, *190*, 33–45. [CrossRef] [PubMed]
25. Mwitari, P.G.; Ayeka, P.A.; Ondicho, J.; Matu, E.N.; Bii, C.C. Antimicrobial Activity and Probable Mechanisms of Action of Medicinal Plants of Kenya: Withania somnifera, Warbugia ugandensis, *Prunus africana* and Plectrunthus barbatus. *PLoS ONE* **2013**, *8*, e65619. [CrossRef] [PubMed]
26. Kadu, C.A.C.; Parich, A.; Schueler, S.; Konrad, H.; Muluvi, G.M.; Eyog-Matig, O.; Muchugi, A.; Williams, V.L.; Ramamonjisoa, L.; Kapinga, C.; et al. Bioactive constituents in *Prunus africana*: Geographical variation throughout Africa and associations with environmental and genetic parameters. *Phytochemistry* **2012**, *83*, 70–78. [CrossRef]
27. Nyamai, D.; Mawia, A.; Wambua, F.; Njoroge, A.; Matheri, F.; Lagat, R.; Kiambi, J.; Ogola, P.; Arika, W.; Cheseto, X.; et al. Pharmacognosy & Natural Products Phytochemical Profile of *Prunus africana* Stem Bark from Kenya. *J. Pharmacogn. Nat. Prod.* **2015**, *1*, 1–8. [CrossRef]
28. Catalano, S.; Ferretti, M.; Marsili, A.; Morelli, I. New constituents of *Prunus africana* bark extract. *J. Nat. Prod.* **1984**, *47*, 910. [CrossRef]
29. Maximillian, J.R.; O'Laughlin, J. Toward sustainable harvesting of Africa's largest medicinal plant export (*Prunus africana*): A case study in Tanzania. *South. For.* **2009**, *71*, 303–309. [CrossRef]
30. Hall, J.B.; O'Brien, E.M.; Sinclair, F.L. *Prunus africana*: A monograph. *Sch. Agric. For. Sci. Publ. Number Univ. Wales Bangor* **2000**, *18*, 104.

31. Anderson, P.K.; Cunningham, A.A.; Patel, N.G.; Morales, F.J.; Epstein, P.R.; Daszak, P. Emerging infectious diseases of plants: Pathogen pollution, climate change and agrotechnology drivers. *Trends Ecol. Evol.* **2004**, *19*, 535–544. [CrossRef] [PubMed]
32. Fashing, P.J. Mortality trends in the African cherry (*Prunus africana*) and the implications for colobus monkeys (*Colobus guereza*) in Kakamega Forest, Kenya. *Biol. Conserv.* **2004**, *120*, 449–459. [CrossRef]
33. Cunningham, A.; Anoncho, V.F.; Sunderland, T. Power, policy and the *Prunus africana* bark trade, 1972–2015. *J. Ethnopharmacol.* **2016**, *178*, 323–333. [CrossRef]
34. Chupezi, T.J. Critical Study of Guidance for A National Prunus africana Management Plan-Cameroon. Under the Supervision of Dr Jean Lagarde BETTI, Regional Coordinator of the ITTO–CITES Program in Africa. Ph.D. Thesis, University of KwaZulu-Natal, Durban, South Africa, 2010; pp. 1–33.
35. Reichel, A.; De Schoenmakere, M.; Gillabel, J. *Circular Economy in Europe-Developing the Knowledge Base (European Environment Agency Report No 2/2016)*; European Environmental Agency: Luxembourg, 2016; ISBN 9789292137199.
36. Ia, W. *Forestry and Forestry Based Industry Implications Digitalisation and Circular Economy: Forestry and Forestry Based Industry Implications*; Union of Scientists of Bulgaria: Sofia, Republic of Bulgaria, 2019; ISBN 9789543970421.
37. Pirc Barčić, A.; Kitek Kuzman, M.; Haviarova, E.; Oblak, L. Circular economy & Sharing collaborative economy principles: A case study conducted in wood-based sector. In *Digitalisation and Circular Economy: Forestry and Forestry Based Industry Implications*; Union of Scientists of Bulgaria: Sofia, Republic of Bulgaria, 2019; pp. 23–28.
38. Schroeder, P.; Anggraeni, K.; Weber, U. The Relevance of Circular Economy Practices to the Sustainable Development Goals. *J. Ind. Ecol.* **2019**, *23*, 77–95. [CrossRef]
39. Magehema, A.; Chang, L.; Mkoma, S. Implication of rainfall variability on maize production in Morogoro, Tanzania. *Int. J. Environ. Sci.* **2014**, *4*, 1077–1086. [CrossRef]
40. Capitani, C.; Van Soesbergen, A.; Mukama, K.; Malugu, I.; Mbilinyi, B.; Chamuya, N.; Kempen, B.; Malimbwi, R.; Mant, R.; Munishi, P.; et al. Scenarios of Land Use and Land Cover Change and Their Multiple Impacts on Natural Capital in Tanzania. *Environ. Conserv.* **2019**, *46*, 17–24. [CrossRef]
41. Dormann, C.F.; Elith, J.; Bacher, S.; Buchmann, C.; Carl, G.; Carré, G.; Marquéz, J.R.G.; Gruber, B.; Lafourcade, B.; Leitão, P.J.; et al. Collinearity: A review of methods to deal with it and a simulation study evaluating their performance. *Ecography* **2013**, *36*, 27–46. [CrossRef]
42. Giliba, R.A. Effects of Climate Change on Potential Geographical Distribuition of *Prunus africana* (African cherry) in the Eastern Arc Mountain Forests of Tanzania. Master's Thesis, Lund University, Lund, Sweden, 2018.
43. Zhang, Q.; Wen, J.; Chang, Z.; Xie, C.; Song, J. Evaluation and prediction of ecological suitability of medicinal plant American ginseng (*Panax quinquefolius*). *Chin. Herb. Med.* **2018**, *10*, 80–85. [CrossRef]
44. Platts, P.J. Spatial Modelling, Phytogeography and Conservation the Eastern Arc Mountains of Tanzania and Kenya. Ph.D. Thesis, University of York, York, UK, 2012.
45. Riley, S.J.; DeGloria, S.D.; Elliot, R. Terrain_Ruggedness_Index.pdf. *Intermt. J. Sci.* **1999**, *5*, 23–27.
46. Fabricius, C.; Coetzee, K. Geographic information systems and artificial intelligence for predicting the presence or absence of mountain reedbuck. *S. Afr. J. Wildl. Res.* **1992**, *22*, 80–86.
47. Telwala, Y.; Brook, B.W.; Manish, K.; Pandit, M.K. Climate-Induced Elevational Range Shifts and Increase in Plant Species Richness in a Himalayan Biodiversity Epicentre. *PLoS ONE* **2013**, *8*, e57103. [CrossRef] [PubMed]
48. Baldeck, C.A.; Harms, K.E.; Yavitt, J.B.; John, R.; Turner, B.L.; Valencia, R.; Navarrete, H.; Davies, S.J.; Chuyong, G.B.; Kenfack, D.; et al. Soil resources and topography shape local tree community structure in tropical forests. *Proc. R. Soc. B Biol. Sci.* **2013**, *280*, 20122532. [CrossRef] [PubMed]
49. Wunder, S. Revisiting the concept of payments for environmental services. *Ecol. Econ.* **2015**, *117*, 234–243. [CrossRef]
50. Tuanmu, M.N.; Viña, A.; Yang, W.; Chen, X.; Shortridge, A.M.; Liu, J. Effects of payments for ecosystem services on wildlife habitat recovery. *Conserv. Biol.* **2016**, *30*, 827–835. [CrossRef]
51. Chen, H.L.; Lewison, R.L.; An, L.; Tsai, Y.H.; Stow, D.; Shi, L.; Yang, S. Assessing the effects of payments for ecosystem services programs on forest structure and species biodiversity. *Biodivers. Conserv.* **2020**, *29*, 2123–2140. [CrossRef]

52. Falcone, P.M.; Tani, A.; Tartiu, V.E.; Imbriani, C. Towards a sustainable forest-based bioeconomy in Italy: Findings from a SWOT analysis. *For. Policy Econ.* **2020**, *110*, 101910. [CrossRef]
53. Gregg, J.S.; Jürgens, J.; Happel, M.K.; Strøm-Andersen, N.; Tanner, A.N.; Bolwig, S.; Klitkou, A. Valorization of bio-residuals in the food and forestry sectors in support of a circular bioeconomy: A review. *J. Clean. Prod.* **2020**, *267*, 122093. [CrossRef]

atmosphere

MDPI

Article

Changes in the Seasonality of Ethiopian Highlands Climate and Implications for Crop Growth

Gashaw Bimrew Tarkegn [1] and Mark R. Jury [2,3,*]

1 Agriculture Department, University of Bahir Dar, Bahir Dar, Ethiopia; gashbimrew@gmail.com
2 Physics Department, University of Puerto Rico, Mayaguez, PR 00681, USA
3 Geography Department, University of Zululand, KwaDlangezwa 3886, South Africa
* Correspondence: mark.jury@upr.edu

Received: 16 June 2020; Accepted: 6 July 2020; Published: 24 August 2020

check for updates

Abstract: Rain-fed agriculture in North-West (NW) Ethiopia is seasonally modulated, and our objective is to isolate past and future trends that influence crop growth. Statistical methods are applied to gauge-interpolated, reanalysis, and satellite data to evaluate changes in the annual cycle and long-term trends. The June to September wet season has lengthened due to the earlier arrival and later departure of rains. Meteorological composites relate this spreading to local southerly winds and a dry-south/wet-north humidity dipole. At the regional scale, an axis of convection over the Rift Valley (35E) is formed by westerly waves on 15S and an anticyclone over Asia 30N. Coupled Model Intercomparsion Project (CMIP5) Hadley2 data assimilated by the Inter-Sectoral Impact Model Intercomparision Project (ISIMIP) hydrological models are used to evaluate projected soil moisture and potential evaporation over the 21st century. May and October soil moisture is predicted to increase in the future, but trends are weak. In contrast, the potential evaporation is rising and may put stress on the land and water resources. A lengthening of the growing season could benefit crop yields across the NW Ethiopian highlands.

Keywords: Ethiopia highlands; seasonal climate; crop impacts

---

1. Background

The effects of rising temperature and changing precipitation affect ecosystems, biodiversity, and people. In both developed and developing countries, climate impacts are reverberating through the economy, from fluctuating water availability to sea-level rise and extreme weather impacts, to coastal erosion and tourism [1] and to disease and pests. Climate change could translate into reduced agricultural performance in Africa where warming of 1 °C in the 20th century and lengthy droughts in recent decades have undermined progress [2–4].

Soil moisture deficits and crop failure undermine livelihoods and need to be offset by local knowledge to enable adaptation [5]. Seasonal precipitation (hereafter 'P') can be forecast to maintain crop yields—with parallel efforts in institutional capacity building and resource management [6–8], but Ethiopia's rainfall occurs before 'maturity' of the El Nino Southern Oscillation and Indian Ocean Dipole, making long-range forecasts less skillful [9]. Understanding changes in the onset and cessation of the growing season could assist coping strategies, particularly if locally tailored to production risks.

World opinion is rightfully pessimistic on the impacts of climate change, but some places may see less harmful trends that could be translated into opportunities. Our study on the NW Ethiopian highlands crop growing season will address the following questions:

1. What temporal and spatial hydro-climate change has occurred in the 20th century and is projected for the 21st century?

2. How will changes in the hydro-climate affect the onset and cessation of the crop growing season?

In NW Ethiopia, the months February–May have high evaporation losses and soil moisture deficit, while the months October–January have cold temperatures. These two factors limit the crop growing season for short-cycle crops [10–13]. Could future climate extend the length of growing season, thus improving yields from rain-fed agriculture?

2. Concepts and Methods

We first analyze Empirical Orthogonal Functions (EOF) for cenTrends P [14] via covariance matrix and time scores. This delineates a 'NW Ethiopia highlands' study area: 8.5–13 N, 35–39.5 E with 1st mode loading pattern covering 73% of variance. We employ statistical techniques to identify the mean annual cycle, measures of association that account for lags between air and land, and linear regression for trends and dispersion [15–18]. Table 1 summarizes the methods of analysis; acronyms are defined following acknowledgements.

Table 1. Sequence of methods applied. CMIP5: Coupled Model Intercomparsion Project.

	Scope	Methods and Variables
1	Determination of homogenous study area	EOF cluster analysis of cenTrends precipitation (P): 8.5–13 N, 35–39.5 E
2	Evaluate potential evaporation (E)	Comparison of observed, reanalysis, model-simulated sensible heat flux (SHF)
3	CMIP5 model validation and selection	Apply criteria to determine annual cycle bias in P, SHF as proxy for E
4	Soil moisture fraction (S)	Compare P–E, latent heat flux (LHF) and NDVI with S
5	Collection of optimal time series	Area-average NW Ethiopian highlands: P, E, S, T; 8.5–13 N, 35–39.5 E
6	Characterization of annual cycle	Calculate annual cycle and percentiles for P, E, S, T, LHF; determine shift/width
7	Meteorological forcing of annual cycle	Composite analysis of reanalysis fields for early-late, wide-narrow LGP
8	Analysis of climate trends	Statistical regression slope and significance; seasonal changes for P, E, S
9	Assess LGP and impact of climate change	Onset and cessation in past (1900–2000) and future (2001–2100)

Atmospheric convection initiates a cascading water cycle of runoff and infiltration that is offset by desiccation due to net radiation and turbulent flux. As our focus is on crop growth, we distinguish between transpiration of moisture via latent heat flux (LHF) and moisture lost by soil via potential evaporation (hereafter 'E'). E can be calculated from station data, measured by A-pan, estimated by satellite, or modeled via sensible heat flux (hereafter 'SHF') [19,20]. We compared the annual cycle of SHF with A-pan data and found a $r^2 = 0.95$, while other proxies such as temperature and LHF exhibited weak relationship and were screened out. The resultant water budget over time produces soil moisture residuals that accumulate to sustain crop growth [19].

Coupled hydrological models estimate the soil moisture fraction in the upper meter (hereafter 'S') via theassimilation of in situ and satellite measurements. These include passive and active microwave radiance and gravity anomalies [21,22], and vegetation color fraction (NDVI). The NDVI represents photosynthetic activity and is used to constrain reanalysis LHF and monitor crop condition [23–32]. The majority of Ethiopia's highlands have an NDVI vegetation fraction > 0.4 (Figure 1) and a mean annual cycle close to soil moisture.

Reanalysis data from NCEP2, ECMWF, and FLDAS (cf. acronym table after Section 4) form an integral part of our study on the evolving atmospheric boundary layer and hydrology [33–36] over the NW Ethiopia highlands [37,38]. We compared multi-station averages with reanalyses and found statistically significant correlations; yet the main reason for parameter choices was due to their availability in the most recent version, underpinned by satellite technology and sophisticated data assimilation. We calculate mean annual cycle percentiles for daily Chirps P [39] and ECMWF LHF and

E. To understand seasonal shifts, we use the gauge-interpolated cenTrends P from 1900 to 2018 and calculate percentage contributions in April–May (early), July–August (narrow), October–November (late), and early + late (wide). Then, we rank those percentages in recent decades (Table 2) and form composite difference fields using NCEP2 reanalysis wind and humidity, and NOAA satellite net outgoing long-wave radiation (OLR), to determine the regional forcing of convection.

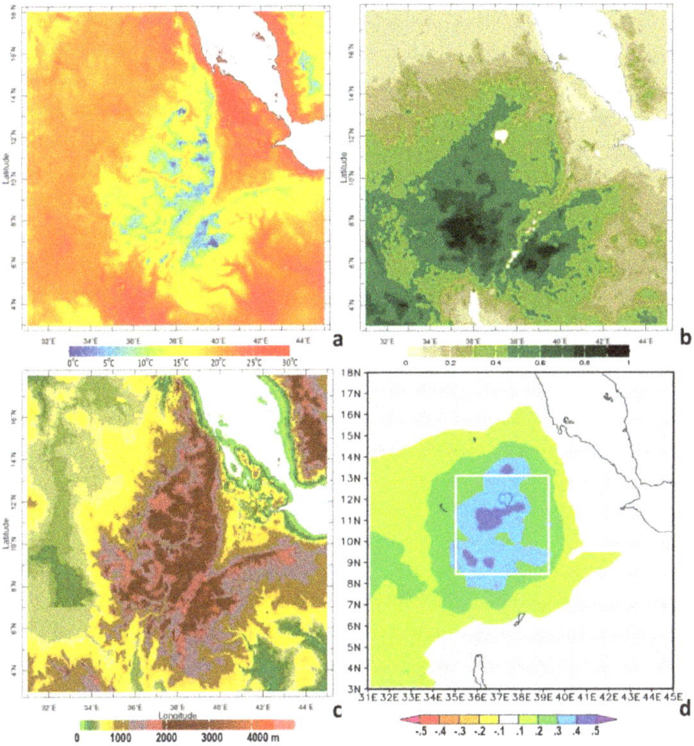

Figure 1. (**a**) Mean nocturnal land surface T and (**b**) NDVI vegetation color fraction of the Ethiopian highlands averaged 2000–2014. (**c**) Topographic map of the study area and (**d**) Empirical Orthogonal Functions (EOF) loading pattern of cenTrends 1st mode P and box for the extraction of time series.

Table 2. Categorization of sub-seasonal rainfall percentages (yellow-least, blue-most).

LEAST	Early	LEAST	Late	LEAST	Wide	LEAST	Narrow
2003	0.06	1984	0.03	2003	0.09	2014	0.38
1990	0.08	1995	0.03	1990	0.11	1997	0.39
2002	0.08	1990	0.03	2002	0.12	2019	0.39
2009	0.08	2003	0.03	2012	0.13	2015	0.40
1988	0.09	1991	0.04	1986	0.14	1987	0.41
2012	0.09	2010	0.04	1991	0.15	1993	0.42
	Apr.–May		Oct.–Nov.		Early + late		Jul.–Aug.
1996	0.18	1982	0.11	2008	0.26	1994	0.50
1993	0.19	2019	0.12	1987	0.26	2013	0.51
1987	0.19	2000	0.12	2016	0.27	2009	0.52
2008	0.19	1992	0.13	2000	0.27	2012	0.52
2016	0.20	1999	0.13	2014	0.29	1981	0.52
2014	0.20	1997	0.16	1997	0.32	1990	0.54
MOST	early	MOST	late	MOST	wide	MOST	narrow

Many of the climate factors that govern our ability to extract resources are seasonal, and thus, we seek ways to determine the annual onset and cessation of crop growing. Most short-cycle crops (e.g., teff) require a length of growing period (LGP) > 100 days [40–45]. At higher elevations in the tropics, the temperature (hereafter 'T') may drop below thresholds (16 °C) that support crop growth, even if soil moisture is available [46]. Figure 1a illustrates cool nocturnal T over the highlands; crop growth tends to slow in October. Crop models use S that depend on cumulative P minus E, conditioned by T and infiltration rates [47–49]. Crop growth is predicted when P accumulates > $\frac{1}{2}$ E, or when S reaches a critical value. Here, we define LGP as the time when area-average S > 15% with T > 16 °C (see Appendix A Figure A2c).

To quantify climate change, we first compare the mean annual cycle of reference P and SHF with all available CMIP5 models (Appendix A Tables A1 and A2) [50–52]. We determined the Hadley2 model [53] as optimal and analyze E and S via ISIMIP 'glowb' and 'watergap' hydrology [54]. Using continuous bias-corrected Hadley2-ISIMIP (Inter-Sectoral Impact Model Intercomparision Project) projected time series with rcp6 scenario [55], we calculate the linear trends and signal-to-noise ratio [56] via the r^2 value and analyze annual cycle differences in past (1900–2000) and future eras (2001–2100). Although much of the analysis uses monthly data, the LGP is detected from daily data.

3. Results

3.1. Historical Trends

The background information reviewed earlier (Figure 1a–d) identified the complex topography of the NW Ethiopian highlands, and climatic responses in T and vegetation that point to orographic rainfall. Most crop production (cf. Appendix A Figure A2b) occurs in the eastern side of our index area, e.g., along 38E, where the NDVI fraction is approximately 0.4. Annual cycle terciles from daily P–E are considered (Figure 2a) based on the area averages of 1980–2018. Surplus conditions begin on day 123-154-190 and end on 285-266-243 (wet-mean-dry). Hence, the season of surplus is 112 days with a tercile range of 162–53. The P–E curve is relatively symmetrical with a crest at the end of July. Upper tercile flood spikes > 10 mm/day extend two months (July–August) and contribute millions of cubic meters to the Blue Nile catchment. The P–E > 0 in dry years is too short for crop production, and the upper–lower spread exceeds 5 mm/day from May through August. Thus, the beginning of the planting cycle is a stressful time for soil moisture and farming practice.

The ECMWF LHF is a useful proxy for vegetation fraction, which satellite NDVI cannot provide at daily intervals due to cloud cover. Its annual cycle terciles in the NW Ethiopian highlands (Figure 2b) exhibit a gradual rise to a plateau in September (approximately day 260), followed by a rapid decline at the end of the year. This asymmetry is quite different than rainfall. Of particular interest is the wide spread between upper and lower terciles in April–May (approximately days 100–130), and limited spread in early July (approximately day 180) and after the peak. Years with low LHF correspond with low NDVI and poor crop yields, and vice versa.

The annual cycle of P–E, LHF, and NDVI guide crop management, but only P has long-term records for analysis of past trends. In Figure 2c, the percentage contribution of sub-seasonal rainfall over the 20th century is calculated. Mean values are: 13% early (April–May), 47% mid (July–August), 7% late (October–November). Linear trends in each sub-season demonstrate that 'late' is becoming prevalent +0.021% yr^{-1}, followed by 'wide' +0.018% yr^{-1} (e.g., early + late), which reduces 'narrow' to0.015% yr^{-1}, leaving 'early' unchanged + 0.003% yr^{-1}. Thus, we see more wet spells at the end of season and ask: what underlies this tendency?

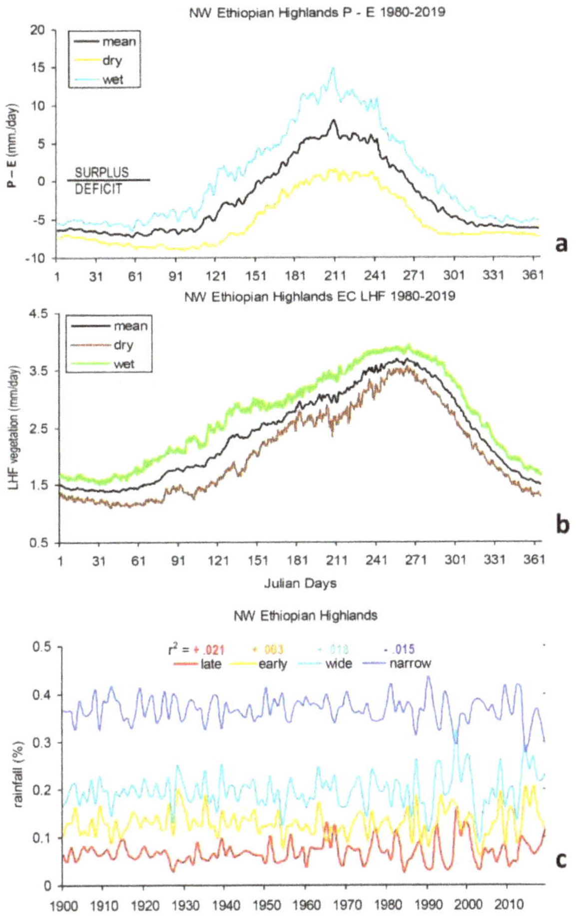

Figure 2. Annual cycle terciles from daily data 1980–2018: (**a**) Seasonal precipitation (P) minus E and (**b**) ECMWF LHF 'vegetation' proxy. (**c**) Seasonal contributions of cenTrends P over the 20th century, where late = Oct.–Nov., early = Apr.–May, wide = early + late, narrow = Jul.–Aug. only. All time series averaged in the study area: 8.5–13° N, 35–39.5° E.

3.2. *Composite Analysis*

To understand the meteorology behind the seasonal changes, we conduct a composite difference analysis (Figure 3a–c) after ranking of 'early', 'late', and 'narrow' and subtracting the five least from the five most. The early composite illustrates that SE wind anomalies from the Turkana Valley push moisture northwestward from Kenya, creating a local humidity dipole. In contrast, the late composite features W wind anomalies from southern Sudan that push moisture northeastward. Again, there is a local humidity dipole corresponding with the source sink. For the narrow composite, we analyze a vertical section and find an S wind anomaly in the 700–600 hPa layer with dry conditions in low latitudes (Kenya). Moisture differences are positive over northern Ethiopia and in the layer 400 hPa. Thus, equatorial convection is 'pushed' northward to the Blue Nile catchment. Yet, [57] find little coherent response of the equatorial trough to global warming.

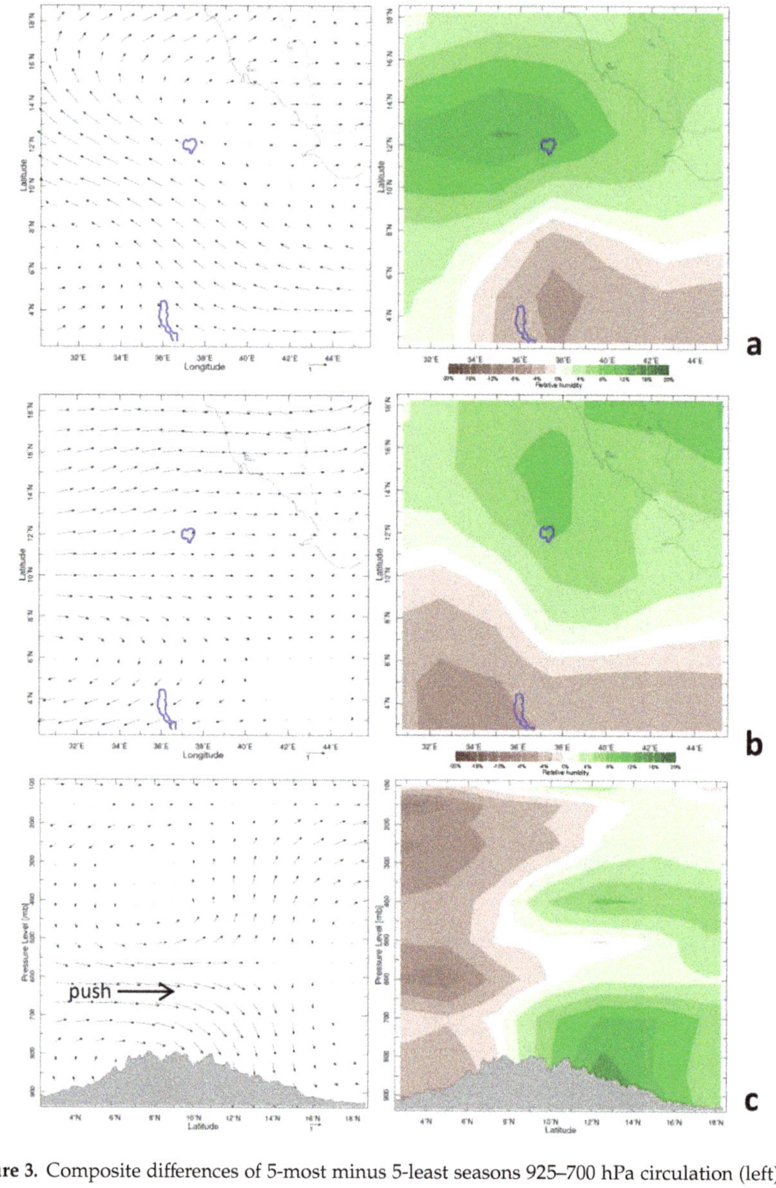

Figure 3. Composite differences of 5-most minus 5-least seasons 925–700 hPa circulation (left) and humidity: (**a**) 'Early' (Apr.–May), (**b**) 'Late' (Oct.–Nov.). (**c**) Meridional circulation and humidity in N-S vertical section with topography for 5-most minus 5-least 'Narrow' (Jul.–Aug.) seasons.

The 'wide' composite differences have mid-latitude influence that require analysis at a larger space scale. Later, we show that CMIP5 hydrological projections support the 'wide' scenario, so here, we establish the underlying process. Figure 4a illustrates that convective differences (−netOLR) over Ethiopia extend southward over the African Rift Valley (35E) and northeastward over the Arabian Peninsula. There are dry zones over the south Indian Ocean [58], Kalahari, and the Mediterraean (+netOLR). Tropospheric wind differences (Figure 4b) are almost absent in the tropics, but there is westerly flow in the southern sub-tropics and a deep anticyclone over southern Asia. The westerly flow along 15S has ridge 10E/trough 35E/ridge 60E features that indicate how anomalies in the sub-tropics lengthen the crop growing season over the NW Ethopian highlands.

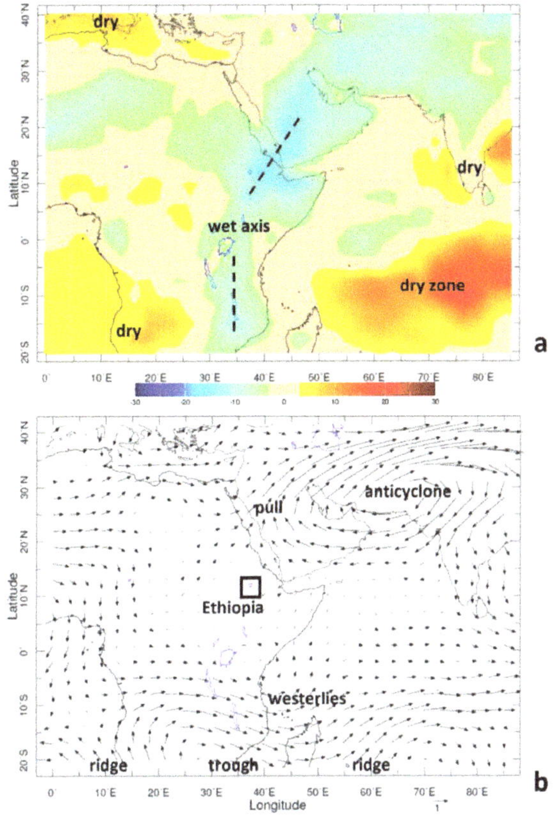

Figure 4. Composite differences of 5-most minus 5-least 'Wide' seasons: (**a**) satellite net outgoing long-wave radiation (OLR) and (**b**) 1000–100 hPa tropospheric circulation vectors with key features, and an index box.

3.3. Annual Cycle

We consider the 1st EOF loading patterns for S and E in Figure 5a,b. There is a center of action over the NW Ethiopian highlands and a sympathetic zone over the White Nile Valley approximately 9N, 33E which identify a unimodal climate. The annual cycle of E reaches an apex in February–April. The mean annual cycle of reanalysis and satellite soil moisture and NDVI in Figure 5c,d reveal that the ECMWF is slightly below FLDAS, which tends to peak later (Sep 31%). The GRACE satellite exhibits dry (March 16%) to wet (August 32%) changes that are relatively sinusoidal. Lag correlations with respect to continuous monthly ECMWF soil moisture (Figure 5e,f) show that P leads by one month and vegetation lags by one month, as expected. Thus, grazing pastures and crops reach peak conditions in September–October. The lag correlation of E is markedly negative and symmetric about zero. The Hadley2 model SHF relates negatively to S in a manner consistent with the reanalysis of E. These serve as references for model projections.

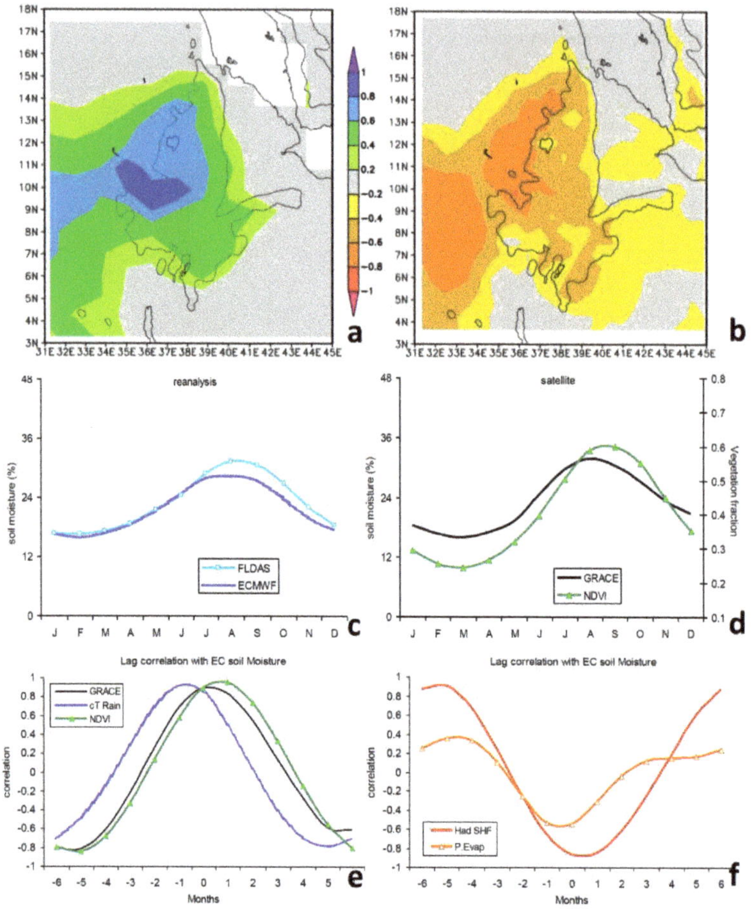

Figure 5. EOF loading patterns of 1st mode ECMWF reanalysis: (**a**) S and (**b**) SHF, identifying zones with unimodal climate. Mean annual cycle of area-averaged: (**c**) reanalysis S and (**d**) satellite S and vegetation NDVI. Lag-correlation of reanalysis S with variables from the NW Ethiopia area: (**e**) cenTrends P, satellite S and NDVI, and (**f**) Hadley2-rcp6 SHF and ECMWF E. Negative months refer to variable leading S.

3.4. Hadley2 Projections

The Hadley2-rcp6 ISIMIP mean annual cycle of soil moisture is given Figure 6a,b. The seasonal range is lower in glowb than watergap: 13% in February–March to 33% in August–September. Both simulations over-deplete S from November–March, but infiltration is near observed from May–August. Changes from the past (1900–2000) to the future (2001–2100) are generally < 1% and retain a unimodal structure consistent with other work [59–63]. There is a seasonal widening of S projected in the future, during May in watergap (1.1%) and during October in glowb (0.7%).

The long-term trend of the Hadley2-rcp6 ISIMIP annual S is slightly downward, with greater multi-year fluctuation in watergap than glowb (Figure 6c,d). Drought conditions may increase slightly during the 21st century. Projected E has a desiccating trend (+0.0022 to +0.0069 mm day^{-1}/yr) and significant signal-to-noise ratio r^2 = 0.60–0.81. Trends in E are initially flat and only become steep in the 21st century, suggesting dependence on the scenario employed. Mapping the past trends (Figure 6e–g), we find that the ECMWF soil moisture is slightly downward over the 20th century around the edges of the Blue Nile catchment < −0.1% yr^{-1}. Both projections show little future trend over the highlands, but the surrounding lowlands become desiccated. While minor adjustments may be needed in agricultural practice and water management to cope with fluctuating soil moisture, greater evaporation will stress the land and reservoirs.

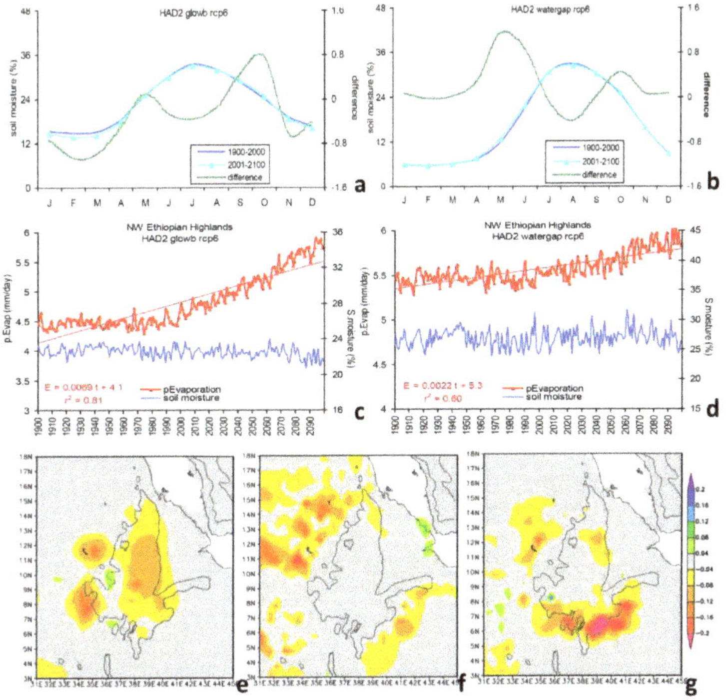

Figure 6. Hadley2-rcp6 annual cycle of S in past, future, and difference: (**a**) glowb and (**b**) watergap. Hadley2-rcp6 projected time series of annual S and E from (**c**) glowb and (**d**) watergap, and E slope (mm day^{-1}/yr) and r^2 fit; scales vary slightly. S trend maps for Jul–Oct wet season: (**e**) ECMWF past (1900–2018), (**f**) glowb future (2001–2100), and (**g**) watergap future (% yr^{-1}). 1500 m elevation contour delineates the highlands.

3.5. LGP Outcome

The LHF annual cycle from Hadley2 rcp6 projection (Figure 7a) has an asymmetrical shape close to NDVI and ECMWF (cf. Figure 2c or Figure 5d). It rises gradually in May–August and reaches a peak in September–October, when crops are harvested. Differences in the future are positive for June–July and otherwise slightly negative. Long-term LHF trends have a small signal-to-noise ratio of approximately 2%. In contrast, we find a considerable increase of minimum T (>2 °C) from the 20th to the 21st century in the Hadley2 rcp6 projection, which is evenly distributed across the annual cycle (Figure 7b). The cool temperatures of October will gradually recede, leaving soil moisture depletion to end the farming season.

Comparing past and future LGP (Figure 7c,d), we determined that the median onset was earlier: day 140 to 138 (trend −0.026 day/yr), cessation was later: day 315 to 317 (+0.035 day/yr), and duration lengthened 175 to 179 days (trend +0.023 day/yr) and exhibited a median range 168–185 days. Appendix A Figure A2c is an example of LGP constraints imposed by daily S and T over the past decade. Variations in duration S > 15% and peak S are evident; in some years, T causes early cessation.

With a longer growing season and adequate minimum temperatures, crop production could shift from temperate to sub-tropical varieties. Alternatively, farming efforts could move gradually upslope to preserve current conditions (−0.7 °C/100 m elevation, Appendix A Figure A2a,b). In any case, the LGP will exceed the 120 days needed for short-cycle crops.

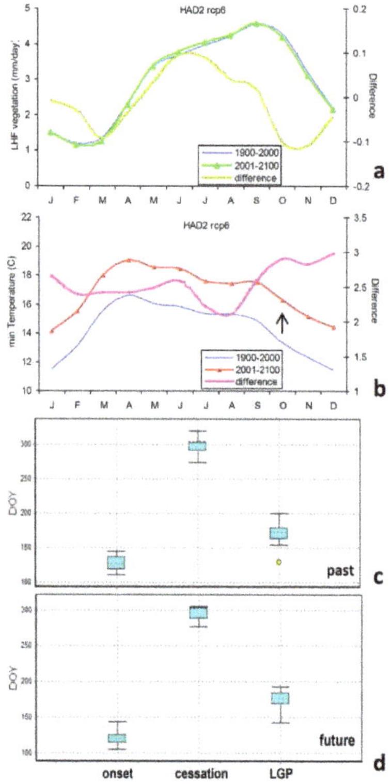

Figure 7. Hadley2-rcp6 annual cycle in past, future, and difference: (**a**) LHF vegetation, and (**b**) minimum T, the arrow points to warming in October. Box and whisker plot of onset, cessation, and LGP of (**c**) past and (**d**) future era: (dashed: median, box: 25/75th percentile, whisker: 10/90th percentile, o: extreme value).

4. Discussion and Conclusions

We have compared hydro-climate change in the 20th century and projections in the 21st century [64], particularly with regard to the seasonal onset and cessation of conditions favoring crop phenology in the NW Ethiopian highlands. Statistical methods were applied to gauge-interpolated, reanalysis, and satellite data to detect the LGP annual cycle. We are motivated to offset climate impacts with more knowledge on cropping cycles that lead to adaptation strategies. Trends in sub-seasonal rainfall over the 20th century show a 'late' season rise $+0.021\%\text{yr}^{-1}$, meaning that conditions favoring crop growth will extend into October. Lag correlations with soil moisture show that P leads by one month and vegetation lags by one month, and that SHF and LHF are valuable proxies via ECMWF reanalysis and Hadley2 model simulation.

To understand the meteorology behind the seasonal changes, we conducted a composite difference analysis. The 'late' composite featured W wind anomalies from southern Sudan that push moisture northeastward from the White Nile to the Blue Nile catchment. Sub-tropical troughs to the north and south that create a meridional axis of convection (–netOLR) that lengthens the crop growing season over the NW Ethopian highlands.

CMIP5 bias-corrected Hadley2 data assimilated by ISIMIP hydrological models gave insights on the unimodal annual cycle of soil moisture in past and future eras. The annual cycle amplitude for S saw a low-point of 13% in February–March and a high point of 33% in August–September. Both hydrology simulations over-deplete S from November–March, but fractional increases in May–August were near observed. The future 'widening' of S was 1.1% during May in watergap and 0.7% during October in glowb.

Projections of both E and S show little future trend over the highlands, but the surrounding lowlands become desiccated. While only minor adjustments are needed in agricultural practice and water management to cope with fluctuating soil moisture, more effort is essential to control stresses from evaporation.

A longer growing season is likely given the rising minimum temperatures in October. Crop production could shift from temperate to sub-tropical varieties, or farming efforts could move gradually upslope to preserve current conditions. Our results show that the LGP will increase from 175–179 days, which is more than adequate for short-cycle crops. Farming efforts could utilize earlier planting and later harvesting with future LGP suitable for longer-cycle crops or double cropping, and they could also employ seasonal forecasts to reduce the risks of climate variability. In a doubled CO_2 future, the number of frost days will decline to zero, meaning that pests and disease may disturb food production.

Author Contributions: This paper is based on the PhD of the first author G.B.T., and was re-analyzed and re-written by the second author M.R.J., who supervised the original thesis. All authors have read and agreed to the published version of the manuscript.

Funding: This research received no external funding.

Acknowledgments: This work is an extension of a PhD thesis by the first author. The second author recognizes ongoing support from the SA Department of Education.

Conflicts of Interest: The authors declare no conflict of interest.

Glossary

cenTrends	centennial trends (precipitation)
Chirps2	satellite-gauge blended rainfall product
CMIP5	coupled model intercomparison project v5
E	potential evaporation (p.Evap)
ECMWF	European community medium-range weather forecasts
EOF	Empirical Orthogonal Function
FLDAS	FEWS land data assimilation system
glowb	hydrological model (ISIMIP)
GRACE	gravity recovery climate experiment (satellite soil moisture)
Hadley2-rcp6	Hadley v2 coupled model with +6 W/m^2 scenario
ISIMIP	inter-sectoral impact hydrological model intercomparison project
LGP	length of (crop) growing period
LHF	latent heat flux (vegetation proxy)
NCEP2	national lefts for environmental prediction reanalysis v2
NDVI	normalized difference vegetation index (colour fraction)
NW	northwest
OLR	(net) outgoing long-wave radiation
P	precipitation
S	soil moisture (0–1 m)
SHF	sensible heat flux
T	temperature
watergap	hydrological model (ISIMIP)

Appendix A

Table A1. Evaluation of NW-Ethiopia highlands CMIP5 model rainfall with gauge-interpolated reference [65] 1981–2010. Correlation of mean annual cycle, Jun.–Sep. seasonal difference between observation and model, model mean mid-summer rain rate (mm/day) and phase/amplitude 'fit' of annual cycle.

No	Model	Rain Correlation	Jun.–Sep. Difference	Jul.–Aug. Value	Annual Cycle
1	bcc-csm1-1	0.65	−3.2	3.8	poor
2	bcc-csm1-1-m	0.90	−1.4	6.3	poor
3	CCSM4	0.76	−2.0	4.3	poor
4	CESM1-CAM5	0.75	−0.9	5.7	poor
5	CSIRO-Mk3-6-0	0.91	−1.1	8.8	moderate
6	FIO-ESM	0.80	−2.1	4.6	poor
7	GFDL-CM3	0.87	−1.5	6.2	poor
8	GFDL-ESM2G	0.86	−0.3	6.7	moderate
9	GFDL-ESM2M	0.87	−0.5	6.4	moderate
10	GISS-E2-H_p1	0.96	−5.3	3.2	poor
11	GISS-E2-H_p2	0.96	−5.5	2.8	poor
12	GISS-E2-H_p3	0.96	−4.7	4.0	poor
13	GISS-E2-R_p1	0.97	−5.6	2.5	poor
14	GISS-E2-R_p2	0.97	−5.7	2.6	poor
15	GISS-E2-R_p3	0.96	−5.3	2.7	poor
16	HadGEM2-AO	0.96	−0.4	8.1	high
17	HadGEM2-ES	0.97	−0.5	8.0	high
18	IPSL-CM5A-LR	0.89	−2.7	6.3	poor
19	IPSL-CM5A-MR	0.90	−3.3	5.3	poor
20	MIROC5	0.98	6.5	16.1	poor
21	MIROC-ESM	0.86	−0.5	7.3	high
22	MIROC-ESM-CHEM	0.88	−0.5	7.4	high
23	MRI-CGCM3	0.93	−2.5	6.0	moderate
24	NorESM1-M	0.66	−2.7	3.4	poor
25	NorESM1-ME	0.62	−2.3	3.3	poor

Table A2. Evaluation of NW-Ethiopia highlands CMIP5 model sensible heat flux (SHF) with ECMWF reanalysis 1981–2010. Correlation of mean annual cycle, Feb.–Apr. seasonal difference between observation and model (mm/day), and phase/amplitude 'fit' of annual cycle.

No	Models	SHF Correlation	Feb.–Apr. Difference	Annual Cycle
1	bcc-csm1-1-m	0.79	0.46	moderate
2	bcc-csm1-1	0.73	0.72	poor
3	CCSM4	0.96	0.63	high
4	CESM1-CAM5	0.78	0.06	moderate
5	CSIRO-Mk3-6-0	0.79	1.04	poor
6	FIO-ESM	0.95	0.81	high
7	GFDL-CM3	0.92	1.26	poor
8	GFDL-ESM2G	0.9	0.92	moderate
9	GFDL-ESM2M	0.92	0.88	moderate
10	GISS-E2-H_p1	0.95	1.48	poor
11	GISS-E2-H_p2	0.96	1.32	poor
12	GISS-E2-H_p3	0.97	1.21	poor
13	GISS-E2-R_p1	0.95	1.29	poor
14	GISS-E2-R_p2	0.97	1.11	poor
15	GISS-E2-R_p3	0.97	1.04	moderate
16	HadGEM2-AO	0.93	0.34	high
17	HadGEM2-ES	0.96	0.24	very high
18	IPSL-CM5A-LR	0.87	1.52	poor
19	IPSL-CM5A-MR	0.74	1.34	poor
20	MIROC5	0.95	−0.95	moderate
21	MIROC-ESM	0.87	0.11	moderate
22	MIROC-ESM-CHEM	0.85	0.05	moderate
23	MRI-CGCM3	0.8	0.31	moderate
24	NorESM1-M	0.95	0.58	high
25	NorESM1-ME	0.91	0.44	high

Table A3. Statistical significance of soil moisture trends per month in the NW-Ethiopian highlands, based on HadGEM2-ES rcp6 projection 1981–2100 and ISIMIP hydrological output, where bold values are significant. Temporal correlation indicating slope of regression line, where negative = drying, and *p*-value with respect to 119 degrees of freedom.

N = 119	Glowb		Watergap	
Months	Time Cor.	*p*-Value	Time Cor.	*p*-Value
Jan	**−0.25**	**0.01**	0.12	0.18
Feb	**−0.28**	**0.00**	0.12	0.19
Mar	**−0.19**	**0.03**	0.12	0.21
Apr	−0.06	0.49	0.14	0.12
May	−0.08	0.38	0.12	0.21
Jun	0.13	0.17	**0.36**	**0.00**
Jul	**−0.43**	**0.00**	−0.07	0.42
Aug	**−0.19**	**0.03**	**−0.42**	**0.00**
Sep	0.15	0.11	−0.10	0.26
Oct	0.17	0.06	0.05	0.57
Nov	**−0.27**	**0.00**	−0.02	0.82
Dec	**−0.23**	**0.01**	−0.03	0.77

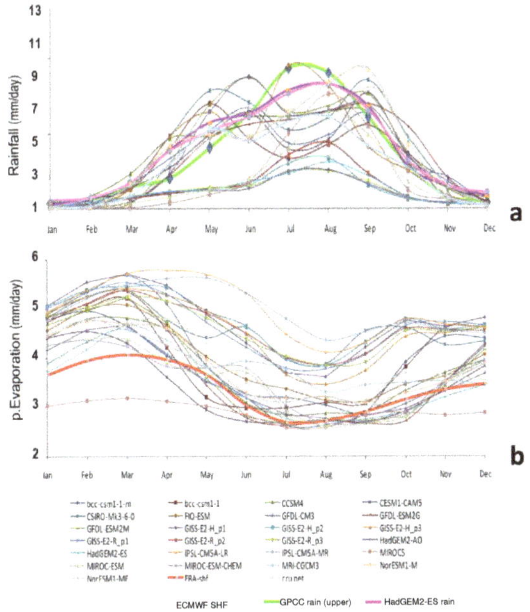

Figure A1. Mean annual cycle of NW Ethiopian highlands area-averaged (**a**) P and (**b**) E derived from SHF; comparing all 25 CMIP5 simulations with observation/reanalysis reference (bold green upper, bold red lower) 1981–2010. HadGEM2-ES rainfall (upper) is bold pink.

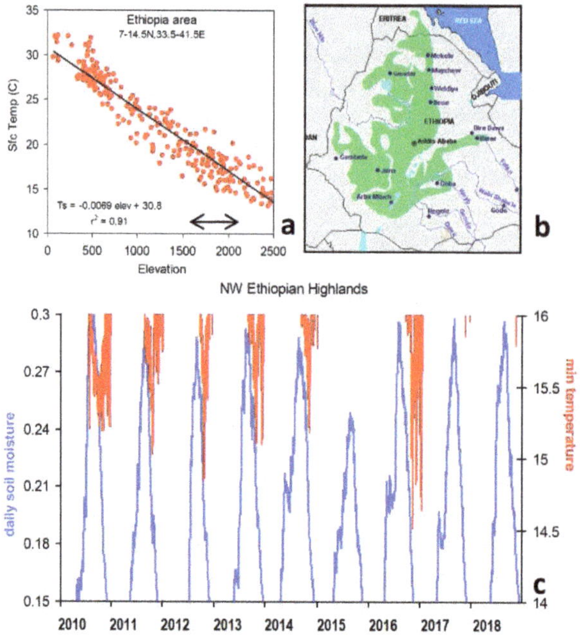

Figure A2. (**a**) Scatterplot of elevation vs. Jul.–Oct. surface T across Ethiopia, arrow highlights the most suitable range, (**b**) main crop-growing areas (green shaded). (**c**) Example of LGP constraints imposed by thresholds of daily soil moisture and minimum T. Note the drought in 2015 and higher T in 2017–2018. The min. T is trimmed above 16C to indicate no threshold exceedance.

References

1. Adenew, B. *Climate Change and Policy Process in Ethiopia: Preliminary Review Results*; Ethiopian Economics Association: Addis Ababa, Ethiopia, 2010.
2. Degefu, W.; Glantz, M. (Eds.) Some aspects of meteorological drought in Ethiopia. In *Drought and Hunger in Africa: Denying Famine a Future*; Cambridge University Press: Cambridge, UK, 1987; pp. 23–36.
3. Davidson, O.; Halsnaes, K.; Huq, S.; Kok, M.; Metz, B.; Sokona, Y.; Verhagen, J. The development and climate nexus: The case of sub-Saharan Africa. *Clim. Policy* **2003**, *3*, S97–S113. [CrossRef]
4. Adem, A.; Bewket, W. *A Climate Change Country Assessment: Report for Ethiopia. Forum for Environment (ECSNCC)*; Epsilon International R&D: Addis Ababa, Ethiopia, 2011.
5. Cheung, W.H.; Senay, G.B.; Singh, A. Trends and spatial distribution of annual and seasonal rainfall in Ethiopia. *Int. J. Climatol.* **2008**, *28*, 1723–1734. [CrossRef]
6. Segele, Z.T.; Lamb, P.J. Characterization and variability of JJAS rainy season over Ethiopia. *Meteorol. Atmos. Phys.* **2005**, *89*, 153–180. [CrossRef]
7. Bewket, W. Rainfall Variability and Crop Production in Ethiopia: Case Study in the Amhara Region. In Proceedings of the 16th International Conference of Ethiopian Studies, Norway, 2009. Available online: http://hpccc.gov.in/PDF/Agriculture/Rainfall%20Variability%20and%20Crop%20production%20in%20Ethiopia%20Case%20study%20in%20the%20Amhara%20Region.pdf (accessed on 16 June 2020).
8. Ayalew, D.; Tesfaye, K.; Mamo, G.; Yitaferu, B.; Bayu, W. Variability of rainfall and its current trend in Amhara region, Ethiopia. *Afr. J. Agric. Res.* **2012**, *7*, 1475–1486. [CrossRef]
9. Shi, L.; Hendon, H.; Alves, O.; Luo, J.-J.; Balmaseda, M.; Anderson, D.L.T. How Predictable is the Indian Ocean Dipole? *Mon. Weather Rev.* **2012**, *140*, 3867–3884. [CrossRef]
10. Henricksen, B.L.; Durkin, J.W. *Moisture Availability, Cropping Period and the Prospects for Early Warning of Famine in Ethiopia*; ILCA: Addis Ababa, Ethiopia, 1987.

11. Belay, S. Agroclimatic analysis in relation to crop production. In Proceedings of the Ethiopian Agricultural Research Organization of Ethiopia, Bahir Dar, Ethiopia, 23 January 2001.
12. Mersha, E. Agroclimatic classification of Ethiopia. *Ethiop. J. Nat. Resour.* **2005**, *2*, 137–154.
13. Tesfaye, M. Water Balance Applications in the Identification and Analysis of Agricultural Droughts in South Wollo Zone, Ethiopia. Ph.D. Thesis, Andhra University, Visakhapatnam, India, 2014.
14. Funk, C.; Nicholson, S.; Landsfeld, M.; Klotter, D.; Peterson, P.; Harrison, L. The centennial trends greater horn of Africa precipitation dataset. *Sci. Data* **2015**, *2*, 150050. [CrossRef]
15. Agnew, C.T.; Chappel, A. Drought in the Sahel. *GeoJournal* **1999**, *48*, 299–311. [CrossRef]
16. Cherkos, T. Intensity–Duration–Frequency Relationships for Northern Ethiopia Rainfall. Master's Thesis, Addis Ababa University, Addis Ababa, Ethiopia, 2002.
17. Jain, S.K.; Kumar, V. Trend analysis of rainfall and temperature data for India. *Curr. Sci.* **2012**, *102*, 37–49.
18. Hemerance, M.A. Addressing Climate Variability in Agricultural Land Evaluation. Case Study for Crop Production in Far North Cameroon. Master's Thesis, Ghent University, Brussels, Belgium, 2013.
19. Allen, R.G.; Pereira, L.S.; Raes, D.; Smith, M. Crop evapotranspiration: A guideline for computing crop water requirements. In *FAO Irrigation and Drainage Paper 56*; FAO Water Resources, Development and Management Service: Rome, Italy, 1998; ISBN 92-5-104219-5.
20. Kebede, B. Analysis of the Variability of Rainfall Distribution and Evapotranispiration over Eastern Amahara Region. Master's Thesis, Arba Minch University, Arba Minch, Ethiopia, 2009.
21. Tapley, B.D.; Bettadpur, S.; Ries, J.C.; Thompson, P.F.; Watkins, M.M. GRACE measurements of mass variability in the earth system. *Science* **2004**, *305*, 503–505. [CrossRef]
22. Liu, Y.Y.; Parinussa, R.; Dorigo, W.; de Jeu, R.; Wagner, W.; van Dijk, A.; McCabe, M.; Evans, J. Developing an improved soil moisture dataset by blending passive and active microwave satellite-based retrievals. *Hydrol. Earth Sys. Sci.* **2011**, *15*, 425–436. [CrossRef]
23. Crippen, R.E. Calculating the vegetation index faster. *Remote Sens. Environ.* **1990**, *34*, 71–73. [CrossRef]
24. Ji, L.; Peters, A.J. Assessing vegetation response to drought in the northern Great Plains using vegetation and drought indices. *Remote Sens. Environ.* **2003**, *87*, 85–98. [CrossRef]
25. Tucker, C.J.; Pinzon, J.E.; Brown, M.E.; Slayback, D.; Pak, E.W.; Mahoney, R.; Vermote, E.; Saleous, N. An extended AVHRR 8-km NDVI dataset compatible with MODIS and SPOT vegetation. *Int. J. Remote Sens.* **2005**, *26*, 4485–4498. [CrossRef]
26. Kaicun, W.K.; Pucai, W.; Zhanqing, L.; Cribb, M.; Michael, S. A simple method to estimate actual evapotranspiration from a combination of net radiation, vegetation index, and temperature. *J. Geophys. Res.* **2007**, *112*, D15107. [CrossRef]
27. United Nations International Strategy for Disaster Reduction. Drought risk reduction framework and practices: Contributing to the implementation of the Kyoto framework for action. In *International Strategy for Disaster Risk*; UNISDR: Geneva, Switzerland, 2009.
28. Mekuria, E. Spatial and Temporal Analysis of Recent Drought Years Using Vegetation Temperature Condition index, Case of Somali Regional State. Master's Thesis, Arba Minch University, Arba Minch, Ethiopia, 2012.
29. Yohannes, Z. Estimation of teff yield using remote sensing and GIS techniques in Tigray region, Northern Ethiopia. In Proceedings of the Research Application Summary, 3rd RU-Forum Biennial Meeting, Entebbe, Uganda, 24–28 September 2012.
30. Tadesse, T.; Demisse, G.B.; Zaitchik, B.; Dinku, T. Satellite-based hybrid drought monitoring tool for prediction of vegetation condition in Eastern Africa: A case study for Ethiopia. *Water Resour. Res.* **2014**, *50*, 2176–2190. [CrossRef]
31. Dodamani, B.M.; Anoop, R.; Mahajan, D.R. Agricultural drought modeling using remote sensing. *Int. J Environ. Sci. Dev.* **2015**, *6*, 326. [CrossRef]
32. Hundera, H.; Berhan, B.; Bewuket, W. Remote sensing and GIS based agricultural drought risk assessment in east Shewa zone, central Rift Valley region of Ethiopia. *J. Environ. Earth Sci.* **2016**, *6*, 48–54.
33. Kanamitsu, M.; Ebisuzaki, W.; Woollen, J.; Yang, S.; Hnilo, J.J.; Fiorino, M.; Potter, G.L. NCEP–DOE AMIP reanalysis R2. *Bull. Am. Meteorol. Soc.* **2002**, *83*, 1631–1644. [CrossRef]
34. Oleson, K.W.; Niu, G.Y.; Yang, Z.L.; Lawrence, D.M.; Thornton, P.E.; Lawrence, P.J.; Stöckli, R.; Dickinson, R.E.; Bonan, G.B.; Levis, S.; et al. Improvements to the community land model and their impact on the hydrological cycle. *J. Geophys. Res.* **2008**, *113*, G01021. [CrossRef]

35. Dee, D.P.; Uppala, S.M.; Simmons, A.J.; Berrisford, P.; Poli, P.; Kobayashi, S.; Andrae, U.; Balmaseda, M.A.; Balsamo, G.; Bauer, D.P.; et al. The ERA-Interim reanalysis: Configuration and performance of the data assimilation system. *Q. J. Roy. Meteorol. Soc.* **2011**, *137*, 553–597. [CrossRef]
36. McNally, A.; Arsenault, K.; Kumar, S.; Shukla, S.; Peterson, P.; Wang, S.; Funk, C.; Peters-Lidard, C.D.; Verdin, J.P. A land data assimilation system for sub-Saharan Africa food and water security applications (FLDAS). *Sci. Data* **2017**, *4*, 170012. [CrossRef] [PubMed]
37. Dinku, T.; Connor, S.J.; Ceccato, P.; Ropelewski, C.F. Comparison of global gridded precipitation products over a mountainous region of Africa. *Int. J. Climatol.* **2008**, *28*, 1627–1638. [CrossRef]
38. Diro, G.T.; Black, E.; Grimes, D.I.F. Seasonal forecasting of Ethiopian spring rains. *Meteorol. Appl.* **2009**, *15*, 73–83. [CrossRef]
39. Funk, C.C.; Peterson, P.J.; Landsfeld, M.F.; Pedreros, D.H.; Verdin, J.P.; Rowland, J.D.; Romero, B.E.; Husak, G.J.; Michaelsen, J.C.; Verdin, A.P. A quasi-global precipitation time series for drought monitoring. *US Geological. Surv. Data Ser.* **2014**, *832*, 1–4. [CrossRef]
40. Masresha, E. Growing season belts of Ethiopia. In Proceedings of the National Workshop on Agrometeorology and GIS, EIAR, Addis Ababa, Ethiopia, 7–11 July 2003; pp. 35–36.
41. Araya, S.D.; Keesstra, L.; Stroosnijder, L. A new agro-climatic classification for crop suitability zoning in northern semi-arid Ethiopia. *Agricul. For. Meteorol.* **2010**, *150*, 1057–1064. [CrossRef]
42. Yemenu, F.; Chemeda, D. Climate resources analysis for use of planning in crop production and rainfall water management in the central highlands of Ethiopia, the case of Bishoftu district, Oromia region, Hydrol. *Earth Syst. Sci. Discuss* **2010**, *7*, 3733–3763. [CrossRef]
43. Dereje, G.; Eshetu, A. *Crops and Agro-Ecological Zones of Ethiopia*; Ethiopian Institute of Agricultural Research: Addis Ababa, Ethiopia, 2011.
44. Gebremichael, A.; Quraishi, S.; Mamo, G. Analysis of seasonal rainfall variability for agricultural water resource management in southern region, Ethiopia. *J. Nat. Sci. Res.* **2014**, *4*, 2224–3186.
45. Sawa, B.A.; Adebayo, A.A.; Bwala, A.A. Dynamics of hydrological growing aeason at Kano as evidence of climate change. *Asian J. Agric. Sci.* **2014**, *62*, 75–78. [CrossRef]
46. Thornton, P.K.; Jones, P.G.; Owiyo, T. *Mapping Climate Vulnerability and Poverty in Africa, Report to the Department of International Development*; UN: Nairobi, Kenya, 2006; 171p.
47. Ati, O.F.; Stigter, C.J.; Olandipo, E.O. A Comparison of methods to determine the onset of the growing season in northern Nigeria. *Int. J. Climatol.* **2002**, *22*, 732–742. [CrossRef]
48. Nata, T. Surface water potential of the Hantebet Basin, Tigray, Northern Ethiopia. *Agric. Eng. Int. CIGRE J.* **2006**, *8*, 5–10.
49. Kebede, K.; Bekelle, E. Tillage effect on soil water storage and wheat yield on the vertisols of north central highlands of Ethiopia. *Ethiop. J. Environ. Stud. Manag.* **2008**, *1*, 49–55. [CrossRef]
50. Taylor, K.E.; Stouffer, R.J.; Meehl, G.A. An overview of CMIP5 and the experiment design. *Bull. Am. Met. Soc.* **2012**, *93*, 485–498. [CrossRef]
51. Intergovernmental Panel on Climate Change. *Climate Change 2014: Synthesis Report. Contribution of Working Group I Fifth Assessment Report*; Pachauri, M., Ed.; IPCC: Geneva, Switzerland, 2014; p. 151.
52. Bhattacharjee, P.S.; Zaitchik, B.F. Perspectives on CMIP5 model performance in the Nile River head waters regions. *Int. J. Climatol.* **2015**, *35*, 4262–4275. [CrossRef]
53. Jones, C.; Hughes, J.K.; Bellouin, N.; Hardiman, S.C.; Jones, G.S.; Knight, J.; Liddicoat, S.; O'connor, F.M.; Andres, R.J.; Bell, C.; et al. The HadGEM2-ES implementation of CMIP5 centennial simulations. *Geosci. Model Dev.* **2011**, *4*, 543–570. [CrossRef]
54. Frieler, K.; Lange, S.; Piontek, F.; Reyer, C.P.; Schewe, J.; Warszawski, L.; Zhao, F.; Chini, L.; Denvil, S.; Emanuel, K.; et al. Assessing the impacts of 1.5 °C global warming—Simulation protocol of the Inter-Sectoral Impact Model Intercomparison Project ISIMIP2b. *Geosci. Model Dev.* **2018**, *10*, 4321–4345. [CrossRef]
55. Van Vuuren, D.P.; Edmonds, J.; Kainuma, M.; Riahi, K.; Thomson, A.; Hibbard, K.; Hurtt, G.C.; Kram, T.; Krey, V.; Lamarque, J.F.; et al. The representative concentration pathways: An overview. *Clim. Chang.* **2011**, *109*, 5–31. [CrossRef]
56. Reda, D.T.; Engida, A.N.; Asfaw, D.H.; Hamdi, R. Analysis of precipitation based on ensembles of regional climate model simulations and observational databases over Ethiopia for the period 1989–2008. *Int. J. Climatol.* **2015**, *35*, 948–971. [CrossRef]

57. Byrne, M.P.; Pendergrass, A.G.; Rapp, A.D.; Wodzicki, K.R. Response of the intertropical convergence zone to climate change: Location, width, and strength. *Curr. Clim. Chang. Rep.* **2018**, *4*, 355–370. [CrossRef]
58. Shanko, D.; Chamberlain, P. The effects of the southwest Indian ocean tropical cyclones on Ethiopian drought. *Int. J. Climatol.* **1998**, *18*, 1373–1388. [CrossRef]
59. Seleshi, Y.; Zanke, U. Recent changes in rainfall and rainy days in Ethiopia. *Int. J. Climatol.* **2004**, *24*, 973–983. [CrossRef]
60. Conway, D.; Schipper, E.L.F. Adaptation to climate change in Africa: Challenges and opportunities. *Glob. Environ. Chang.* **2011**, *21*, 227–237. [CrossRef]
61. Jury, M.R. Ethiopian highlands crop-climate prediction 1979–2009. *J. Appl. Meteorol. Climatol.* **2013**, *52*, 1116–1126. [CrossRef]
62. Jury, M.R. Statistical evaluation of CMIP5 climate change model simulations for the Ethiopian highlands. *Int. J. Climatol.* **2015**, *35*, 37–44. [CrossRef]
63. Ethiopian Panel on Climate Change (EPCC). *First Assessment Report, Working Group I Physical Science Basis*; Ethiopian Academy of Sciences: Addis Ababa, Ethiopia, 2015.
64. Jury, M.R.; Funk, C. Climatic trends over Ethiopia: Regional signals and drivers. *Int. J. Climatol.* **2013**, *33*, 1924–1935. [CrossRef]
65. Harris, I.; Jones, P.D.; Osborn, T.J.; Lister, D.H. Updated high-resolution grids of monthly climatic observations the CRU4 Dataset. *Int. J. Climatol.* **2014**, *34*, 623–642. [CrossRef]

atmosphere

MDPI

Article

Climate and the Global Spread and Impact of Bananas' Black Leaf Sigatoka Disease

Eric Strobl [1,*] and Preeya Mohan [2]

1 Department of Economics, University of Bern, 3008 Bern, Switzerland
2 Sir Arthur Lewis Institute of Social and Economic Studies, University of the West Indies, St. Augustine 82391, Trinidad and Tobago; preeya.mohan@sta.uwi.edu
* Correspondence: eric.strobl@vwi.unibe.ch

Received: 7 August 2020; Accepted: 31 August 2020; Published: 5 September 2020

check for updates

Abstract: While Black Sigatoka Leaf Disease (*Mycosphaerella fijiensis*) has arguably been the most important pathogen affecting the banana industry over the past 50 years, there are no quantitative estimates of what risk factors determine its spread across the globe, nor how its spread has affected banana producing countries. This study empirically models the disease spread across and its impact within countries using historical spread timelines, biophysical models, local climate data, and country level agricultural data. To model the global spread a empirical hazard model is employed. The results show that the most important factor affecting first time infection of a country is the extent of their agricultural imports, having increased first time disease incidence by 69% points. In contrast, long distance dispersal due to climatic factors only raised this probability by 0.8% points. The impact of disease diffusion within countries once they are infected is modelled using a panel regression estimator. Findings indicate that under the right climate conditions the impact of Black Sigatoka Leaf Disease can be substantial, currently resulting in an average 3% reduction in global annual production, i.e., a loss of yearly revenue of about USD 1.6 billion.

Keywords: bananas; Black Sigatoka Leaf Disease; climate; global spread & impact

---

1. Introduction

While early farming hunter-gatherers were probably aware of the existence of fungal crop diseases and their potential impact, given that they depended on a local natural, often diverse, population of plants, the range for gathering was likely easily extended and thus any impact was minimized (Agrios [1] and Scheffer [2]). However, as the domestication of plants and the development and dissemination of techniques for raising them productively increased around 8000 years ago, resulting in larger areas of plantation as well as the reliance on fewer crops, food security became increasingly more vulnerable to disease outbreaks. As a matter of fact, there are ample references in historical documents that make reference to such events and their often devastating impacts (Agrios [1]). Modern globalization and specialization of agricultural production in the 19th century further encouraged the focus on fewer crop varieties, leading to further susceptibility to crop diseases. In some instances crop disease outbreaks have even been argued to have changed history, as, for example, through massive migration following the 1845 potato famine in Ireland (Gráda and O'Rourke [3]), the near downfall of the wine industry during the Downy Mildew of Grapes outbreak in the Mediterranean for wine in 1865 (Simms [4]), or the switch from drinking coffee to tea in the British Empire as a result of the coffee leaf rust in the 1890s in Sri Lanka (Money [5]). Finally, with the agricultural green revolution in the 1960s, which involved breeding and encouraging specific varieties that had higher yield potential, monocropping became firmly established across the globe (Hunter et al. [6]).

The link between crop monoculture systems and crop diseases is straight forward: by focusing on fewer varieties with higher genetic uniformity, many crops have become substantially more susceptible to both old and newly arising varieties of fungi (Wolfe [7] and Garrett and Mundt [8]). Moreover, with increasing globalization and specialization the transmission of diseases has also increased on a much greater geographical scale (Gergerich et al. [9]). Measures to prevent or dampen crop disease outbreaks through the increasing use of existing and the development of new pesticides, as well as strictly controlling the import of plant related products into countries, have only been partially effective. In terms of the former, many fungi have shown increased resistance to existing pesticides over time (Lucas et al. [10]) (As a matter of fact, despite a clear increase in pesticide use, crop losses have not significantly decreased during the last 40 years (Oerke [11])). For the latter, not only are there arguably inefficiencies in current legislation implementation (Perrings [12]), but for many crop fungi even strictly enforced physical borders may not be effective as these can still spread through the atmosphere over long distances (Brown and Hovmøller [13]). As a matter of fact, rough estimates suggest that currently losses of major crops due to fungal diseases amount to enough to feed 8.5% of the global population (Fisher et al. [14]), and between 10 to 40% of global production (Savary et al. [15]). These losses continue to occur despite the fact that many countries have implemented integrated disease management, including the biological control of many pests and diseases (HE et al. [16]).

Within the context of the important role that fungal diseases play in the evolution of many major crops, bananas are perhaps the most exemplary. More specifically, 160 years ago few people outside banana growing countries would have even known the taste of a banana (Marin et al. [17]). However, after 1870, with the first commercialization of banana exports, the introduction of refrigerated shipping, the growing taste for the tropical fruit, and the expansion of organized cultivation into Central America, bananas became one of the most important crops globally (Abbott [18] and Koeppel [19]). Today it is the most exported fruit, and the fourth most imported crop globally. Bananas earn approximately US$ 8 billion annually from the production of 114 million tonnes on 5.6 million hectares of land (Authors' own calculations using data from FAOSTAT), and are produced in more than 100 countries in tropical and subtropical regions, including Africa, Asia, the Pacific islands, Latin America and the Caribbean (Churchill [20]). However, bananas have also been a crop decidedly marked by disease. More precisely, early exports were dominated by a single banana cultivar, the Gros Michel banana, but the appearance of Panama fungal disease in Central America, the main global exporting region of bananas, wiped out vast tracts of plantations (Koeppel [19]). Consequently, most banana exporting plantations replaced the Gros Michel with the Cavendish cultivar, which is resistant to Panama disease. However, in the early 1960s a new fungal disease, Black Sigatoka Leaf Disease (BSLD), to which the Cavendish cultivar is extremely susceptible, started spreading across the banana growing world. It has now been detected in nearly half of all banana producing countries and is likely to further spread through natural and human induced channels (Brown and Hovmøller [13]). The disease can reduce yields by up to 90% and induce early ripening, the latter being an important drawback for a fruit that is usually shipped unripe and then artificially ripened in transport or industrial greenhouses (de Bellaire et al. [21] and Alamo et al. [22]). BSLD is now considered one of the world's main crop diseases, and while chemical treatment can partially help control it, such measures have increased production costs substantially and fungicide resistance appears to be increasing(Jones et al. [23]).

Despite the potential historical importance of fungi diseases for many crops, apart from rough figures, direct evidence on the actual quantitative impact at a global and long-term scale is virtually non-existent (For example, Savary et al. [15] conduct a survey among crop health experst in order to derive their estimates). For example, in terms of the quantitative impact of BSLD on bananas specifically, there is, as far as we are aware, no existing empirical study, not even at a local scale (The only two econometric studies related to the topic, Edmeades et al. [24] and Kenneth et al. [25], both investigate the impact of the perceived risk of black sigatoka and the adaption of resistant varieties, respectively, on farmers' banana planting decisions in Uganda). Rather, a few papers have used simulations to predict the economic impact of the disease and possible preventive measures on banana production. For instance, Alamo et al. [22] use an equilibrium displacement model for Puerto Rico and find that even with import prohibition measures and assistance from the government, the introduction of the disease would result in loss of yields of 10%. In their partial equilibrium model for Australia, Cook et al. [26] show that under import restrictions expected damage to the banana industry due to Black Sigatoka would be around USD 40 million dollars, and would increase to USD 130 million, which is about a third of the gross annual value of production, if all quarantine restrictions were removed.

In this paper we, to the best of our knowledge for the first time, estimate the historical global impact of a major crop disease, using the case study of bananas and BSLD. To this end we first construct the history of first time infections across the globe. We then simulate the long distance wind dispersal and local diffusion of the disease using gridded ($\approx$50 km) climate data and known optimal conditions relevant to BSLD. These data allow us to empirically model the risk factors related to the spread of the disease across countries, as well as the impact on banana production once a country becomes infected. The fact that we model variations in the disease spread only based on optimal climatic conditions, while controlling for climatic factors in general, allows us to arguably identify true causal effects. Our results show that trade in agricultural products has played the main role in diffusion across countries, while long distance wind dispersal has only played a minor part. The analysis also demonstrates that once countries are infected, climatic conditions conducive to the local diffusion of BSLD can cause considerable losses, currently on average about a 2–3% reduction in global banana production.

2. Black Sigatoka Leaf Disease & Climate

BSLD is caused by the pathogens *Myscosphaerella fijiensis* in their sexual state, and by *Paracercospora fijiensis* in their asexual form. Infection can occur via both ascospores (sexual) and conida (asexual). The evolution of the disease occurs mainly from the top to the bottom of the banana plant, where aerospores first affect the stomata and then ultimately the leaves (de Bellaire et al. [21]). This can lead to the production of conida which further infects the tree. The first symptoms are reddish brown streaks which grow to form large darker lesions. This leaf spotting has two types of impacts on banana yield. Firstly, because it affects the photosynthetic area of the leaves and diminishes the leaf area, which consequently has a strong effect on bunch weight (Ramsey et al. [27]). Secondly, it reduces the greenlife, i.e., the time between harvest and climacteric rise, of harvested fruits from diseased plants, and thus the ability to export the fruit over long distances (de Bellaire et al. [21]).

Importantly spore germination of BSLD crucially depends on the micro-climate and it is this feature that is used in this study to model the aerial dispersion and local diffusion of the disease (de Bellaire et al. [21] and Bebber [28]). More specifically, in order to germinate and infect the leaf spores *Myscosphaerella fijiensis* requires very high relative humidity or a wet leaf surface, and once these conditions are present the rate of germination and infection will also depend on the temperature. In terms of the spread of the disease, both conida and ascospores seem to play a role, again subject to the right climatic conditions. For conida the principal agents of dispersal appear to be rain wash and rain splash. In contrast, while rainfall still plays a role in the release of ascospores, wind appears to be the primary carrier, where its speed and degree of turbulence appear to be important drivers (Marin et al. [17], Norros et al. [29]). It is common to distinguish between gradual local disease spread (LDS) and much rarer, stochastic long distance dispersal (LDD) (Golan and Pringle [30]).

3. Results

3.1. Descriptive Results

Figure 1 shows the global spatial distribution of areas suitable for banana production that is used as the basis of allocating climate and modelling disease spread for the analysis. As can be seen, these are located in the tropical and sub-tropical regions across Asia, Africa, the South Pacific and the Americas.

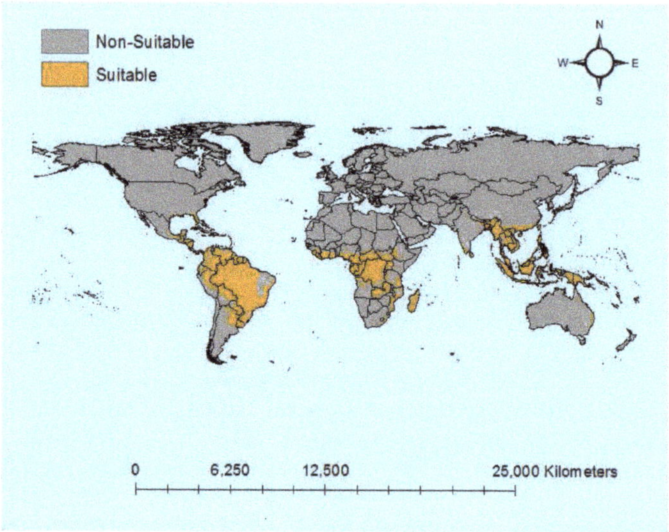

Figure 1. Bananas Growing Suitability Area. Note: This figure depicts banana suitability areas.

Next the percentage of banana producing countries infected with BSLD is depicted in Figure 2 over the sample period of the analysis (1961–2016). Accordingly, at the start BSLD was present in 4.4% of these, but this rose steadily to over 53% by 2016. The change in geographical distribution of this spread is shown in Figure 3. Accordingly, in 1961 it was essentially only in the United States (Hawaii) and small parts of Asia and the South Pacific that BSLD had been detected. By 1980 Black Sigatoka spread more widely across Asia, and began to additionally appear in Africa. As of 1999, it had further spread to the South American Continent and the Caribbean, as well as expanded more into Africa. At the end of the sample period the Caribbean and Africa had been further affected.

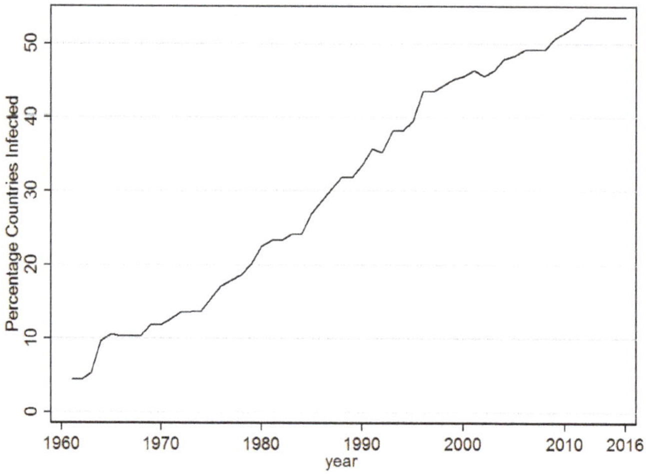

Figure 2. Black Sigatoka Detected-% Countries. Note: This figure shows the percent of banana producing countries that have been infected.

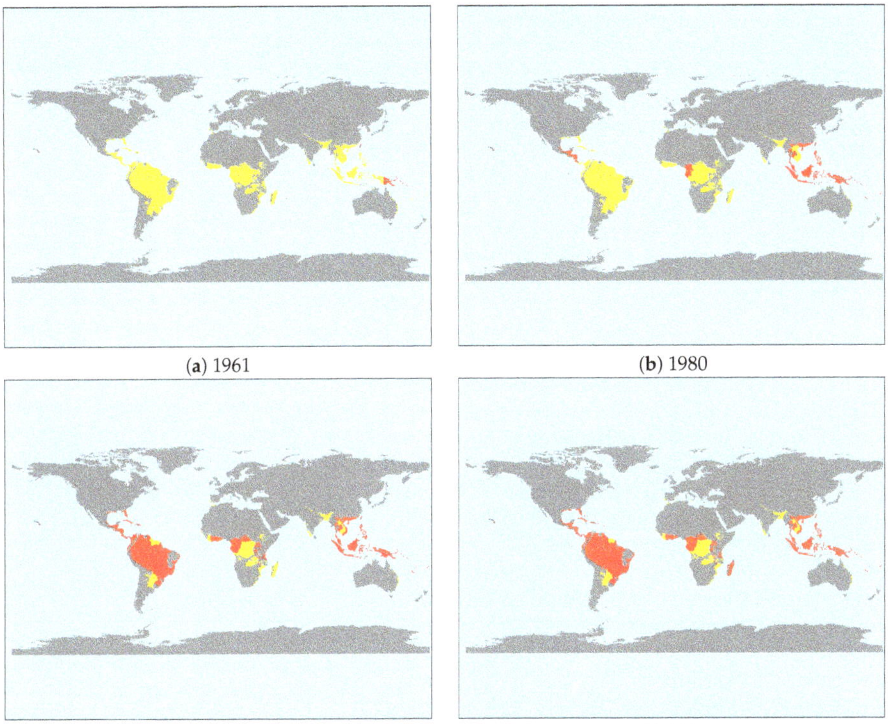

(**a**) 1961 (**b**) 1980 (**c**) 1999 (**d**) 2016

Figure 3. Temporal Spread of Black Sigatoka. Note: (i) This figures shows the distribution of countries where BSLD has been detected over time. (ii) Detected; Not Detected.

Table 1 provides summary statistics of all the variables used in the analysis. The average annual country level production of bananas is about 465,000 tons, but with considerable variation. This output is comes from harvested areas of on average 3000 hectares. In the data sample the mean first time BSLD infection is 31.7%. The climate optimal for disease diffusion (F) exists about 2.6% of the time, where this is slightly lower in countries that have already been infected. The potential infection rate through long distance dispersal is on average very small (4.09×10^{-6}), but with a large standard deviation. One may also want to note that for more than 48% of the time countries did not have the optimal amount of water available for banana production.

3.2. Regression Results

3.2.1. First Time Infection

For model assessment of Equation (10) the Cox-Snell are plotted against the cumulative Hazard in Figure 4. These are relatively close to the reference line and thus indicate a satisfactory fit.

The estimates from Equation (10) are given in the first column of Table 2. As can be seen, agricultural imports have a significant positive impact on a country becoming infected by BSLD. In contrast, banana imports play no play significant role. One also finds that the long distance dispersal of the fungus from other infected countries under the right climatic conditions is a positive risk factor in becoming infected. All other control variables do not constitute significant risk factors for first time infection. The results of additionally including $DWBSLD$ is included in the last column of Table 2. The coefficient on this variable is significantly negative and increases the estimated coefficient on LDD and $AIMP$.

Table 1. Summary statistics.

Variable	Definition	Mean	Std. Dev.
PROD	Production (tons)	464,628	1,621,986
HAREA	Area Harvested (Ha)	3,0087	79,202
BSLD	Detection Indicator	0.3169	0.4653
F	Disease Diffusion rate	0.0259	0.0526
F (BSLD = 1)	F once Infected	0.021	0.039
LDD	Long Distance Dispersal Probability	4.09×10^{-6}	0.0003
RAIN	Rainfall (mm/day)	4.0223	2.5112
EVAPO	Evapotranspiration (mm/day)	2.6141	1.1214
HUMID	Relative Humidity (%)	74.9114	11.7134
CMOIST	Moisture Storage on Canopy	2.2819	2.513
CTEMP	Canopy Temperature (°C)	24.0699	2.4552
WIND	Wind (m/s)	2.9039	1.7498
WSTRESS	% Days Soil Water Stressed	0.4804	0.3589
CTEMP8	% Days CTEMP $<$ 8 °C	0.0053	0.0188
CTEMP38	% Days CTEMP $>$ 38 °C	0.0001	0.0014
HUMID60	% Days HUMID $>$ 60%	0.166	0.219
WIND4	% Days WIND $>$ 4 m/s	0.403	0.3092
BIMPORT	Import of Bananas (tons)	24,006	113,401
AIMPORT	Import of Agr. Products (tons)	1,720,699	4,777,477
BSUIT	Area of Banana Suitability (Ha)	1691	4409
DWBS	Distance Weight. Detection	0.037	0.202
DWAEXP	Distance Weight. logged Agricultural Exports	0.717	2.544
DWBEXP	Distance Weight. logged Bananas Exports	0.406	1.785

Note: This table provides summary statistics for all variables used in the analysis.

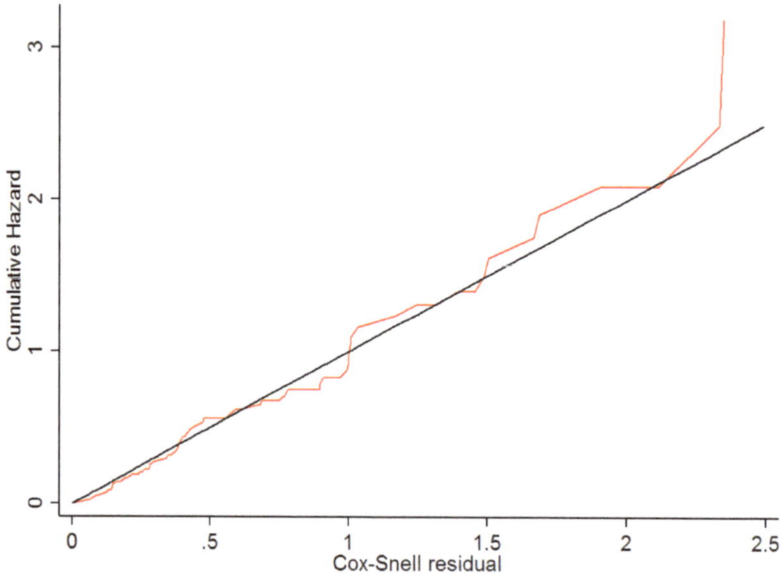

Figure 4. Cox-Snell Residuals Model Fit Assessment. Note: This figure provides the Cox-Snell Residual Model Fit Assessment of the Cox Proportional Hazard Model from Equation (10).

Converting the estimated coefficients of the Cox model in column 2 to hazard ratios for agricultural imports indicates that a standard deviation increase in $log(AIMP)$ raises the (relative) hazard of being infected by over 450%. When the probability of LDD is at its mean value, this would increase the (relative) risk by 18.3%.

The estimated coefficients of the logit model explicitly estimating the baseline hazard (t) are given in the last column of Table 2 provide the basis for conducting the counterfactual analysis of setting LDD and $AIMP$ alternatively to zero and comparing the predicted hazards to using their actual values. Noteworthy is that the estimated coefficients of this logit specification are very similar to the estimates of the Cox proportional hazard in terms of all common variables. The coefficients on t and t^2 suggest that the baseline hazard takes on an inverted u-shaped function, where it first increases, then after reaching an optimum decreases. Calculating marginal effects from the coefficients suggest a turning point of around 33 years.

Table 2. First-Time Infection.

	(1)	(2)	(3)
LDD	0.032 * (0.013)	0.0397 ** (0.012)	0.0411 ** (0.0116)
log(AIMP)	0.6039 ** (0.1666)	0.6735 ** (0.1684)	0.6328 ** (0.1554)
log(BIMP)	−0.0031 (0.0928)	−0.0275 (0.0991)	−0.0356 (0.0954)
DWBSLD		−1.4895 * (0.6779)	−1.0984 (0.7962)
FT	−0.2109 (2.5275)	−0.3030 (2.5928)	−01.0721 (2.6881)
RAIN	0.13 (0.1384)	0.1269 (0.1399)	0.1619 (0.1499)
EVAPO	−0.8477 (0.6737)	−0.8978 (0.6778)	−0.7071 (0.7295)
HUMID	0.0759 (0.1153)	0.0974 (0.1116)	0.0327 (0.1347)
CMOIST	−0.3582 (0.2181)	−0.4000 (0.216)	−0.3787 (0.2319)
CTEMP	0.1954 (0.1513)	0.2236 (0.1467)	0.199 (0.1605)
WIND	−0.8690 (1.0180)	−01.1943 (1.0125)	−0.9247 (0.9748)
WSTRESS	−4.9328 (2.8768)	−5.2234 (2.8953)	−4.3036 (3.1743)
CTEMP8	2.5629 (13.0638)	2.7856 (12.6117)	2.9621 (12.5182)
CTEMP38	−2968.0320 (5605.0180)	−2750.2310 (5690.3960)	−5314.0590 (8580.5970)
HUMID60	0.2426 (6.6567)	1.3038 (6.5067)	−1.8477 (8.0296)

Table 2. *Cont.*

	(1)	(2)	(3)
WIND4	2.9073 (5.8018)	4.7726 (5.8627)	3.4434 (5.6950)
log(HAREA)	0.0598 (0.1316)	0.0494 (0.1268)	0.0122 (0.1534)
log(BAREA)	−0.3020 (0.2067)	−0.3588 (0.2156)	−0.3106 (0.2671)
t			0.2261 ** (0.073)
t^2			−0.0034 ** (0.0011)
MODEL: Obs.	COX 4137	COX 4137	LOGIT 4137

Notes: (a) This table provides the estimates from the Cox Proportional Hazard model in Equation (10), as well as the logit model of survival; (b) * and ** indicate 1 and 5 per cent significant levels, respectively; (c) All regressions include 12 sub-regional indicator variables and measures of the logged area and the logged banana suitable areas. (d) Standard errors clustered by country in parentheses. (e) COX: Cox Proportional Hazard.

The predicted actual and counterfactual probability of hazard for an average banana producing country are shown in Figure 5. Accordingly, for *LDD* there is little difference in the actual and counterfactual average probability of being infected. As a matter of fact, by 2016 the actual probability is only about 0.8 per centage points higher than under the counterfactual of no long distance dispersion. The same counterfactual exercise but setting *AIMP* to zero is depicted in Figure 6. With no agricultural imports the average probability of being infected by 2016 is just a little over 2%, compared to 71% when *AIMP* takes on its observed values.

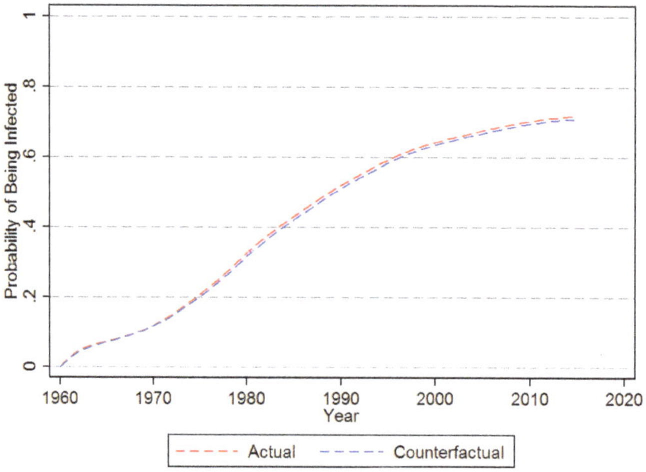

Figure 5. Probability of First Time Infection—No Long Distance Dispersal. Note: This figure provides the counter-factual prediction of the impact assuming of no LDD as estimated from the Cox Proporational Hazard Model in Equation (10).

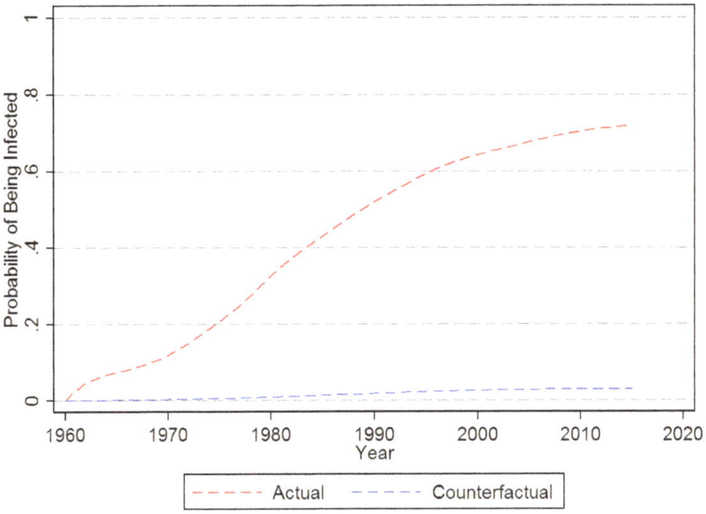

Figure 6. Probability of First Time Infection—No Agricultural Imports. Note: This figure provides the counter-factual prediction of the impact assuming no agricultural imports as estimated from the Cox Proporational Hazard Model in Equation (10).

3.2.2. Impact of Disease Diffusion on Banana Production

The results of estimating Equation (11), i.e, the impact on country level banana production once they become infected, are provided in Table 3. The R^2 statistic indicates that the model explains around 68% of the variation in banana production. The estimated coefficients show that having first reported BSLD in the past does not per say reduce banana production, as the coefficient on $BSLD$ on its own is not significant. Rather it is only its interaction with F that produces a significant (negative) predictor of banana production. Alternatively using (logged) harvested area rather than production as the dependent variable in Equation (11), as shown in the last column, produces the same findings, although the model fit is slightly lower (0.66%). In terms of the other control variables, only the number days that the temperature was below 8 degrees is a significant (negative) predictor of banana production across all three specifications.

Taking at face value the size of the coefficients in the fourth column of Table 3 suggests that when a country is already infected and the diffusion probability, i.e., $F(BSLD = 1)$, is at the mean of the sample (0.021), banana production falls by nearly 3.8%. In Figures 7 and 8 this estimated coefficient on $F \times BSLD$ was used to predict the counterfactual implied losses as a total in tons, and as a percentage of annual production, respectively, over the sample period. Accordingly, since the year 2000 average annual losses have been at least 2, and since 2010 close to 3 million tonnes. As a percentage of total potential productions this translates into annual losses of over 2% since 1998. If one takes the average level of F over the sample time period for each country and assumes that $BSLD = 1$, i.e., that all countries have been at least once infected, then annual expected losses would nearly double to about 4.2% of total global banana production.

Table 3. Banana Production.

	(1)	(2)
BS	0.076 (0.047)	0.057 (0.041)
F	−1.315 (1.089)	−0.732 (0.904)
F × BS	−1.846 ** (0.545)	−2.717 ** (0.611)
RAIN	−0.025 (0.016)	−0.012 (0.014)
EVAPO	−0.061 (0.148)	−0.005 (0.108)
HUMID	0.011 (0.009)	0.01 (0.008)
CMOIST	0.009 (0.019)	0.008 (0.017)
CTEMP	−0.051 (0.027)	−0.046 (0.025)
WIND	−0.005 (0.043)	0.042 (0.042)
WSTRESS	−0.082 (0.244)	0.101 (0.214)
CTEMP8	−4.916 ** (1.778)	−4.845 ** (1.458)
CTEMP38	−0.127 (4.859)	1.202 (3.754)
HUMID60	0.531 (0.431)	0.718 (0.37)
WIND4	−0.471 (0.409)	−0.488 (0.293)
Dep. Var:	PROD	BAREA
Obs.	6793	6793
R^2	0.677	0.66

Notes: (a) This table provides the estimates from the linear regression model in Equation (11), as well as the logit model of survival; (b) ** indicates 1 per cent significant levels, respectively; (c) All regressions yearly indicators as well as country specific time trends. (d) Driscoll and Kraay [31] standard errors allowing for cross-sectional and serial correlation in parentheses.

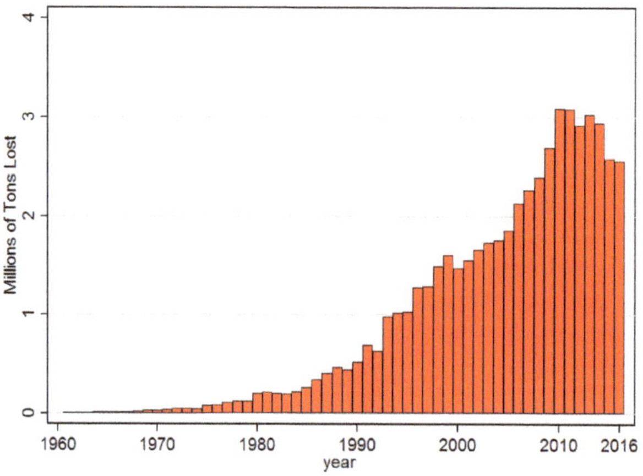

Figure 7. Potential Losses. Note: This figure provides the total predicted losses as estimated from the linear regression model in Equation (11).

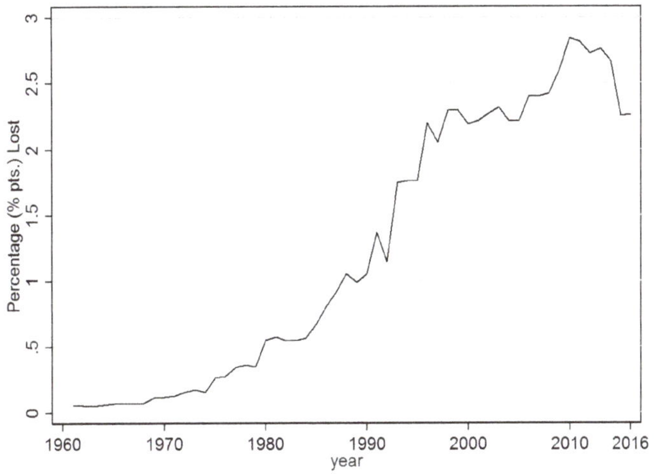

Figure 8. Potential Losses (%). Note: This figure provides the percentage of predicted losses as estimated from the linear regression model in Equation (11).

4. Discussion

Global agricultural losses of major crops due to crop disease pathogens are believed by experts to be considerable (Savary et al. [15]). As a matter of fact, rough estimates suggest that currently losses of major crops due to fungal diseases amount to enough to feed 8.5% of the global population (Fisher et al. [14]), and between 10 to 40% of global production (Savary et al. [15]). Yet there is no quantitative estimate of what risk factors increase their dispersion, nor what their impact is once regions become infected. Using the case study of Bananas and BSLD-the most important pathogen affecting the industry—this study models the risk factors associated with the historically observed global cross-country spread of the disease using hazard models, as well as the impact on country level banana production once a country is infected using panel linear regression models.

Our results showed that the main driver of first time infection was the import of agricultural products. While long distance diffusion based on climatic factors also played a role, it was small compared to the trade channel. One should note that these two findings support the current literature in that BSLD is likely to have spread over long distances rather than through slow local diffusion. For instance, such claims have not only been substantiated by the sequence of first reporting within and across continents (Jones et al. [23]), but also by genetic evidence. More specifically, Robert et al. [32] showed that the genetic drift between samples across countries and continents was large enough to serve as evidence that the introduction of BSLD in several locations had been over long distances. In other words, rather than a steady diffusion of an epidemic frontier, the evidence is consistent with stochastic spread of the disease [33].

Nevertheless, there is some disagreement as to the exact causes of such spread. Most studies would agree that human drivers, such as the importation of plant material into a country, are likely to have played the most important role- see, for instance, Robert et al. [32] and Burt [34]-and this is certainly supported by our result that the degree of agricultural imports into a country is quantitatively an important predictor of first time infection. However, there is still considerable disagreement if wind dispersal on its own could explain some of the stochastic introductions within and across continents. More specifically, it has been shown that some fungi can be spread over several thousands of kilometers by wind under the right conditions, as, for example, wheat leaf rust (Aylor [35]). In this regard, Burt [34] notes that under optimal conditions it would have taken about 37 days for BSLD to have been carried from Australia to the Caribbean and that perhaps the South Easterly trade winds might have brought

BSLD from Africa to the Caribbean. However, importantly *Myscosphaerella fijiensis* ascospores, unlike many others, are killed by ultraviolet radiation and are thus much less likely to survive being carried across long distance over long time periods by wind unless they happen to be protected by clouds (Parnell et al. [36]). As a matter of fact, some studies, such as Rivas et al. [33], have argued that for BSLD wind dispersal is likely to be limited to just a few hundred of kilometers and thus probably occurred within rather than across continents. The results of our study indicate that long distance dispersal through climatic factors might have still played a role, albeit only a very small one.

Local climate, apart from that incorporated in the long distance dispersal probability measures, was not a significant predictor of first time infection, again pointing to stochastic dispersal as the main cross-country spread channel. We also find that the import of bananas did not increase a country's likelihood of first time infection. This may not be surprising given that banana fruits themselves, unlike the leaves on their plant, are possibly not infected by the fungus (Robert et al. [32]). Leaves, typically used as packing material for other goods, are unlikely to be captured in the FAO data of banana imports, which strictly refers to imports of the fruit. Nevertheless, any more reliable conclusions in the role of banana imports would require bilateral trade data, so that we would need to be able to trace imports from infected countries. Such trade data would also allow us to more accurately estimate the role of agricultural imports in introducing the disease.

The fact that a distance weighted measure of first time infection of other countries (*DWBSLD*) had a negative impact on first time infection and that its inclusion as a control variable reduced the estimates on agricultural imports and LDD suggests that it is indeed possibly capturing the role that policy may play. More specifically, the presence of BSLD in neighboring countries may have induced uninfected nations to implement greater preventive measures and legislation. For instance, in many banana producing countries in the Caribbean agricultural quarantine precautions became fairly strict after the first outbreaks in the region and may have prevented further spread (Burt [34]). The possible benefits of such measures, even at the cost of reduced trade, should not be overlooked. Using an equilibrium displacement model for Puerto Rico, Alamo et al. [22] show that not having trade restrictions would cause a net loss in welfare. However, in actuality not only are there arguably inefficiencies in current legislation implementation (Perrings [12]), but for many crop fungi even strictly enforced physical borders may not be effective as these can still spread through the atmosphere over long distances (Brown and Hovmøller [13]), as evidence here also indicates. Nevertheless any accurate assessment on the efficacy of trade restrictions and legislation to reduce infection would require detailed historical construction of policies implemented across countries and time.

Our findings from the banana production model revealed that once a country is infected and the right climatic conditions prevail, losses in banana production due to the disease can be substantial. Currently these average about 2-3% of total production a year. If we take the average current producer price of bananas from available FAOSTAT data, i.e., USD 630 per tonne, then this would imply annual expected losses currently of about USD 1.6 billion. This would more than double if the remaining banana producing countries become infected. While this arguably demonstrates that the effects of BSLD on the banana industry are economically important, it is difficult compare this to other major crops since no other comparable global study exists, rather, as noted above, just qualitative estimates by experts (Savary et al. [15]).

Climatic conditions were found to be crucial in terms of the impact of BSLD on banana production. More specifically, even if BSLD has already been reported in a country, it is only when climate is optimal for diffusion that this will have an impact on aggregate banana production of a country. This result is echoed in Yonow et al. [37] who find that there is a strong relationship between climate suitability of BSLD and export ratings for disease pressure. In this regard, Bebber [28], on whose biophysical model our local disease diffusion framework is heavily based, showed that the risk of a disease outbreak has increased by a median of 44.2% in Latin America and the Caribbean since the 1960s. Moreover, using different climate dependent predictive factors of BSLD under two climate change scenarios, Júnior et al. [38] calculate that while the areas favourable to the disease will decrease, extensive areas will continue to be favourable to BSLD.

Surprisingly climate stress factors specific to bananas, as taken FAO [39], were not found to be significant predictors of country level production, except days when the temperature was below 8 °C. Assuming that most of them should be, this suggests that there is considerable error in their measurement, leading to attenuation bias and thus leading to Type II errors in our estimated coefficient hypothesis testing; see Wooldridge [40]. One obvious reason is that we only know the banana suitability areas and not actual banana growing areas, so that our climate proxies are not only capturing the relevant local climate variations. Another explanation is that since banana are potentially grown year round except for the sub-tropical regions, but we have no information as to the local growing cycles, this may again introduce some measurement error in our climate variables. Importantly, such measurement error in our climate variables would also have implications for our ability to estimate the impact of the probability of long distance and local disease diffusion, which are also based on climate factors, so that we may be underestimating the true impact these as well.

Although we were not able to investigate this specifically due to data availability, one way to control the spread of BSLD is through chemical control. In this regard, the high susceptibility of the Calvendish crop to the disease necessitates the use of both protectant and systemic fungicides at relatively high frequencies (Marin et al. [17], de Bellaire et al. [21]). However, the costs of such treatment are substantial, so much so that Cavendish cultivars are among the top global inputs of agricultural fungicides (Churchill [20]), and thus making this treatment option not feasible for many smaller banana producers. Moreover, the disease has shown over time to develop increased resistance to the treatment (Jones et al. [23]. For example, experience in Costa Rica has shown that within 20 years of use the amount of fungicide needed to control BSLD increased by around two thirds (de Bellaire et al. [21]).

5. Materials and Methods

5.1. Methods

Local Disease Spread (LDS)

The approach in this study follows Bebber [28] closely and employs a local infection diffusion model based on micro-climate to simulate the spread of BSLD once a country i is infected. Consider a set of localities $m = 1, ..., M$ in country i during days $d = 1, ..., D$. It is assumed that the diffusion rate of spores at locality m during day d, F_{imd}, follows a probabilistic survival process of spores transitioning to infections, which depends on the number of days, s_{imd}, that passed since the outbreak and temperature, T_{imd}. Moreover, local diffusion is dependent on the occurrence of a sufficiently wet period, $\mathbf{1}_{C_{imd}}$. Thus F_{imd} is determined by:

$$F_{imd} = (1 - e^{-H(s_{imd},T_{imd})}) \times \mathbf{1}_{C_{imd}} \tag{1}$$

where H is a Weibull hazard function such that:

$$H_{imd} = r(T_{imd}) \left(\frac{s_{imd}}{\alpha}\right)^{\gamma} \tag{2}$$

The temperature dependent rate r depends on T_{imd}'s value relative to given thresholds of minimum (T_{min}), optimum (T_{opt}), and maximum (T_{max}) temperatures:

$$r(T_{imd}) = \left(\frac{T_{max} - T_{imd}}{T_{max} - T_{opt}}\right) \left(\frac{T_{imd} - T_{min}}{T_{opt} - T_{min}}\right)^{\frac{T_{opt}-T_{min}}{T_{max}-T_{opt}}} \tag{3}$$

The incidence of a sufficiently wet period, $\mathbf{1}_{C_{imd}}$, is contingent on minimum wetness (WET^{thresh}) and humidity (RH^{thresh}) thresholds:

$$\mathbf{1}_{C_{imd}} = \begin{cases} 1 & (WET_{imd} > WET^{thresh}) \cup (RH_{imd} > RH^{thresh}) \\ 0 & (WET_{imd} \leq WET^{thresh}) \cap (RH_{imd} \leq RH^{thresh}) \end{cases} \tag{4}$$

As in Bebber [28], in order to parameterize Equations (1)–(4), $\alpha = 32.6$, $\beta = 37.6$, $T_{min} = 16.6$ °C, $T_{opt} = 27.2$ °C, $T_{max} = 30.3$ °C, $WET^{threshold} = 0$, and $RH^{threshold} = 98\%$. (These parameters were estimated by Bebber [28] using Brazilian data on Black Sigatoka and temperature) One should note when $\mathbf{1}_{C_{imd}} = 0$, s_{imd} is reset to 0 and will start to sequence again only once $\mathbf{1}_{C_{imd}} = 1$.

Since the empirical analysis is at the country level, LDS is measured as the country level annual average daily diffusion rate:

$$F_{it} = \frac{\sum_m^M \sum_d^D F_{imd}}{M \times D} \tag{5}$$

where D is the total amount of banana growing days in year t and M the total number of banana growing localities in country i.

5.2. Long Distance Dispersal (LDD)

To model long distance aerial dispersal of BSLD a simplified version of the LDD model of Aylor [35] is employed, where it is 'simple' in the sense that it does not model the time between release of spores in the source region and target region, but rather simply assumes that the transport happens within the same year. This is in part because only the first infection date by year is known and the banana production data is only annual, but also because climatic data to model the waiting time between release and potential infection is not available. Moreover, there is no data to model the potentially important impediment of ultra-violet radiation along any transport route, and this aspect is thus ignored.

It is assumed that the amount of possible spores released from location m on day d in source country j at time t, Q_{jt}, depends on the local diffusion rate F_{jmd}, as defined in Equation (5), and appropriate climate conditions conducive to aerospore release $\mathbf{1}_{A_{jmd}}$:

$$Q_{jt} = \frac{\sum_m^M \sum_d^D F_{jmd}}{M \times D} \times \mathbf{1}_{A_{jmd}} \tag{6}$$

$$\mathbf{1}_{A_{jmd}} = \begin{cases} 1 & (RAIN_{jmd} > 0) \cap (0.2 \leq TU_{jmd} \leq 0.5) \\ 0 & (RAIN_{jmd} = 0) \cup (TU_{jmd} \leq 0.2) \cup (0.5 \leq TU_{jmd}) \end{cases} \tag{7}$$

where $RAIN$ is the amount of daily local rainfall and TU the local wind turbulence measured as the ratio of the standard deviation relative to the mean wind speed. These optimum spore release climatic conditions were chosen since rainfall is necessary for spore release (Burt [34]), and Norros et al. [29] found that LDD of small spores, like those of *Myscosphaerella fijiensis*, increased within the wind turbulence range between 0.2 to 0.5 m/s.

Along the route to possible destinations it is assumed that there are constant and favourable transport conditions in that the rainfall rate is zero, there is no spores mortality due to ultraviolet radiation or other reasons. This is done since the necessary data, particularly over water bodies, is not available for most the time period of the analysis. The concentration of viable spores located at i at a distance x_{ij} downwind from source countries, $j = 1, ..., J; j \neq i$, can be described by a Gaussian puff as:

$$LDD_{it} = \sum_{j \neq i}^{J} Q_{jt} \times e^{-x_{ij}} \times RAINRATE_{it} \tag{8}$$

where the Equation (8) is made dependent on the $RAINRATE$, i.e., the amount of rainy days during the growing season in target country i at time t, since it is well known that the deposition velocity of spores is dependent on the rate of rainfall (Aylor [35]). In order to measures x_{ij} the shortest distance between banana growing areas between countries is used.

5.3. BSLD Presence

A country level indicator of BSLD presence, $BSLD_{it}$, takes on the value of zero until the first infection appears in country i in year t and then from a value of one. A distanced weighted measure of BSLD presence in neighboring countries $j = 1, ..., J; \&j \neq i$ for country i is generated as:

$$DWBSLD_{it} = \sum_{j \neq i}^{J} exp(-x_{ij}) \times BSLD_{jt} \tag{9}$$

where x_{ij} is defined as in Equation (8).

5.4. Empirical Modeling

5.4.1. First Time Infection Model

A country's hazard of becoming infected by BSLD for the first time is modelled using a Cox Proportional Hazards Model where time varying and time invariant covariates are allowed to influence a country's hazard function $h_i(t)$ from the disease free state as:

$$\begin{aligned} h_i(t) = h_0(t) exp\Big(& \alpha_{LDD} LDD_{it} + \alpha_{AIMP} log(AIMP_{it}) \\ & + \alpha_{BIMP} log(BIMP_{it}) + \alpha_{log(HAREA)} log(HAREA)_{it-1} \\ & + \alpha_{BSUIT} log(BSUIT_i) + \alpha_C C_{it} + \alpha_{CS} CS_{it} \Big) \end{aligned} \tag{10}$$

where $BSUIT$ is the total area suitable for banana growing, C is a vector of the disease spread optimal climatic conditions (rainfall, evapotranspiration, relative humidity, canopy moisture, canopy temperature, and wind speed), and the vector CS consists of productivity reducing climatic stress factor thresholds in terms of water availability, temperature, humidity, and wind. Additionally, twelve sub-regional dummy indicators are included to capture regional time invariant differences. In order to make the estimated coefficient on LDD more readable in the estimation output its value is multiplied by 10^6. In order to assess model fit we examined the Cox-Snell residuals, as suggested by Grambsch and Therneau [41].

One should note that the Cox Proportional Hazard model is semi-parametric in that the baseline hazard $h_0(t)$ is estimated non-parametrically and only depends on time t, but the risk factors are estimated parametrically, where this parametric function takes on an exponential form. The hazard of country i becoming infected is thus multiplicatively proportional to the baseline hazard h_0. The drawback of this non-parametric flexibility is that the hazard for any country is a fixed proportion of the hazard for any other country, and thus only the relative, and not the absolute, hazard of countries can be estimated. Thus, when the estimates from Equation (10) are used to make counterfactual predictions, as in Zheng and Cai [42] a logit model version of Equation (10) of infection is employed, where the baseline hazard $h_0(t)$ is specified as a linear and quadratic function of time elapsed. The estimated coefficients from this logit model are then used to construct counterfactual predictions by setting alternatively LDD or $AIMP$ equal to zero, but keeping all other controls at their observed values, and predicting the probability of a representative country being infected. This is then compared to doing the same prediction but using the observed values of all control variables.

5.4.2. Impact of Disease Diffusion Model

To determine how, once a country is infected, diffusion of BSLD has impacted banana production is estimated using the following regression equation:

$$\begin{aligned} log(BANANAS_{it}) &= \beta_0 + \beta_{BSLD}BSLD_{it} + \beta_F F_{it} \\ &+\beta_{F\times BSLD}F_{it} \times BSLD_{it} + \beta_C C_{it} + \beta_{CS}CS_{it} \\ &+\beta_{TR4}TR4_{it} + trend_{it} + \pi_t + \mu_i + \epsilon_{it} \end{aligned} \tag{11}$$

where $trend$ is a vector of country specific time trends, π is a vector of year specific indicator variables. In order to purge the vector of country specific effects μ from Equation (11) a panel fixed effects linear estimator is employed (Wooldridge [40]). To take account of serial correlation due to growing seasons spanning across calendar years and cross-sectional dependence we calculate standard errors as recommended by Driscoll and Kraay [31]. Model fit is assessed by examining the R-squared of the residuals, where a good fit is if these are close to the diagonal intersecting the origin. The estimated coefficients from Equation (11) are also used to predict what banana production would be if there was no local disease spread by setting the interaction term of $F_{it} \times BSLD$ to zero, but keeping all other control variables at their observed values, and calculating out predicted production. This is then comparing to the predicted banana production when all variables are at their observed values.

5.5. Data

5.5.1. BSLD Presence Data

To construct the history of the BSLD spread across the countries and time a number of sources were resorted to, most prominently Stover et al. [43], Pasberg-Gauhl et al. [44], Jones et al. [23], Jacome [45], Jones et al. [23], Rivas et al. [33], de Bellaire et al. [21], and Blomme et al. [46]. Additionally, the list of first infection reports provided by the European and Mediterranean Plant Protection Organization (EPPO) disease database and the CABI International's Invasive Species Compendium were consulted. For all banana producing countries that were not on these lists the internet was extensively searched for any information on first time outbreaks and the years of these, if any.

5.5.2. Climatic Data

To construct climatic variables the Japanese 55-year Reanalysis (JRA55) climate reanalysis data set from the National Center for Atmospheric Research (NCAR) Research Data Archive was used, which consists of griddded data with a spatial resolution of TL319 (about 55 km). Unfortunately there is no data set that allows one to know the exact location of banana growing areas within a country at any point in time. In order to nevertheless capture the climate likely to be relevant to local production of bananas, the local banana growing areas are proxied by the local areas suitable for banana production, as given in the FAO's Global Agro-Ecological Zones (GAEZ) database at the 0.5 degree resolution. This set of cells provides set of (potentially) banana growing localities $m = 1, ..., M$ and thus for which climate data from the JRA55 is extact. To this end 6-hourly data over the time period 00:00 h UTC on 1/1/1960 to 21:00 h UTC on 31 December 2016 on measures of canopy moisture (CMOIST), canopy temperature (CTEMP), relative humidity at 2m (HUMID), rainfall (RAIN), temperature (T), and the u- and v-wind components, which are used to calculate wind speed (WIND), were downloaded. Daily mean values were then calculated for those pixels that fell within the banana growing suitability areas, as derived from the GAEZ map described above. Since the banana growing season may be throughout the year for tropical areas, all daily values within a year for this region where used, while for the sub-tropical region (deliminated by being outside the -34° and 42° latitude zone) we restrict the climatic data to fall within the March to December window. While all the derived climatic variables are employed as potential climatic factors for banana production, canopy moisture and canopy temperature are specifically used to measure WET and T in order to construct F in Equation (1). The v- and w-wind component data to generate wind turbulence ($TURB$).

The climatic stress indicators that are likely to retard banana growth for inclusion in Equations (10) and (11) were also constructed, following FAO [39]. More specifically, these were the percentage of days in a year (or the 10 month growing period for sub-tropical regions) that mean wind speed was above 4 m/s (WIND4), canopy temperature was below 8 °C (CTEMP8), canopy temperature was above 38 °C (CTEMP38), and relative humidity was above 60% (HUMID60). In order to estimate the incidence of water stress, following Allen et al. [47], the daily soil water balance using the appropriate parameters for bananas, and calculated as the percentage of days that the water balance was below absolute optimum level (WSTRESS).

5.5.3. Banana and Agricultural Products Data

Data on banana production, area harvested, exports, and imports are taken from the FAOSTAT database. More specifically, these data provide annual country level banana production in tons ($BANANAS$), area harvested area in Ha ($HAREA$), exports in tons ($BEXP$), and imports in tons ($BIMP$) from 1961 to 2016 for all 129 major banana producing countries. Additionally information on total agricultural (plant based) imports ($AIMP$) and exports ($AEXP$) was used. Combining those countries that changed name over the period, and summing production and area harvested for those that split into several territories, left a total sample of 129 nations that produce bananas.

6. Conclusions

Our study of the global spread and impact of Bananas's Black Sigatoka Leaf Disease highlights the danger and potential cost of relying on just a few varieties with genetic uniformity for production of a specific crop on a global scale. In particular the results show that agricultural trade may play an important role in spreading the disease across countries. In this regard, while strict import restrictions and securities measures may be hypothetically able to prevent the influx of crop diseases across countries due to the transfer of diseased material by humans, there is still nevertheless the chance that the disease is transmitted over long distances from elsewhere under the right climatic conditions. Countries should thus plan for the likely arrival of a debilitating disease at some stage. Of course, once such a disease arrives chemical treatment may be able to keep it partially under control, although costs of such treatments may be prohibitive for some farmers and their effectiveness is likely to fade over time. Hence, international efforts to look for disease resistant crop varieties, such as FAO's Technical Cooperation program (The Technical Cooperation Programme (TCP) was created to enable FAO to make its know-how and technical expertise available to member countries upon request), should be continuously supported. Nevertheless, if anything, the history of banana crop diseases has shown that, while disease resistant varieties or treatments are eventually likely to be discovered or developed, new or mutations of existing fungi also continuously emerge, thus potentially restarting the vicious circle.

Author Contributions: Conceptualization, E.S.; methodology, E.S.; investigation, E.S. and P.M.; data curation, E.S. and P.M.; writing—original draft preparation, E.S. and P.M.; writing—review and editing, E.S. and P.M.; visualization, E.S. All authors have read and agreed to the published version of the manuscript.

Funding: This research received no external funding.

Conflicts of Interest: The authors declare no conflict of interest.

References

1. Agrios, G.N. *Plant Pathology*; Academic Press: New York NY, USA, 2005.
2. Scheffer, R.P. *The Nature of Disease in Plants*; Cambridge University Press: Cambridge, UK, 1997.
3. Gráda, C.Ó.; O'Rourke, K.H. Migration as disaster relief: Lessons from the Great Irish Famine. *Eur. Rev. Econ. Hist.* **1997**, *1*, 3–25. [CrossRef]
4. Simms, C. The grape depression. *New Sci.* **2017**, *236*, 60–62. [CrossRef]
5. Money, N.P. *The Triumph of the Fungi: A Rotten History*; Oxford University Press: Oxford, UK, 2006.
6. Hunter, D.; Guarino, L.; Spillane, C.; McKeown, P.C. *Routledge Handbook of Agricultural Biodiversity*; Taylor & Francis: New York, NY, USA, 2017.
7. Wolfe, M. The current status and prospects of multiline cultivars and variety mixtures for disease resistance. *Annu. Rev. Phytopathol.* **1985**, *23*, 251–273. [CrossRef]
8. Garrett, K.; Mundt, C. Epidemiology in mixed host populations. *Phytopathology* **1999**, *89*, 984–990. [CrossRef]
9. Gergerich, R.C.; Welliver, R.A.; Gettys, S.; Osterbauer, N.K.; Kamenidou, S.; Martin, R.R.; Golino, D.A.; Eastwell, K.; Fuchs, M.; Vidalakis, G.; et al. Safeguarding fruit crops in the age of agricultural globalization. *Plant Dis.* **2015**, *99*, 176–187. [CrossRef]
10. Lucas, J.A.; Hawkins, N.J.; Fraaije, B.A. The evolution of fungicide resistance. In *Advances in Applied Microbiology*; Elsevier: Amsterdam, The Netherlands, 2015; Volume 90, pp. 29–92.
11. Oerke, E.C. Crop losses to pests. *J. Agric. Sci.* **2006**, *144*, 31–43. [CrossRef]
12. Perrings, C. Options for managing the infectious animal and plant disease risks of international trade. *Food Secur.* **2016**, *8*, 27–35. [CrossRef]
13. Brown, J.K.; Hovmøller, M.S. Aerial dispersal of pathogens on the global and continental scales and its impact on plant disease. *Science* **2002**, *297*, 537–541. [CrossRef]
14. Fisher, M.C.; Henk, D.A.; Briggs, C.J.; Brownstein, J.S.; Madoff, L.C.; McCraw, S.L.; Gurr, S.J. Emerging fungal threats to animal, plant and ecosystem health. *Nature* **2012**, *484*, 186. [CrossRef]
15. Savary, S.; Willocquet, L.; Pethybridge, S.J.; Esker, P.; McRoberts, N.; Nelson, A. The global burden of pathogens and pests on major food crops. *Nat. Ecol. Evol.* **2019**, *3*, 430. [CrossRef]
16. HE, D.C.; ZHAN, J.S.; XIE, L.H. Problems, challenges and future of plant disease management: From an ecological point of view. *J. Integr. Agric.* **2016**, *15*, 705–715. [CrossRef]
17. Marin, D.H.; Romero, R.A.; Guzmán, M.; Sutton, T.B. Black Sigatoka: An increasing threat to banana cultivation. *Plant Dis.* **2003**, *87*, 208–222. [CrossRef] [PubMed]
18. Abbott, R. *A Socio-Economic History of the International Banana Trade, 1870–1930*; European Union University: Fiesole, Italy, 2009.
19. Koeppel, D. *Banana: The Fate of the Fruit that Changed the World*; Penguin: Harmondsworth, UK, 2007.
20. Churchill, A.C. Mycosphaerella fijiensis, the black leaf streak pathogen of banana: Progress towards understanding pathogen biology and detection, disease development, and the challenges of control. *Mol. Plant Pathol.* **2011**, *12*, 307–328. [CrossRef] [PubMed]
21. de Bellaire, L.d.L.; Fouré, E.; Abadie, C.; Carlier, J. Black Leaf Streak Disease is challenging the banana industry. *Fruits* **2010**, *65*, 327–342. [CrossRef]
22. Alamo, C.; Evans, E.; Bruguer as, A.; Nalampang, S. Economic impact and trade implications of the introduction of Black Sigatoka (Mycosphaerella fijiensis) into Puerto Rico. *J. Agric. Appl. Econ.* **2007**, *39*, 5–17. [CrossRef]
23. Jones, D. The distribution and importance of the Mycosphaerella leaf spot diseases of banana. In *Mycosphaerella Leaf spot Diseases of Bananas: Present Status and Outlook*; INIBAP: San José, Costa Rica, 2003; pp. 25–41.
24. Edmeades, S.; Phaneuf, D.J.; Smale, M.; Renkow, M. Modelling the crop variety demand of semi-subsistence households: bananas in Uganda. *J. Agric. Econ.* **2008**, *59*, 329–349. [CrossRef]
25. Kenneth, A.; Gerald, O.; Edilegnaw, W.; Wilberforce, T. *Ex-Ante Adoption of New Cooking Banana (Matooke) Hybrids in Uganda Based on Farmers' Perceptions*; Technical Report; International Association of Agricultural Economists: Foz Do Iguacu, Brazil, 2012.
26. Cook, D.C.; Liu, S.; Edwards, J.; Villalta, O.N.; Aurambout, J.P.; Kriticos, D.J.; Drenth, A.; De Barro, P.J. Predicted economic impact of black Sigatoka on the Australian banana industry. *Crop. Prot.* **2013**, *51*, 48–56. [CrossRef]

27. Ramsey, M.; Daniells, J.; Anderson, V. Effects of Sigatoka leaf spot (Mycosphaerella musicola Leach) on fruit yields, field ripening and greenlife of bananas in North Queensland. *Sci. Hortic.* **1990**, *41*, 305–313. [CrossRef]
28. Bebber, D.P. Climate change effects on Black Sigatoka disease of banana. *Philos. Trans. R. Soc. B* **2019**, *374*, 20180269. [CrossRef]
29. Norros, V.; Rannik, Ü.; Hussein, T.; Petäjä, T.; Vesala, T.; Ovaskainen, O. Do small spores disperse further than large spores? *Ecology* **2014**, *95*, 1612–1621. [CrossRef]
30. Golan, J.J.; Pringle, A. Long-distance dispersal of fungi. *Microbiol. Spectr.* **2017**, *5*. [CrossRef]
31. Driscoll, J.C.; Kraay, A.C. Consistent covariance matrix estimation with spatially dependent panel data. *Rev. Econ. Stat.* **1998**, *80*, 549–560. [CrossRef]
32. Robert, S.; Ravigné, V.; Zapater, M.F.; Abadie, C.; Carlier, J. Contrasting introduction scenarios among continents in the worldwide invasion of the banana fungal pathogen Mycosphaerella fijiensis. *Mol. Ecol.* **2012**, *21*, 1098–1114. [CrossRef] [PubMed]
33. Rivas, G.G.; Zapater, M.F.; Abadie, C.; Carlier, J. Founder effects and stochastic dispersal at the continental scale of the fungal pathogen of bananas Mycosphaerella fijiensis. *Mol. Ecol.* **2004**, *13*, 471–482. [CrossRef] [PubMed]
34. Burt, P.J.A. Windborne dispersal of Sigatoka leaf spot pathogens. *Grana* **1994**, *33*, 108–111. [CrossRef]
35. Aylor, D.E. Spread of plant disease on a continental scale: Role of aerial dispersal of pathogens. *Ecology* **2003**, *84*, 1989–1997. [CrossRef]
36. Parnell, M.; Burt, P.J.; Wilson, K. The influence of exposure to ultraviolet radiation in simulated sunlight on ascospores causing Black Sigatoka disease of banana and plantain. *Int. J. Biometeorol.* **1998**, *42*, 22–27. [CrossRef]
37. Yonow, T.; Ramirez-Villegas, J.; Abadie, C.; Darnell, R.E.; Ota, N.; Kriticos, D.J. Black Sigatoka in bananas: Ecoclimatic suitability and disease pressure assessments. *PLoS ONE* **2019**, *14*, e0220601. [CrossRef]
38. Júnior, J.; Valadares Júnior, R.; Cecílio, R.A.; Moraes, W.B.; Vale, F.X.R.d.; Alves, F.R.; Paul, P.A. Worldwide geographical distribution of Black Sigatoka for banana: Predictions based on climate change models. *Sci. Agric.* **2008**, *65*, 40–53. [CrossRef]
39. FAO. 2019. Available online: http://www.fao.org/3/y5102e/y5102e04.htm (accessed on 22 March 2020).
40. Wooldridge, J.M. *Econometric Analysis of Cross Section and Panel Data*; MIT Press: Cambridge, MA, USA, 2002.
41. Grambsch, P.M.; Therneau, T.M. Proportional hazards tests and diagnostics based on weighted residuals. *Biometrika* **1994**, *81*, 515–526. [CrossRef]
42. Zheng, Y.; Cai, T. Augmented estimation for t-year survival with censored regression models. *Biometrics* **2017**, *73*, 1169–1178. [CrossRef] [PubMed]
43. Stover, R. Distribution and probable origin of Mycosphaerella fijiensis in southeast Asia. *Trop. Agric. Trinidad Tobago* **1978**, *55*, 65–68.
44. Pasberg-Gauhl, C.; Gauhl, F.; Jones, D. Black leaf streak: Distribution and economic importance. *Dis. Banan. Abaca Enset* **2000**, 37–44.
45. Jacome, L. *Mycosphaerella Leaf Spot Diseases of Bananas: Present Status and Outlook*; Bioversity International: Rome, Italy, 2003.
46. Blomme, G.; Ploetz, R.; Jones, D.; De Langhe, E.; Price, N.; Gold, C.; Geering, A.; Viljoen, A.; Karamura, D.; Pillay, M.; et al. A historical overview of the appearance and spread of Musa pests and pathogens on the African continent: Highlighting the importance of clean Musa planting materials and quarantine measures. *Ann. Appl. Biol.* **2013**, *162*, 4–26. [CrossRef]
47. Allen, R.G.; Pereira, L.S.; Raes, D.; Smith, M. Crop evapotranspiration-Guidelines for computing crop water requirements-FAO Irrigation and drainage paper 56. *Fao Rome* **1998**, *300*, D05109.

atmosphere

Article

Climate Change and the Future Heat Stress Challenges among Smallholder Farmers in East Africa

Genesis Tambang Yengoh [1,*] and Jonas Ardö [2]

[1] Lund University Centre for Sustainability Studies—LUCSUS, Lund University, Biskopsgatan 5, SE 223 62 Lund, Sweden

[2] Department of Physical Geography and Ecosystem Science, Lund University, Sölvegatan 12, SE 223 62 Lund, Sweden; jonas.ardo@nateko.lu.se

* Correspondence: yengoh.genesis@lucsus.lu.se; Tel.: +46-46-222-0690

Received: 26 May 2020; Accepted: 13 July 2020; Published: 16 July 2020

Abstract: Agricultural production in sub-Saharan Africa remains dependent on high inputs of human labor, a situation associated with direct exposure to daylight heat during critical periods of the agricultural calendar. We ask the question: *how is the Wet-Bulb Globe Temperature (WBGT) going to be distributed in the future, and how will this affect the ability of smallholder farmers to perform agricultural activities?* Data from general circulation models are used to estimate the distribution of WBGT in 2000, 2050 and 2100, and for high activity periods in the agricultural calendar. The distribution of WBGT is divided into recommended maximum WBGT exposure levels (°C) at different work intensities, and rest/work ratios for an average acclimatized worker wearing light clothing (ISO, 18). High WBGTs are observed during the two periods of the East African. In February to March, eastern and coastal regions of Kenya and Tanzania witness high WBGT values—some necessitating up to 75% rest/hour work intensities in 2050 and 2100. In August to September, eastern and northern Kenya and north and central Uganda are vulnerable to high WBGT values. Designing policies to address this key challenge is a critical element in adaptation methods to address the impact of climate change.

Keywords: climate change; farm work; heat stress; WBGT; mitigation; East Africa

1. Introduction

Climate change is already adversely affecting the health of populations around the world, with the greatest impacts in low-income countries [1,2]. As a result of climate change, mean annual temperatures and the intensity and frequency of heat waves are expected to increase [3]. An increase in average temperatures, as well as the frequency, duration, and intensity of heat waves, has already been reported in some regions, with significant adverse impacts on local economies, agriculture, water resources and public health [1,4–6]. Scenario-based projections forecast that average global surface temperatures will increase by 1.4 to 5.8 °C from 1990 to 2100 [7]. This is bound to have substantial implications for human health, with the potential of contributing to an increase in future heat-related morbidity and mortality [1,8–10]. This study sought posits that Wet Bulb Globe Temperature (WBGT) (the concept of Wet Bulb Globe Temperature (WBGT) is further defined and operationalized in the 'Methods' section of this paper) in East African croplands in 2000 remains unchanged in 2050 and in 2100. The heat-related human health impacts of climate change are therefore expected to become more widespread and profound in the future. The future health impacts of climate change will vary spatially and temporally and will depend on changing socioeconomic and environmental conditions, as well as the preparedness of communities and health systems to avoid preventable health outcomes. Populations that are particular vulnerable to heat-related conditions include the elderly, children, the chronically ill, the socially isolated, and at-risk occupational groups. Some experts see climate change as "the biggest global health threat of the 21st century" [11].

Research on the impact of climate change on agriculture in Africa has expanded in some fields. These include studies on the impact of severe events associated with a changing climate, on the implications of changes in precipitation amounts and frequencies on cropping cycles, yields, and production [12–15]. Efforts have also been made to understand the implication of rising sea levels on agriculture in coastal regions, the role of climate change in the distribution of agricultural pests, and outcomes for animal production, including fisheries [16–19]. Changes in production systems and the spatial distribution of resources that support agricultural production are bound to have implications on social and economic systems dependent on or supported by agriculture [20]. This dimension of climate change impacts has also been widely investigated.

Data on the human element of climate change impacts, especially regarding labor for agricultural production, remain scarce, and our understanding of implications of the impact of climate change on the ability to work in labor-driven economic systems (such as agriculture) remains scarce. Initiatives have been undertaken to understand the implications of climate change on increasing heat impacts on labor productivity [2,21,22]. Kjellström et al. focused on estimating populations exposed to heat stress resulting from climate change [1]. Others have assessed human productivity under conditions of heat stress [6], and compared heat stress and its impacts on the health of workers from different occupational sectors [10].

The potential health impacts of climate change can be relieved through a combination of strategies, including strengthening key health system functions and improving the management of associated risks. To achieve robust health systems and manage human health risks, there is a need for a better understanding of the geography and scale of its potential impacts, especially impacts associated with the most common economic activities of populations.

In this study, we seek to examine whether environmental determinants of heat stress in the croplands of East Africa vary between the periods 2000–2100. East Africa is a diverse environment, with elevations rising from sea level to 5825 m above sea level, and a variety of agroecological zones (Figure 1). This study focuses on three countries of this region: Kenya, Tanzania, and Uganda. Together, these three countries comprise a total land area of over 1.6 million Km^2—Kenya (569,140 Km^2); Uganda (200,523 Km^2); and Tanzania (885,800 Km^2). To put in perspective the combined land area of the study area is larger than that of western Europe at 1.4 million Km^2 [23].

This study sought to assess the hypothesis that Wet Bulb Globe Temperature (WBGT) (the concept of Wet Bulb Globe Temperature (WBGT) is further defined and operationalized in the 'Methods' section of this paper) in East African croplands in 2000 remains unchanged in 2050 and in 2100. This study therefore contributes to assessing the geographical distribution of heat stress. We attempt to associate the geographical distribution of this challenge with the most important economic activity in East Africa, i.e., agriculture. By so doing, we examine the spatial distribution of heat stress, as a current and future challenge to agricultural productivity in East Africa. This initiative constitutes to strengthening the case for planning and investment in health protection within the context of climate change challenges.

1.1. Heat Waves and Heat Stress

Until recently, the severity of heat waves has been largely ignored. The Intergovernmental Panel on Climate Change (IPCC) now emphasizes the risk of drastically increasing incidences of heat waves with severe consequences for human health, livelihoods, agriculture, ecosystems, and societies at large [24]. The IPCC special report on extreme events, Risks of Extreme Events and Disasters to Advance Climate Change Adaptation (SREX), concluded that the length or number of heat waves have increased in many parts of the world and will virtually certainly increase further in the 21st century. Under a medium warming scenario, Coumou, Robinson [25] used a global $2 \times 2°$ grid with roughly 12,500 grid points with monthly data and predicted the number of monthly heat records to be over 12 times more common by the 2040s. This equates, on average, to roughly one record-breaking warm month per year in the tropics, including East Africa. A recent report estimated that mortality in Europe due to heat waves would increase by over 5550% by 2100 in a medium warming scenario [26]. If the global mean temperature increases to +7 °C or more, the habitability of large parts of the tropics and mid-latitudes will be at risk [27].

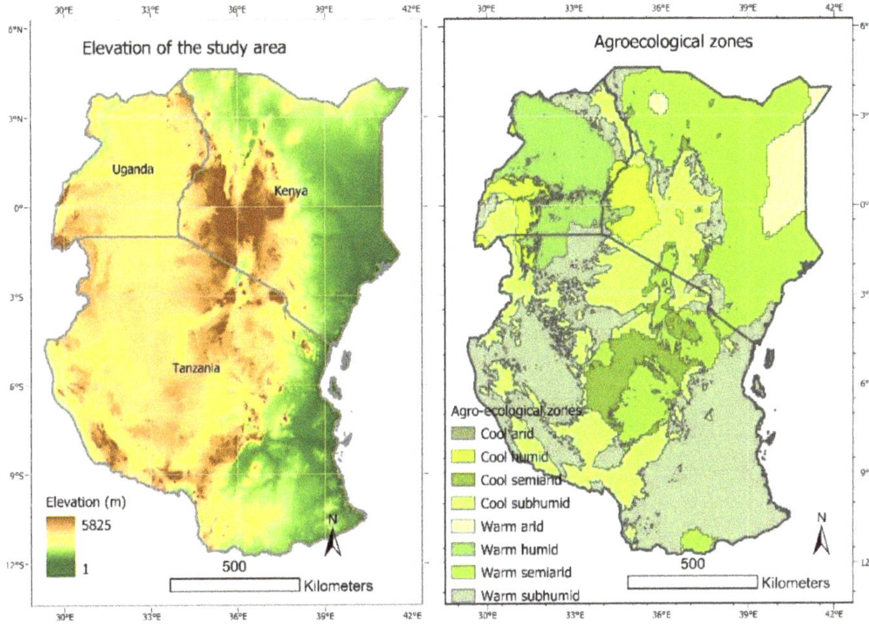

Figure 1. Location of the countries studied in East Africa, Kenya, Tanzania, and Uganda—showing the varied altitudinal ranges as well as the diversity of agro-ecological zones.

1.2. Impacts of Heat on Human Health

Excessive heat exposure affects natural and human systems, directly and indirectly, often resulting in severe losses of lives, assets and resources, and even social unrest, and may trigger tipping points in both natural and social systems [1,6,22,28,29]. Even if knowledge of climate change impacts on health has increased markedly in the last decade, research has mainly focused on direct physical rather than indirect mental health impacts. The physical impacts include: (1) mortality and morbidity from extreme weather events; (2) physical health impacts of extreme heat stress and heat waves; an increased intensity and spread of vector-borne disease; (3) effects of air quality on respiratory disorders; climate-induced changes in food and water quality and availability; (4) impacts on hygiene as effects of changed access to water [30]. When subject to heat stress, our cognitive ability is affected. These cognitive tasks may involve decision making, problem solving, memory, attention and judgement. Numerous studies have investigated the impact of heat stress on cognition-related outcomes [31], including performance responses and protective strategies [32], and even its outcomes on labor productivity and the economic implications [33]. Table 1 shows the conditions and symptoms of some common heat-related illnesses. Individuals show varying adaptive and maladaptive responses [34]. Heat stroke is another serious direct impact of heat stress. The role of contributing factors and the pathway to heat stroke from occupational heat stress has been mapped by Seichi Horie [35]. A combination of strenuous physical activity, a hot and humid atmosphere, continuous work without rest, and the use of inappropriate clothing for the heat environment such as some protective clothes and masks as the main contributing factors to elevated body temperatures. This can lead to a reduction of sodium concentration in the blood and dehydration. Elevated body temperature can also cause circulatory disturbance from cutaneous vasodilation, as well as cause elevated brain temperatures. These together have the potential of contributing to heat stroke characterized by symptoms such as heat cramps, muscle weakness, nausea, oliguria, fainting, headaches and high body temperatures [35]. Poverty may act as a barrier preventing adaptive behavior if poor people dependent on outdoor hard physical work consider resting. In a review of barriers to climate change adaptation among natural resource-dependent communities and livelihoods, financial constraints on agricultural production and rural development was identified as a major factor [36]. The inability to afford alternative means of production limits the potential for smallholder communities to easily change from current production practices to practices that limit their exposure to high WBGT.

Table 1. Some heat related illnesses (condition and symptoms)—modified from the Centres for Disease Control (CDC).

Condition	Symptoms
Heat stroke ▪ The body's response to loss of water and salt from heavy sweating.	• High body temperature (103°F or higher) • Hot, red, dry, or damp skin • Fast, strong pulse • Headache • Dizziness • Nausea • Confusion • Losing consciousness (passing out)
Heat exhaustion ▪ Develops when a person is working or exercising in hot weather and does not drink enough liquids to replace those lost liquids.	• Heavy sweating • Cold, pale, and clammy skin • Fast, weak pulse • Nausea or vomiting • Muscle cramps • Tiredness or weakness • Dizziness • Headache • Fainting (passing out)
Heat cramps ▪ Caused by the loss of body salts and fluid during sweating. Low salt levels in muscles cause painful cramps.	• Heavy sweating during intense exercise • Muscle pain or spasms
Sunburn ▪ A painful sign of skin damage from spending too much time outdoors without wearing a protective sunscreen	• Painful, red, and warm skin • Blisters on the skin
Heat rash ▪ Caused by sweat that does not evaporate from the skin.	• Red clusters of small blisters that look like pimples on the skin (usually on the neck, chest, groin, or in elbow creases)
Heatstroke (also known as sunstroke) ▪ Occurs when the body fails to regulate its own temperature and body temperature continues to rise, often to or above 40.6 °C.	• Unconsciousness for longer than a few seconds. • Confusion, severe restlessness, or anxiety. • Convulsion (seizure). • Symptoms of moderate to severe difficulty breathing. • Fast heart rate. • Sweating that may be heavy or may have stopped. • Skin that may be red, hot, and dry, even in the armpits. • Nausea and vomiting.

1.3. *Impacts of Heat on Society*

The economic effects of heat stress are huge, primarily regarding lower labor productivity [2], higher demands on health care, and increasing welfare costs. In all countries, heat stress is associated with social consequences, such as increasing violence, emotional problems and low life satisfaction including various secondary social stressors [24]. Impacts are highly differential with disproportionate burdens on people (often women and children) who toil daily under a scorching sun [24]. Small-scale farmers and fisher folks are particularly exposed to high temperatures and are already reporting excessive heat as a major burden [24]. From a livelihood perspective, food production is a particular concern. Since around 2010, knowledge about the sensitivity of crops to extreme heat has increased substantially, showing that heat stress is a major reason for productivity declines and crop failures, particularly in the tropics. For tropical systems where moisture availability or extreme heat will limit the length of the growing season, it is likely that the growing season and overall suitability for crops will decline due to heat stress. By 2050, most African countries will experience temperatures—over at least half of their current crop area—that lie outside the currently experienced range. As a further threat to small-scale farming, heat stress affects the health and productivity of livestock in meat and dairy production. As the most predictable, widespread and severe climate change impact on human societies, heat stress is already affecting, directly and indirectly, millions of people every year—and that the situation will get increasingly worse is 'virtually certain' according to the IPCC [24].

2. Smallholder Farmers and the Climate Change Context

The definition of smallholder farmers varies across geographies, agroecological zones, and even contexts. More holistic definitions incorporate elements of farm size, education level, knowledge of farming practices, land tenure situation, household demographics, and farming assets (which include access to financial resources, and technologies). Generally, however, smallholders tend to be defined based on the size of their farm holdings and levels of technology integration in the practice of agriculture. In sub-Saharan Africa, farmers operating less than 2 hectares of cropland are commonly categorized as smallholders [37,38]. Data on the distribution of farmland sizes and factors associated have been analyzed in a recent FAO/UNCTAD study [38]. In Ethiopia and Egypt, farms with an area of 2 hectares or less constitute nearly 90% of all farms. While in Kenya, if the classification of smallholders is taken to be those farmers with farmlands of 2 hectares or less is applied, it will cover nearly the entire land. In Tanzania, it will account for nearly 80% of all farms [38]. The focus on smallholder farmers is important because this group of agricultural producers makes up the vast majority of actors in the food production sector in the developing world [39]. An estimated two-thirds of the developing world's 3 billion rural people live in about 475 million smallholder farming households, working on land plots smaller than 2 hectares. Kenyan and Tanzanian smallholder farmers produce 63% and 69% of the food in the country, respectively [38].

Land management systems of smallholder agriculturalists rely heavily on human (often, family) labor [38]. Manual labor is a key feature of activities such as farm clearing, tillage, planting, weeding, harvesting, as well as traditional processing or farm products. The use of machinery or other labor-saving technologies is minimal [40]. Smallholder agriculture is therefore characterized by high labor inputs [41].

Agricultural practices among smallholder farmers in most of sub-Saharan Africa puts most agriculturalists under the category of outdoor workers. Outdoor workers refer to any workers who spend a substantial portion of the shift outdoors. For these groups of workers, their sources of heat exposure and potential for overheating can come in two ways:

1. The environmental conditions in which they work: most of which are already very warm in many parts of sub-Saharan Africa.
2. The internal heat generated by physical labor: smallholder farming practices depend heavily on manual labor for many strenuous farming activities, such as farm preparation, planting, weeding, harvesting, etc. Levels of technology use in agriculture remain very low.

Heat-related illnesses occur when the body is not able to lose enough heat to balance the heat generated by physical work and external heat sources. Weather conditions are the primary external heat sources for outdoor workers. Smallholder agricultural workers in sub-Saharan Africa are therefore at risk of heat-related illness when the heat index (WBGT) is high.

Four main on-farm activities substantially expose smallholder farmers in our study region to outdoor heat stress. These include (1) farm preparation, which involves clearing, tilling, and the preparation of planting surfaces (mounds or ridges); (2) planting; (3) farm maintenance (weeding); and (4) harvesting [42]. The maize production calendar is taken as a crop of choice for this study because of the importance of maize in the agricultural and food system in the East African region. The maize-mix farming system is the most important food production system in Eastern Africa [38].

2.1. Farm Preparation

Farm preparation is a dominantly manual activity for smallholder farmers in sub-Saharan Africa. In 2006, for example, human muscle power accounted for 65% of the energy used for land preparation in sub-Saharan Africa [43]. This is compared to 40% in East Asia, 30% in South Asia, and 25% in Latin America and the Caribbean [43]. Manual clearing involves the shearing of trees and bush vegetation with a cutting blade for new or fallowed land. In continuously cultivated land, it involves cutting of grasses that have colonized the farmland since the last cropping season. In some cases, the burning of vegetation is used as a means of clearing the land [44]. Tilling refers to turning the soil over so some of the lower soil comes up and some of the upper soil goes down aerating is and, in some cases, burying plant material that will eventually decompose in the process. Manual tilling is still a dominant agricultural practice among most smallholder farmers in sub-Saharan Africa [45]. A hoe with a large blade is the universal tool for tilling, even though there are variations in the design of this tool preferred in different agroecological zones [45]. Tilling and planting are done at the same time. Preparation of mounds or ridges is sometimes done during the tilling process. In more fertile farmlands, farmers may decide to eliminate the use of mounds and ridges altogether. The preparation of the planting surface (mounds and ridges) has been a well-established practice in the agricultural history of East Africa [46,47]. C.G. Knight observed the system of "nkule" in Tanzania in 1980 [48]. Planted crops take advantage of the take advantage of the nutrients provided by decomposing grass under the mounds and released by the process of burning [48]. In Zambia, these ridging systems are referred to as "ibala" [49], and as "ankara" in the North West Region of Cameroon. Common tools used for land preparation include hoes; machete; axes; forks; rakes; spades; grass hooks.

2.2. Planting

Planting for most crops is not done according to precisely measured distances; rather, it is based on estimates of required distances between crops. With a heavy reliance on manual labor, there is little need for adherence to precise measurements and geometrical patterns during most farming practices (including planting). Maize may be planted on mounds, ridges, or on flat, tilled, or untilled fields. Common tools used for planting are dibbers, hoes, and machetes. Better-off farmers occasionally make use of semi-mechanized tools such as jab planters, push-pull seed drills, and manual rotary injection planters [50].

2.3. Weeding

Manual weeding is another backbreaking work intensity in smallholder farming practice. Most weeding is done with the hands and hoe. Weeds are pulled out with the hands while the hoe is used to till and soften the soil, as well as to cover the dislodged vegetation with soil to ease the process of decomposition. As with tilling, the preparation of ridges and planting, the worker undertakes this process in the crouched position as he/she moves from one row to another.

2.4. Harvesting

As with other activities, harvesting is a manual activity for the smallholder farmer. The process of harvesting maize usually involves plucking the maize cob from the standing plant, collecting the harvested crop and transporting the harvested crop. The problem of transportation of harvested produce stems chiefly from the poor development of roads in rural sub-Saharan Africa [51]. This, together with the limited resources of smallholder farmers to afford the high transportation costs makes the arduous task of transportation a practice that relies heavily on human muscle power. The use of human muscle power limits the amount of crop that can be transported per person and trip. This means that smallholder farmers require more trips to transport their crops. Hence, smallholder farmers would be exposed to potential heat stress during the process of harvesting and the transportation of crops.

3. Materials and Methods

3.1. Wet Bulb Globe Temperature

Wet Bulb Globe Temperature (WBGT) is the most common index used for assessing heat stress in occupational health. It was developed by the US Army two decades ago [52,53] to guide military and civilian health care providers and allied medical personnel on understanding, identifying and managing heat stress among troops. Early studies investigated total heat stress imposed on military personnel in three camps by physical training, temperature, radiation, humidity and wind [54]. This index considers air temperature, radiant temperature, humidity and air movement, and is the reference for time limitations of work under different heat exposure conditions (Table 2).

Table 2. Recommended maximum Wet Bulb Globe Temperature (WBGT) exposure levels (°C) at different work intensities and rest/work ratios for an average acclimatized worker wearing light clothing. Source: compiled by Kjellström et al. 2009 from the International Organization for Standardization (ISO) 18 and the National Institute for Occupational Safety and Health (NIOSH) (criteria for a recommended standard: occupational exposure to hot environments. NIOSH Publication No. 86113. Atlanta, GA: National Institute of Occupational Health; 1986).

Metabolic Rate Class (Work Intensity)	1 (Light)	2 (Medium)	3 (Heavy)	4 (Very Heavy)
Continuous work, 0% rest/hour	31	28	27	25.5
25% rest/hour	31.5	29	27.5	26.5
50% rest/hour	32	30.5	29.5	28
75% rest/hour	32.5	32	31.5	31
No work at all (100% rest/hour)	39	37	36	34

3.2. Meteorological Data

Meteorological data for the year 2000, 2050 and 2100 were downloaded from the Earth System Grid Federation repository (https://esg-dn1.nsc.liu.se/) and used for the calculation of WBGT. This data originates from CORDEX—Coordinated Regional Climate Downscaling Experiment, http://www.cordex.org/ [55]. The data have a spatial resolution of 0.44 x 0.44 degrees latitude/longitude and 3 h temporal resolution.

Near-surface relative humidity (RH,%), surface down welling shortwave radiation (Rs, W m^{-2}) and surface air temperature (Ta, °C) at 2 m were derived from historical and RCP4.5 simulations with the Second Generation Canadian Center for Climate Modelling and Analysis Earth System Model (CanESM2) [56] (Data ID's listed in Appendix A). RCP 4.5 was selected as a reasonable lower range scenario with peaking emissions around 2040—a middle scenario among the available ones (RCP2.6, RCP4.5, RCP6, and RCP8.5).

3.3. Calculation of WBGT

To compute WBGT, we used the Natural wet-bulb temperature (combined with dry-bulb temperature indicates humidity, *Tw*); the Globe thermometer temperature (measured with a globe thermometer, also known as a black globe thermometer, *Tg*); and the Dry-bulb temperature (actual air temperature, *Td*).

$$WBGT = 0.7Tw + 0.2Tg + 0.1Td \quad (1)$$

where:

Tw = natural wet-bulb temperature (combined with dry-bulb temperature indicates humidity, °C);
Tg = globe thermometer temperature (measured with a globe thermometer, also known as a black globe thermometer, °C);
Td = dry-bulb temperature (actual air temperature, °C).

We calculated Tg according to Hajizadeh et al. [57] as:

$$Tg = 0.01498 * Rs + 1.184 * Ta - 0.0789 * RH - 2.739 \tag{2}$$

Tw was calculated according to Stull [58] (see units and abbreviations above):

$$\begin{aligned} Ta * atan(0.151977 * (RH + 8.313659)^{0.5}) + atan(Ta + RH) - atan(RH\text{-}1.676331) + \\ 0.00391838 * (RH)^{\wedge 1.5} * atan(0.023101 * RH) - 4.686035 \end{aligned} \tag{3}$$

For each day during the planting period (February–March) and the harvest period (August–September), the maximum WBGT was calculated and used for further analysis. This resulted in about 60 daily maximum WBGT values per grid cell that were further used for analysis.

3.4. Uncertainty and Bias Correction

When using climate simulation data for prognostic studies it is important to quantify the bias of observed versus simulated data [59,60] as this bias can cause systematic errors [61]. In the supplement (Appendix C), we compare observed surface air temperature from a set of climate stations to the estimated surface air temperature in order to quantify this bias.

Farm tasks such as manual clearing, hoeing, planting, weeding, and harvesting involve substantial inputs of labor and energy. Given that these energy-demanding activities are carried out mainly in the outdoors with exposure to elements of weather, WBGT is judged to be a suitable index for assessing risks for heat stress among this population of workers. Kohut [62] identifies some activities associated with military exercises in hot, dry climates that fall under the metabolic rate. These activities are matched to activities of comparable categories carried out by smallholder farmers in a typical farming cycle (Table 3).

This research is one study in a project whose goal is to understand the implications of environmental changes on human welfare and health (see *Funding* for details). In this project, the environmental determinants of human wellbeing, including on the burden of tropical diseases will be investigated. The study area is in the Lake Victoria region, hence the choice of Kenya, Tanzania, and Uganda as locations of interest.

To assess the sensitivity of our computations of WBGT, we used the software (WBGT) developed by Strategic Security Sciences Division, Argonne National Laboratory [53]. For this analysis, we examined the variability of WBGT for the years 2000, 2050 and 2100 (Appendix B Figure A1) as well as assessed the sensitivity of WBGT to Air Temperature (Ta) and Relative humidity (Rh) (Appendix B Figure A2).

3.5. Annual Calendar of Agricultural Activities

Farming activities associated with maize production are used to assess the potential of heat stress on food crop production. Maize is used as the reference crop for assessing farmers' activities as it is the main food crop in the east African region. In many parts of the region, maize is cultivated in two cycles within the year. The timing of agricultural activities was derived from the database of the Food and Agriculture Organization of the United Nations (FAO) (Found here: www.fao.org/agriculture/seed/cropcalendar/). This database gives the onset and end dates of planting and harvesting cycles for major food crops, including maize (see Appendix D, Table A2). While the database contains planting and harvesting dates, it does not have information on weeding periods.

Table 3. Examples of activities within metabolic rate categories (from Naval Medical Command, 1988, Manual of Naval Preventive Medicine, Washignton DC, 20372–5120, https://www.med.navy.mil/directives/CanPublications/5010-3.pdf) equated to farming practices undertaken by smallholder farmers. (Estimated from Monica Dungarwal and Maya Choudhry, 2003, Energy Balance of Farm Labourers. J. Hum. Ecol., 14(1): 51–55. Starred activities are estimates of the metabolic category of farm activities by authors.)

Physical Activity	Average Metabolic Rate Kcal/hr	Comparative Farm Activities	Average Metabolic Rate Kcal/hr
(a) Sitting ▪ Moderate arm and trunk movement (e.g., typing, drafting, driving a car in light traffic)	68		
▪ Moderate arm and leg movement (e.g., general laboratory work, slow movement about an office)	82		
▪ Heavy arm and leg movement (e.g., driving a car in moderate traffic)	99		
(b) Standing ▪ Light work at machine or bench, mostly arms	82	▪ Threshing ▪ Weeding ▪ Harvesting maize in a standing position	95
▪ Light work at machine or bench, some moving about (e.g., using a table saw, driving a truck in light traffic)	99		
▪ Moderate work at machine or bench, some walking about (e.g., replacing tires, driving a car in heavy traffic)	119	▪ Manual planting of maize using machete, dibber or hoe	109
(c) Walking About, with Moderate Lifting or Pushing (e.g., driving a truck in moderate traffic, scrubbing in a standing position)	164	Hoeing	179.6
(d) Intermittent Heavy Lifting, Pushing or Pulling (e.g., sawing wood by hand, callisthenic exercise, pick and shovel work)	238	Bunding—Ridging and mound formation during land preparation	205
(e) Hardest Sustained Work	300		

3.6. *Cropping Intensity*

We used the European Space Agency's prototype high-resolution LC map over Africa based on 1 year of Sentinel-2A observations from December 2015 to December 2016 (Found here: http://2016africalandcover20m.esrin.esa.int/) to identify areas of cropland. This "Prototype land cover map of Africa" v1.0 dataset was downloaded from http://2016africalandcover20m.esrin.esa.int/. This dataset divides land cover in Africa into 10 generic classes that describe the land surface at a 20 m × 20 m spatial resolution: "trees cover areas", "shrubs cover areas", "grassland", "cropland", "vegetation aquatic or regularly flooded", "lichen and mosses/sparse vegetation", "bare areas", "built-up areas", "snow and/or ice" and "open water". From the 10 land cover classes included, cropland was extracted and resampled to spatially fit the WBGT data, whereas the percentage of cropland per grid cell was calculated. This data was then used for the stratifications of cropping/cropland intensity (Figure 2).

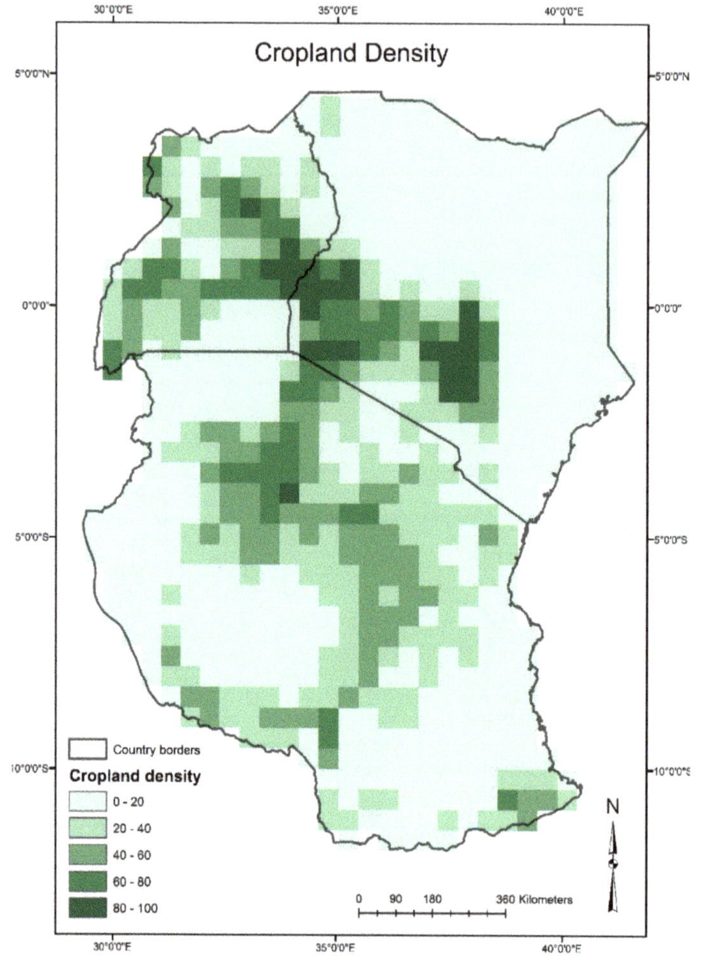

Figure 2. The distribution of cropland density—computed as a percent of cropland per 0.44° × 0.44° grid cell. This is input data for the stratification of cropping intensity.

4. Results and Discussion

The hypothesis advanced by this study that the Wet Bulb Globe Temperature (WBGT) in East African croplands in 2000 remains unchanged in 2050 and in 2100 is rejected, as high WBGT are observed during the two periods of the East African farming calendar studied. We also asked the question: *How is the Wet-Bulb Globe Temperature (WBGT) going to be distributed in the future, and how will this affect the ability of smallholder farmers to perform agricultural activities?* Generally, in February to March, the eastern and coastal regions of Kenya and Tanzania witness high WBGT values—some necessitating at least 50%, and up to 75% in some cases, rest/hour work intensities in 2050 and 2100. In August to September, the eastern and northern regions of Kenya, as well as the north and central regions of Uganda are vulnerable to high WBGT values. During this maize harvesting period, rest/work intensities with up to 50% rest/hour are expected in 2050 and 2100. Planning to understand and craft policies to address this key challenge is a critical element in adaptation methods to address the impact of climate change.

4.1. The Geographical Distribution of Maximum WBGT

The geographical distribution of maximum WBGT generally shows high values along coastal regions of of Kenya and Tanzania (Figure 3). WBGT values in coastal communities remain relatively high (generally above 25 °C), especially for the February–March season in all time segments. In Tanzania, regions with currently high maximum WBGT values and whose condition is going to be sustained in the future include Tanga, Pwani, Lindi, Mtwara, Ruvuma, and sections of Morogoro. In 2000, high maximum WBGT values were also observed in Shinyanga, Tabora, Kigoma, and Rukwa—a condition that becomes more widespread in the southeastern regions in 2100 (Figure 3c).

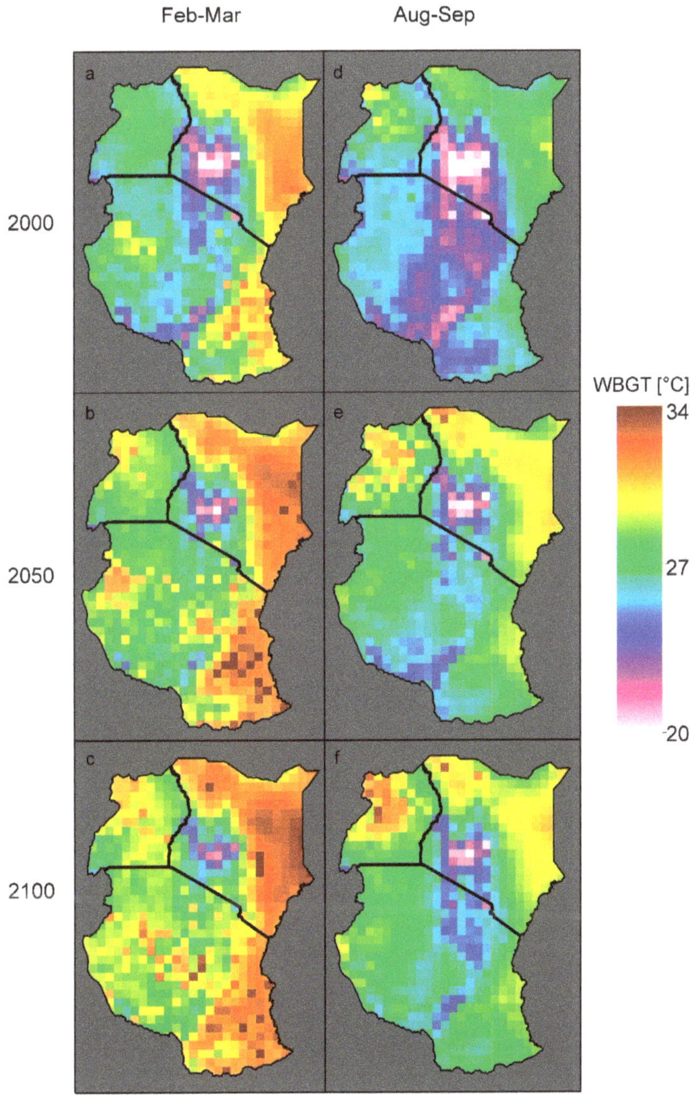

Figure 3. Distribution of maximum WBGT for the months of February–March (**a**–**c**), and August–September (**d**–**f**) in the years 2000, 2050 and 2100.

In August–September, maximum WBGT values affect the eastern and north eastern regions of Kenya more profoundly—with values reaching well above 30 °C in most of these areas (Figure 3e). High values are also observed north of the Rift Valley Region. Comparatively lower maximum WBGT values are observed in the central egion, Nyanza, and the southern portions of the Rift Valley region in all years and seasons. The central regions of Kenya have areas of high cropland density, many above 60%. In many of these areas, observed maximum WBGT values tend to be high, more than 28 °C (Figure 3b–f). In 2000, Uganda had a comparatively lower maximum WBGT (Figure 3a). By 2050, however, higher values are evident in the north-western (north of Gulu) and southern parts of the country (around Kampala) (Figure 3b,e). While generally lower maximum WBGT values are observed in Uganda in February–March, higher values are more widespread in August–September, affecting the central and north-western regions (especially in 2050, Figure 3e).

The distribution of high WBGT values illustrated by Figure 3 above is a cautionary tale of experiences that are already being observed in other parts of the Sub-Saharan Africa region. In a study of heat exposure on farmers in northeast Ghana, Frimpong, Eddie Van Etten [63] suggested that farmers were a population at risk and noted that the farming methods (using rudimentary tools and labor-intensive methods) were at the center of their vulnerability to high WBGT. In a study of the impact of heat stress and farmers' adaptation to it, Frimpong, Odonkor [64] found that heat stress associated with farming activities was a challenge for small-holder farmers in the region. They also recognized that, even though there were adaptation strategies for coping with heat stress in Baku East in Northern Ghana, these strategies were ineffective. In their study of the impact of heat on health and productivity among maize farmers in the Gombe province of Nigeria, Sadiq, Hashim [65] found that farmers were frequently exposed to heat stress—a condition that was contributing to heat exhaustion and productivity decreases among small-holder producers.

4.2. WBGT Frequncies and Farming Practices

Studies of labor inputs (in man-hours per hectare per year) into smallholder African agriculture reveals that farm preparation is the most labor-intensive activity in the cropping cycle. This has been reported in the study of pure maize cultivation in seven locations in Malawi [66]. Harvesting is the next most labor-intensive activity, then weeding, then planting. This distribution of labor inputs across crop types was also true for millet, legumes, manioc, and peanuts. While up-to-date data on labor use for different crops and different activities is hard to get, the mean amount of time spent on farms by smallholder farmers can give an idea of the potential for their exposure to potential elements of heat stress risk. Table 4 summarizes mean national data for the average amount of days spent on-farm by smallholder farmers. Family labor days on the farm supplied over a day refer to the total number of person-days that family members spend on the farm during one working day [67]. The FAO computes it by taking the total family labor day supplied on the farm over a day period, which is the total number of days at household level divided by the number of working days in a year—here, 300 days. In our case, the yearly labor inputs are presented [67].

Table 4. Mean national summaries of on-farm labour (days per year).

	Family on-Farm Labour (Days per Year)
Kenya, 2005	231
United Republic of Tanzania, 2013	189
Uganda, 2012	192

Drawing from values presented in Table 4, farmers will be putting in between 189 to 231 days of labor on farms annually. This is well above half the number of days of each year working in farms. If we draw on the Malawi [66] example to examine the distribution of activities between tasks, we find a total of 341 man-hour ha^{-1} year^{-1} during the garden preparation and planting seasons in February and March. (We use Malawi for a breakdown of farming activities as we do not have this data for the study area. Malawi has the same agroecological zones as found in portions of the study area (such as Tanzania) and the practice of smallholder agriculture is quite comparable in methods.) These activities connect each other closely regarding the periods in which they are practices (see Appendix D). Assuming an eight-hour work day, this translates to 42.6 days ha^{-1} of exposure to mean maximum WBGT (i.e., the mean value of all maximum WBGT for February-March 2000) ranging from 23.5 to 24.7 °C in 2050, with maximum reaching 31.3 °C. In 2100, the mean maximum WBGT will be 25.4 to 26.6 °C, with maximums reaching 30.3 °C (Table 5). Based on our classification of the manual planting of maize and associated crops using machete, dibber or hoe as heavy work, it follows that there are regions in which high WBGT will warrant at least 50% rest/hour work intensity (Table 2). Farm preparation tasks are classified as very heavy work (Table 3). These include clearing vegetation using a machete and a grass hook; ridging and mound formation during land preparation; and hand weeding and tilling at a crouched position. Performing these tasks will warrant work intensities with up to 75% rest/hour (see Table 2). A high number of workdays in conditions of high WBGT in February–March will be affecting the eastern half of Kenya, the east and southern regions of Tanzania as well as pockets of regions in Tanzania's center and north-west in 2050. In 2100, the geography of maximum WBGT values does not change much from the 2050 situation (Figure 4).

During the harvesting season in August–September, labor inputs of 141 person-hours ha^{-1} year^{-1} translate to 17.6 days of eight-hour workdays (Table 3). Maximum daily WBGT values during this period have a mean of 26.6 to 28.5 °C in 2050, with the maximum reaching 34.0 °C (Table 4). In 2100, maximum daily WBGT have a mean of 27.1 to 29.2 °C, with the maximum reaching 31.9 °C.

Harvesting maize in a standing position is classified as medium work (Table 2). At WBGTs of up to 30.3 °C in August–September (Table 5), there are regions in which maize harvesting will warrant work intensities with up to 50% rest/hour (Table 2). Areas in the north and northeastern Kenya, including the coastal regions, will be particularly vulnerable to high WBGT in 2050 (Figure 3). The central and northwestern regions of Uganda will also be vulnerable. While the eastern coastal regions of Tanzania will see increased WBGT values compared to other parts of the country, the country as a whole will not be as vulnerable as its northern neighbors in August to September (Figure 5).

Table 5. Maximum daily WBGT per grid cell, disaggregated to areal cropland percentage for the months of February–March and August–September for the years 2000, 2050 and 2100. The values hence describe the variability of the max daily WBGT (n, min mean, max and standard deviation, (stdev)).

AUG-SEP WBGT			**2000**					**2050**					**2100**		
Cropland%	n	Min	Mean	Max	Stdev	n	Min	Mean	Max	Stdev	n	Min	Mean	Max	Stdev
>0%	1131	16.74	25.19	32.58	2.57	1131	18.89	26.72	34.00	2.23	1131	18.50	27.14	31.99	2.57
>50%	193	17.41	25.01	29.10	3.03	193	19.85	26.63	29.88	2.37	193	19.57	26.73	30.32	2.84
>75%	107	19.26	26.23	28.81	2.76	107	21.74	27.57	29.88	2.05	107	20.90	27.85	30.18	2.67
>90%	71	21.57	27.68	28.81	1.49	71	23.74	28.55	29.88	1.17	71	23.07	29.23	30.07	1.39
FEB-MAR WBGT			**2000**					**2050**					**2100**		
Cropland%	n	Min	Mean	Max	Stdev	n	Min	Mean	Max	Stdev	n	Min	Mean	Max	Stdev
>0%	1131	15.30	22.94	27.4§9	2.61	1131	16.42	23.49	31.30	2.90	1131	18.43	25.38	30.33	2.27
>50%	193	16.30	23.13	27.48	2.27	193	17.11	23.72	30.26	2.38	193	19.22	25.40	28.88	2.03
>75%	107	18.64	24.13	25.43	1.59	107	19.52	24.47	30.26	1.55	107	21.82	26.18	28.73	1.40
>90%	71	21.28	24.81	25.42	0.65	71	22.17	24.70	27.14	0.95	71	23.53	26.57	27.46	0.74

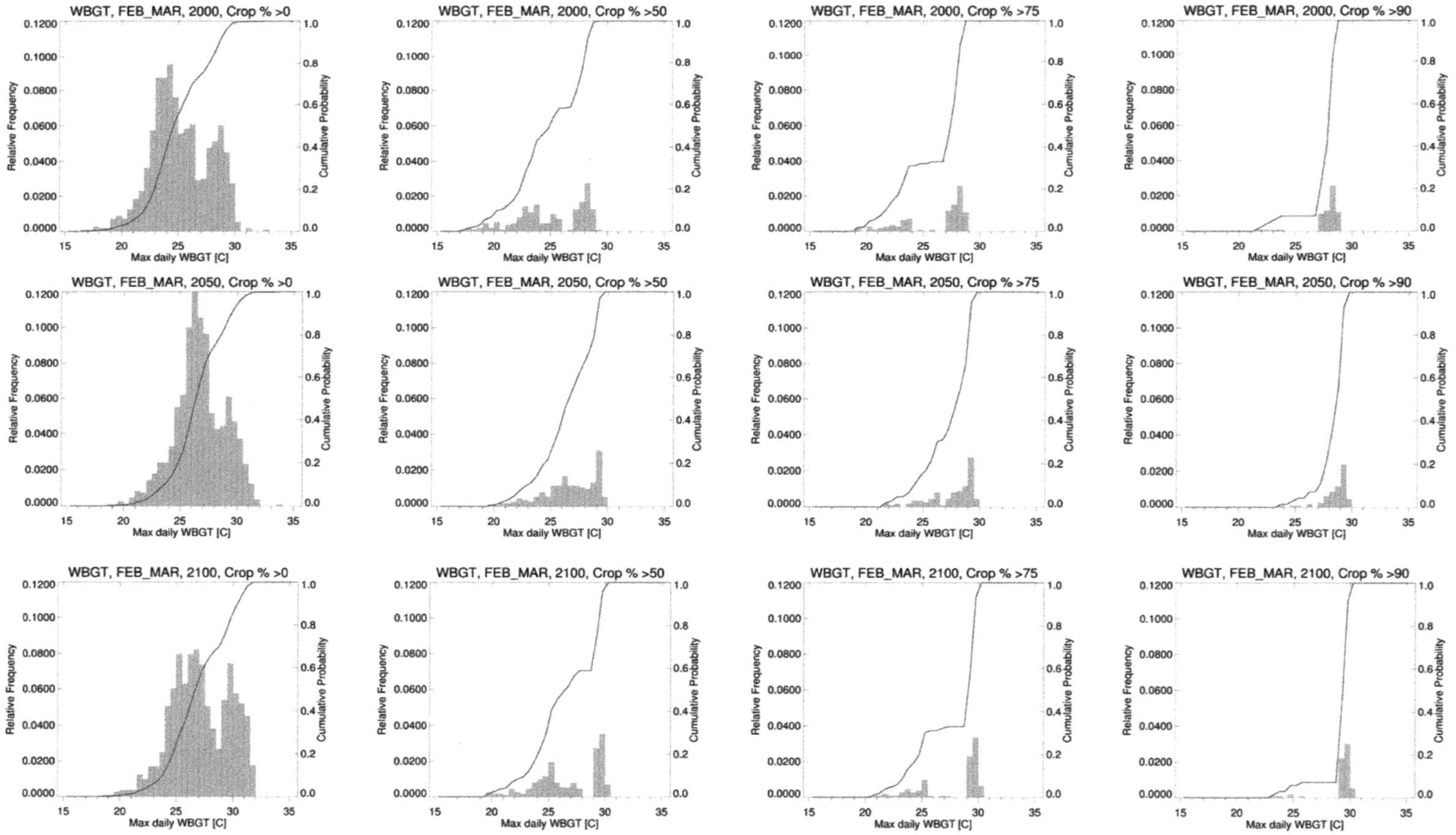

Figure 4. Frequency distribution and cumulative probability of WBGT and varying cropping intensities for February–March 2000, 2050, and 2100.

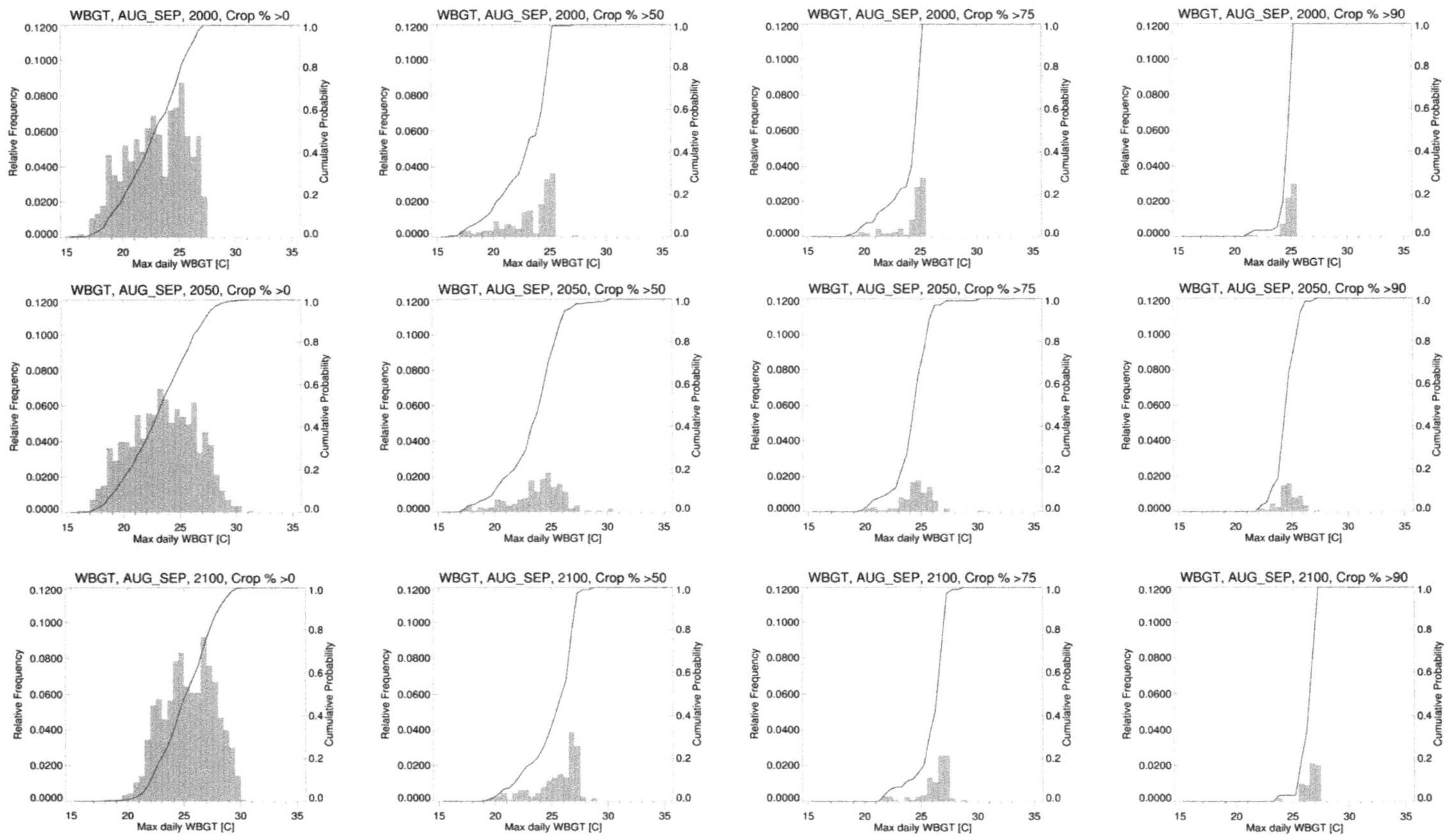

Figure 5. Frequency distribution and cumulative probability of WBGT and varying cropping intensities for August–September 2000, 2050, and 2100.

4.3. Implications for Climate Change Planning in the Agricultural Sector

There is consensus on the observation that African countries will suffer serious health consequences due to impacts of climate change. The rapidly growing populations in many African countries are among the most vulnerable to climatic changes in the world [30,68]. Beyond the direct negative health impacts, the impact of climate change stands to affect key socio-economic sectors such as agriculture. We demonstrate in this study that the impact of changing climates on the availability and productivity of labor will affect many geographies in East Africa. This study demonstrates that key sectors such as agriculture are particularly vulnerable to the effects of climate change. In many parts of the continent, this vulnerability is due in part to existing problems of poverty, weak institutions, political unrest, and the activities of some international financial institutions, which limit the capacity of some countries to deal with the challenges posed by a changing climate [69,70]. This vulnerability poses threats to human health, well-being and the economic productivity of agriculture-dependent countries in sub-Saharan Africa. Such threats warrant the need for mitigation measures to adapt smallholder agriculture to a warming world and policy engagement on occupational health programs that protect individuals at risk of heat-related morbidity and mortality.

4.4. Mitigation Measures to Adapt Smallholder Agriculture to A Warming World

Mitigation measures that adapt agriculture to changing conditions brought about by climate change. One is inclined to propose a more concerted drive towards practices of agroecological farming and ecological intensification to achieve the goals of sustainable food production and viable food systems in sub-Saharan Africa. In this case, a focus should be on management approaches that reduce human exposure to elements that contribute to heat stress within the agricultural environment. Examples may include:

(a) Reduction in exposure to heat stress in farm preparation activities (Figure 4 and Table 4). This may include the diffusion of best practices in no-till farming that eliminate the need to spend time and energy clearing, tilling and ridging the land. No-till farming usually involves (a) sustaining the availability of mulch or crop residue, or a careful section of cover crops for maintaining soil cover at all times (including off-farming seasons), and (b) using suitable crop rotations [50]. Within the context of human exposure to the elements of heat stress, the labor savings of no-till farming is especially important. In some cases, it has been observed that planting in a no-till system can reduce labor input by as much as 60% [71]. Notwithstanding the potential to reduce exposure to heat stress be adopting no-till farming, there are constraints associated with making it work in East Africa. Constraints include the small sizes of farms, which make farmers less willing to set aside portions of it to experiment with new approaches; problems of land tenure that decrease incentives for long-term investments in no-till practices; and access to information on the best practices for no-till farming. In addition, the highly degraded soils of the region mean that the transition period to achieving viable no-till farming systems is longer and may not be appreciated by farmers with restricted economic margins.

(b) Reduction in exposure to heat stress in planting activities (Figure 5 and Table 4). Planting is the activity that has experienced the most diverse innovations in mechanization among smallholder farming practices. Grain planting, in particular, has seen substantial innovation in small-scale mechanization that can reduce the work intensity of the practices [50]. Constraints to accessing and using these planting aids remain at the level of affordability (because many smallholder farmers may not have the financial resources to purchase this machinery) and organization (because, at the level of the institutional framework, it makes it possible for these tools to reach smallholder farmers, farming communities and farming organizations in the first place [72]). In the absence of mechanization, farming activities can be planned to reduce the risk of excessive exposure to heat, through a modification of the timing of practice of some activities. In a study of heat exposure among non-harvest sugarcane workers in Costa Rica, one of the conclusions drawn was that changes in the attitudes of employers that involve creative ways of organizing work shifts, among other things, can reduce worker exposure to heat stress [73].

(c) Reduction in exposure to heat stress during harvesting. It is challenging identifying what can be done to reduce the work intensity or harvesting for smallholder farmers. The mechanization of the harvesting process for crops such as maize and potatoes, for example, seems to be challenging for a variety of reasons, namely the small size of farms means that they are often not suited to the large-scale mechanization of the harvesting activity. In addition, the haphazard geometry of planted crops also reduces the suitability of mechanizing the process. However, an important component of crop harvesting is its transportation to homes or markets, since the purpose of harvesting is to get produce to where they can be sold or consumed. In this regard, there is a lot that can be done to reduce the long distances over which farmers transport agricultural produce, as well as the number of times that the harvesting of a single farm has to be done because the family can transport only so much at a time. There is also potential for reducing heat exposure through the smart planning of farming activities using existing intervention programs already in use. For example, interventions that make use of the Occupational Safety and Health Administration (OSHA)'s Water–Rest–Shade program (WRS) have been evaluated in El Salvador and found to contribute to reductions in symptoms associated with heat stress and with dehydration [74].

4.5. Policy Engagement on Occupational Health Programs

Policy engagement on occupational health programs that protect individuals at risk of heat-related morbidity and mortality is an essential part of mitigation and adaptation planning [1]. This falls within a recommendation proposed by the 2015 Lancet Commission to scale-up financing for climate resilient health systems worldwide [75]. As warmer temperatures become a reality in many parts of sub-Saharan Africa, redesigning medical services to meet and address the emerging challenges of outdoor occupational heat stress are indispensable. Among some of the areas of concern is the need to include heat-related morbidity and mortality into training programs for personnel in the health sector [76]. Such training will enable emergency medical personnel, clinicians, and doctors to respond to an increase in incidences of heat-related emergencies in a warming world. Just as important is the need to invest in infrastructure and equipment to facilitate the ability of these medical practitioners to meet the challenge. In the countries of Europe, these policy engagements culminate in what is referred to as "heat-health action plans" [77]. A policy engagement plan that responds to the emerging challenges of heat stress would address three key issues:

(a) Develop an illness prevention plan for outdoor work based on the heat index that is appropriate for specific agro-ecological zones in each country. The goal of such a prevention plan would be to prevent heat-related illnesses and deaths by raising awareness among agricultural practitioners, support personnel, and policymakers about the health risks associated with working in hot environments.

(b) Train workers in the agricultural sector (practitioners, agricultural extension workers, members of agricultural common initiative groups) how to recognize and prevent heat-related illness. To be fully effective, an interdisciplinary approach that engages stakeholders at different levels of the agricultural production and distribution chain, as well as on associated agricultural support services, would be essential [78].

(c) Define protective measures for dealing with outdoor work conditions for smallholder farmers. These measures may include work/rest schedules, clothing choices under different heat stress conditions, techniques for keeping cool, the importance of hydration during working hours, as well as how to deal with heat-related emergencies. Many examples of such protective measures have been examined by previous studies [1]. These measures can contribute to addressing heat-related morbidities.

4.6. Uncertainty in the Distribution of Future Agricultural Areas

The African climate is determined by three main processes. Two of these are local processes that determine the regional and seasonal patterns of temperature and rainfall-tropical convection, and the alternation of the monsoons. The third, El Niño-Southern Oscillation is foreign to the continent, but strongly influences interannual rainfall and temperature patterns in Africa. Some of the most reported impacts of anthropogenic global warming and climate change in Africa are higher sea and land surface temperatures, and an increase in the incidence and severity of droughts, floods and other extreme weather events. It is forecast that over the next 100 years, mean temperatures across Africa will rise faster than the global average, exceed 2 °C, and may reach as high as 3 °C to 6 °C greater than 20th century levels [79]. Drier subtropical regions are expected to warm more than the moister tropics, with northern and southern Africa becoming much hotter by as much as 4 °C. It is also expected to become drier, with precipitation falling by as much as 15% or more [79]. In East Africa, climate change is projected to increase temperature and precipitation variability as well [80]. These changes in key factors that determine the suitability of rain-fed agriculture are bound to have an effect on the distribution of suitable areas for smallholder agriculture on the African continent. Indeed, even if rainfall remains constant, existing water stress will be amplified as a result of increased temperatures, putting even more pressure on agricultural systems on the continent, especially in arid and semiarid areas [14]. Climate change is projected to decrease the yields of cereals overall in Africa through shortening growing season length, amplifying water stress and increasing the incidence of diseases, pests and weed outbreaks [80]. In East Africa, cereal mixes, especially the maize mixed cropping system, covers over 40% of the area [12].

This study therefore acknowledges that suitable areas for food crops and agriculture in general may therefore change by the middle and end of the century, meaning that some of the areas that host agriculture today may not be hosting these activities in the future. We also acknowledge that there is a possibility that improved technologies may reduce the burden of human labor in smallholder agricultural systems. However, if we draw from experience over the last three decades, these changes have neither been fast nor widespread enough to expect that substantial radical changes may have changed the agricultural landscape before the middle of the century. Poor agricultural performance in sub-Saharan Africa has led to a situation of stagnating real incomes of farmers, stagnating and often increasing rural poverty, and a farming landscape whose methods and productivity have not changed substantially over the last three decades [81]. Indeed, the Alliance for a Green Revolution in Africa (AGRA) notes that, despite the positive outlook on the role of agriculture and plans for its development in the sub-region, "there remains significant need for improvement to achieve an inclusive agricultural transformation: (i) agricultural growth is still too slow and yield increase too marginal; (ii) food security is not yet sustainable in most places; (iii) new challenges such as climate change, pests and diseases threaten progress, etc. [81]". Nonetheless, our findings point to the need to take human labor and its vulnerabilities in the face of climate change into consideration when examining or exploring adaptation policies.

4.7. Uncertainty of the CORDEX Data

The results of the comparison of measured surface air temperature versus the CORDEX estimated surface air temperature (Appendix C) show no systematic bias. Approximately 80% of the observations (n = 63,027 daily mean temperature observations) showed a mean absolute error of 1 °C or less, whereas the percent bias was 5% or less for about 60% of the 252 station years studied.

5. Conclusions

We find that heat stress is already affecting regions of East Africa. This condition is set to continue in the middle of the century and beyond. Not all areas of the East African region or all areas inside national boundaries are affected equally. Different regions of each country are affected at different degrees and at different times of the year. While Kenya and Tanzania experience large portions of their national land mass affected by high WBGT values, a neighboring country (Uganda) is relatively less affected in the two seasons of the agricultural calendar examined. High WBGT has implications on the rest/work cycles of smallholder farmers whose use of machinery for many farming practices remains very limited. There is therefore a need to design and implement mitigation measures to adapt smallholder agriculture to a warming world. These could be measures that target exposure to heat stress in different farming cycle activities, such as land preparation, planting, weeding, and harvesting. There is also a need for policy engagement to protect from the risks of heat-related morbidity and mortality. Reduced work capacity in heat-exposed jobs constitutes one of the important effects of climate change. This has implications for the attainment of key social and economic goals for societies in which economic production relies on high inputs of manual labor and high levels of exposure to climatic elements during key production periods.

Author Contributions: G.T.Y. and J.A. contributed equally in the following tasks: conceptualization of the study; methodology development; investigation, data analysis; writing—original draft preparation; as well as the review and editing of the final draft. All authors have read and agreed to the published version of the manuscript.

Funding: This research was funded by Vetenskapsrådet (Swedish Research Council), Grant No: 2016-04884 and the Swedish International Development Cooperation Agency through the BREAD project.

Acknowledgments: We are grateful for the support provided by James C. Liljegren of the Strategic Security Sciences Division, Argonne National Laboratory, Illinois USA for sharing his software WBGT with us. We used this software to perform sensitivity analysis for our computations.

Conflicts of Interest: The authors declare no conflict of interest. The funders had no role in the design of the study; in the collection, analyses, or interpretation of data; in the writing of the manuscript, or in the decision to publish the results.

Appendix A

Identification Values for CORDEX Data Used:
Below are the ID's for data sets used:

Historical (1951–2005)
cordex.output.AFR-44.SMHI.CCCma-
CanESM2.historical.r1i1p1.RCA4.v1.3hr.tas.v20180109|esg-dn1.nsc.liu.se
cordex.output.AFR-44.SMHI.CCCma-
CanESM2.historical.r1i1p1.RCA4.v1.3hr.rsds.v20180109|esg-dn1.nsc.liu.se
cordex.output.AFR-44.SMHI.CCCma-
CanESM2.historical.r1i1p1.RCA4.v1.3hr.hurs.v20180109|esg-dn1.nsc.liu.se

Simulations (2006–2100):
cordex.output.AFR-44.SMHI.CCCma-
CanESM2.rcp45.r1i1p1.RCA4.v1.3hr.tas.v20180109|esg-dn1.nsc.liu.se
cordex.output.AFR-44.SMHI.CCCma-
CanESM2.rcp45.r1i1p1.RCA4.v1.3hr.hurs.v20180109|esg-dn1.nsc.liu.se
cordex.output.AFR-44.SMHI.CCCma-
CanESM2.rcp45.r1i1p1.RCA4.v1.3hr.rsds.v20180109|esg-dn1.nsc.liu.se

Appendix B

Sensitivity of WBGT to Air Temperature (Ta) and Relative humidity (Rh). (WBGT used for the sensitivity analysis is derived using the software "WBGT", developed by Liljegren JC, Carhart RA, Lawday P, Tschopp S, Sharp RJJoo, hygiene e (2008), who model the wet bulb globe temperature using standard meteorological measurements. 5:645–655).

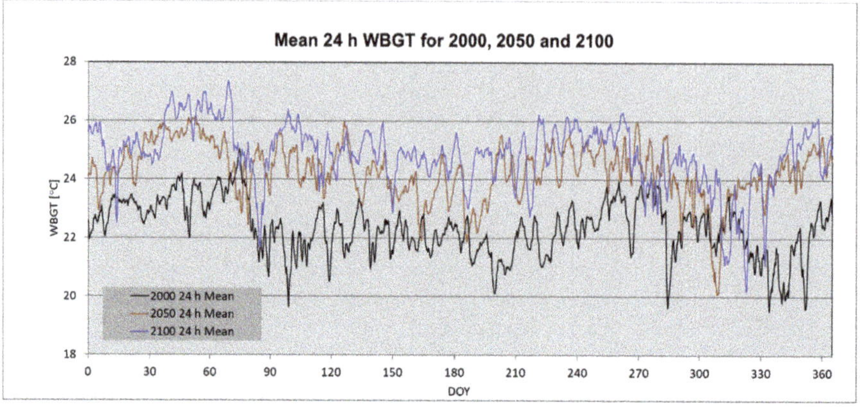

Figure A1. Mean 24 h WBGT (°C) variability for one grid-cell in the year 2000, 2050 and 2100. WBGT calculated using near-surface relative humidity (RH, %), surface down welling shortwave radiation (Rs, W m^{-2}) and surface air temperature (Ta, °C) at 2 m were derived from historical (year 2000) and RCP4.5 simulations (years 2050 and 2100) with the Second-Generation Canadian Center for Climate Modelling and Analysis Earth System Model (CanESM2) [56].

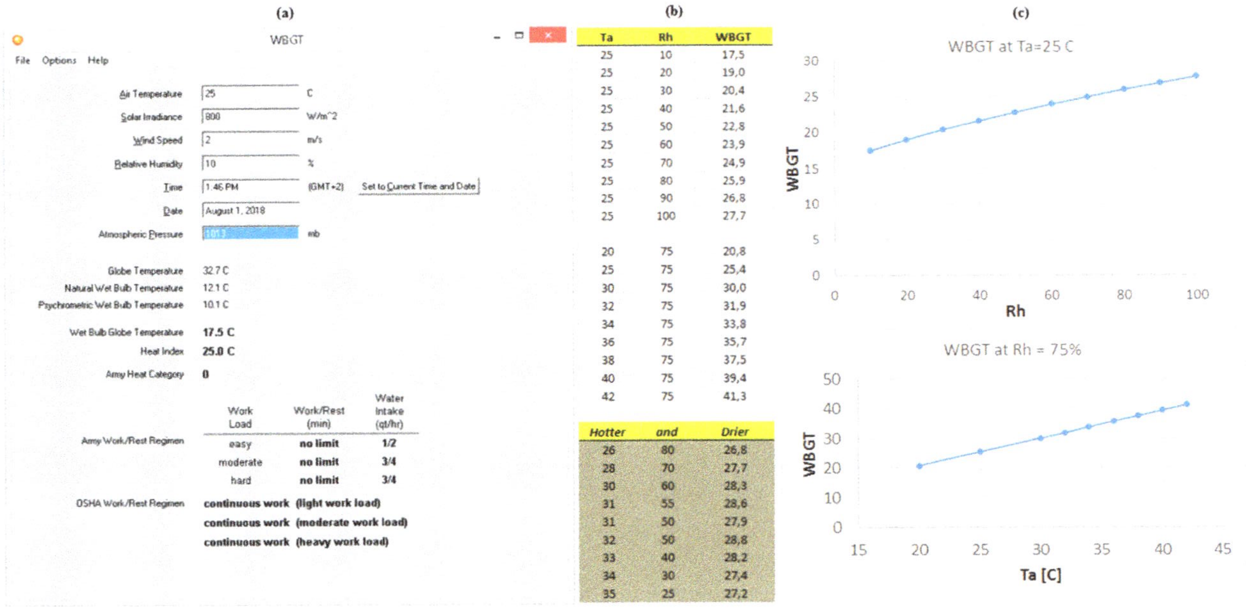

Figure A2. Assessing the sensitivity of WBGT to air temperature (Ta) and relative humidity (Rh): (**a**) the interface for the software "WBGT" developed by Liljegren et al. (2008) [53]; (**b**) input data, air temperature (Ta) and relative humidity (Rh) used for the assessment of WBGT; (**c**) plots of the WBGT response to relative humidity (Rh) and to air temperature (Ta).

Appendix C

Appendix C.1 Bias of CORDEX RCM Surface Temperature Data as Compared to Observations from Meteorological Stations

Appendix C.1.1 Introduction

When using climate simulation data for prognostic studies, it is important to quantify the bias between observed and simulated data. Either in order to do a bias correction or to get an estimate of the uncertainty of the prognoses done and the potential effects on downstream calculations. Here, we briefly described such a comparison between observations and model estimates, including potential effects on calculated WBGT.

Appendix C.1.2 Data and Methodology

Data

From the 3h CORDEX, surface air temperature data was a daily mean temperature calculated for the locations of 18 meteorological stations within the study area for the period 2006–2019 (Table A1). In total, 63027 daily mean temperature observations were used. These temperatures were compared to daily mean temperature data for the corresponding stations that were downloaded from the Global Surface Summary of the Day (GSOD) database provided by the National Oceanic and Atmospheric Administration (NOAA) at https://www.nodc.noaa.gov/ (Table A1).

Table A1. Meteorological stations used. United State Air Force number (USAF), station name, country, latitude, longitude and altitude.

USAF	Station Name	Country	Latitude [Degrees]	Longitude [Degrees]	Altitude [m]
636020	ARUA	Uganda	3.05	30.917	1211
636300	GULU	Uganda	2.806	32.272	1069
636310	LIRA	Kenya	2.283	32.933	1189
637720	LAMU MANDA	Kenya	−2.252	40.913	6
636710	WAJIR	Kenya	1.733	40.092	234
639620	SONGEA	Tanzania	−10.683	35.583	1067
637260	KABALE	Uganda	−1.25	29.983	1869
636020	ARUA	Uganda	3.05	30.917	1211
636120	LODWAR	Kenya	3.122	35.609	522
639710	MTWARA	Tanzania	−10.339	40.182	113
636120	LODWAR	Kenya	3.122	35.609	522
636410	MARSABIT	Kenya	2.300	37.900	1345
636610	KITALE	Kenya	0.972	34.959	1850
636860	ELDORET	Kenya	0.483	35.300	2120
637170	NYERI	Kenya	−0.500	36.967	1759
637230	GARISSA	Kenya	−0.464	39.648	148
637400	NAIROBI JKIA	Kenya	−1.319	36.928	1623
638700	ZANZIBAR	Tanzania	−6.222	39.225	16

Methodology

For each station and year was the corresponding goodness of fit (R2), Mean Absolute Error, $MAE = \frac{\sum_{i=1}^{n}|Poi - Psi|}{n}$ and the Percent Bias $PBIAS = \frac{\sum_{i=1}^{n}(Poi - Psi)}{\sum_{i=1}^{n} Poi}$ suggested by Luo et al. (2018).

Results

In total, 252 station years and 63027 daily mean temperature observations were used. R^2 ranged from 0.0 to 0.42, and PBIAS ranged from −13.9% to 36.5% with a mean of 0.37%. MAE ranged from −2.3 °C to 6.5 °C with a mean of 0.04 °C. Approximately 80% of the observations (Figure A3a) show a MAE of 1 °C or less, whereas the percent bias is 5% or less for about 60% of the 252 station years studied (Figure A3b). Station 636410 (MARSABIT, Kenya) stands out with PBIAS ranging from 23 to 36% and MEA ranging from 1 to 6.5 °C. This indicate that, for most stations, the systematic error is low when described on an annual basis. This does not exclude larger unsystematic errors during shorter periods and individual days (Figure A4 gives an example for one year and one station.).

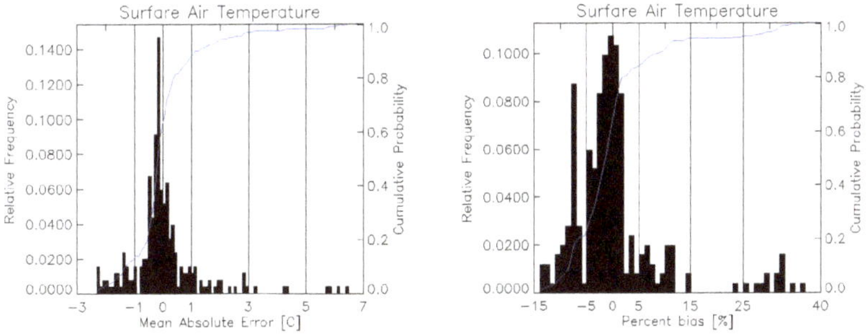

(a) (b)

Figure A3. Distribution describing the relative and cumulative distributions of the mean absolute error (**a**) and percent bias (**b**) for 252 station years of Global Surface Summary of the Day (GSOD) daily man temperature compared to surface air temperature from a regional climate model for CORDEX-Africa. Both the error and the bias are centered around 0 indicating no systematic deviations.

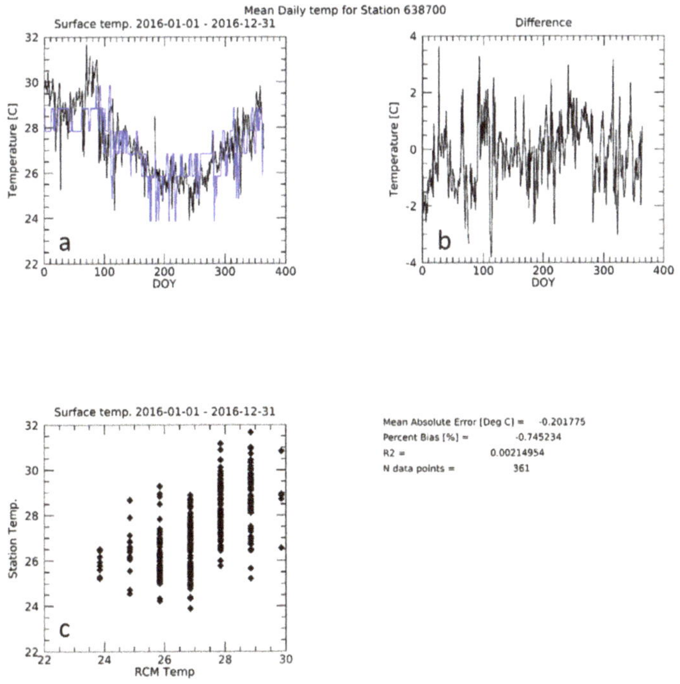

Figure A4. Comparison of mean daily surface temperature from Zanzibar (Station Temp. In graph) versus the mean daily surface temperature from CORDEX (RCM Temp) for the year 2016. (**a**) The mean daily observed temperature is in black and the CORDEX temperature is in blue. (**b**) Daily temperature difference (observed-CORDEX) and (**c**) scatterplot of observed versus estimated (RCM) temperature.

Appendix C.1.3 Implications for WBGT

The effect of the uncertainty or bias of CORDEX-predicted surface air temperature (Ta, the most important input when calculating WBGT) was quantified by recalculation of WBGT for February-March

2050 (corresponding to Figure 2b in the main paper) using Ta − 2 °C and Ta + 2 °C. We consider ±2 °C as a reasonable approximation of the uncertainty of the CORDEX-predicted surface air temperature based on Figure A3b. Figure A5 describe the spatial distribution of the outputs as well as the histograms for WBGT calculated with Ta − 2, Ta, Ta + 2. The increase in WBGT (February–March, 2050) when using Ta + 2 compared to using Ta (Figure A6a) range from about 1.5° to 2.1° The decrease in WBGT (February–March, 2050) when using Ta−2 compared to WBGT calculated with Ta range from approximately −2.1° to −1.5° (Figure A6b).

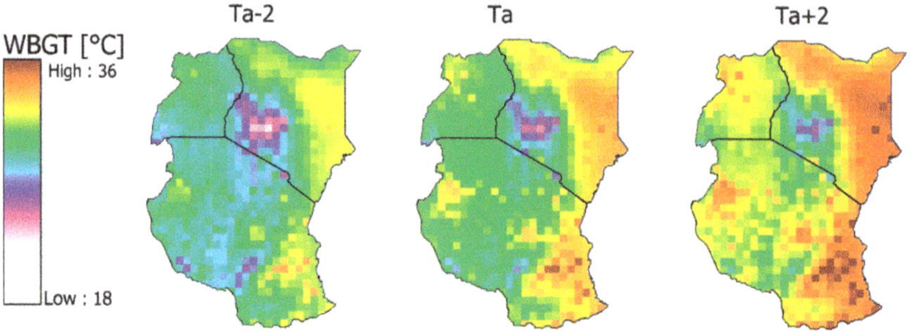

(**a**)

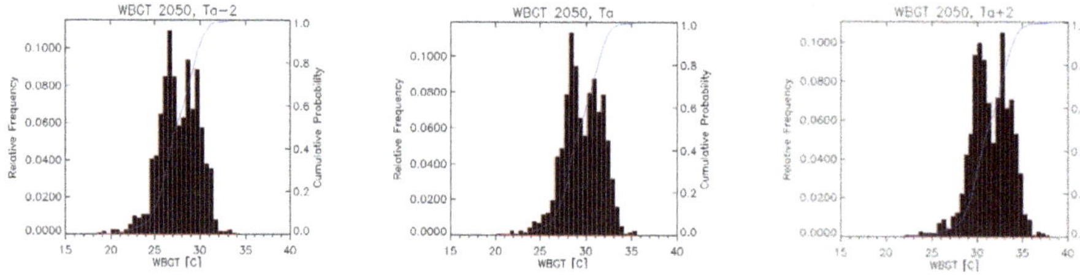

(**b**)

Figure A5. Maximum WBGT for February-March in 2050 assuming a data uncertainty of the surface air temperature of ±2 °C. Spatial distribution (**a**) and histograms with cumulative distributions in blue of maximal WBGT (**b**).

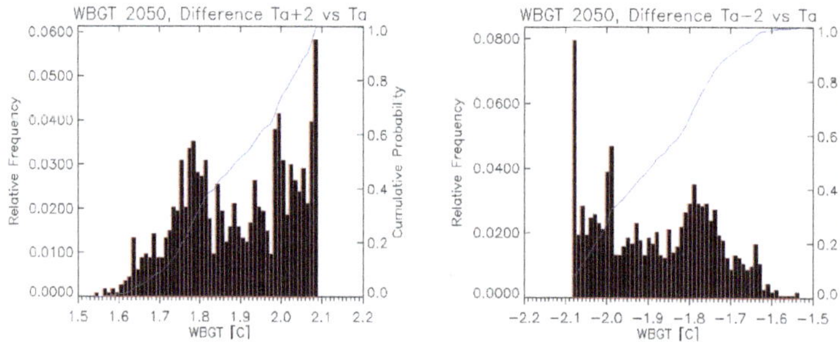

(**a**) (**b**)

Figure A6. Relative and cumulative (blue line) distributions of the effects of an uncertainty of ±2 °C in air surface temperature on maximum WBGT for February-March 2050. WBGT[C] (*x*-axis) is the difference ($WBGT_{Ta+2}$–$WBGT_{Ta}$) assuming Ta to be CORDEX Ta + 2 °C (**a**) and ($WBGT_{Ta-2}$–$WBGT_{Ta}$) assuming Ta to be CORDEX Ta − 2 °C (**b**).

Appendix D

Table A2. Agricultural calendar showing planting and harvesting periods, as well as crop cycles within each agro-ecological zone in Kenya, Uganda and Tanzania. Source: FAO.

Country	Agro-Ecological Zones	Cropping Cycle	Planting Period-Onset	Planting Period-End	Crop Cycle (Days)	Harvest Period-Onset	Harvest Period-End
Kenya	Lower Highland Zone 1 (LH1)	First	28 February	31 March	180-270	01 September	20 December
Kenya	Lower Highland Zone 1 (LH1)	Second	01 September	30 September	180–270	01 February	30 June
Kenya	Lower Highland Zone 2 (LH2)	First	01 March	31 March	180–270	01 September	20 December
Kenya	Lower Highland Zone 2 (LH2)	Second	01 August	15 October	180–270	01 January	15 May
Kenya	Lower Highland Zone 3 (LH3)	First	15 March	31 March	180–270	15 September	20 December
Kenya	Lower Highland Zone 3 (LH3)	Second	01 October	31 October	180–270	01 May	30 June
Kenya	Lower Highland Zone 4 (LH4)	First	15 March	31 March	180–270	15 September	20 December
Kenya	Lower midland zone 1 (LM1)	First	15 March	15 April	110–150	01 August	15 September
Kenya	Lower midland zone 1 (LM1)	Second	01 September	15 October	110–150	01 January	15 March
Kenya	Lower midland zone 2 (LM2)	First	15 March	15 April	110–150	01 August	15 September
Kenya	Lower midland zone 2 (LM2)	Second	01 August	15 October	110–150	01 December	15 March
Kenya	Lower midland zone 3 (LM3)	First	01 March	31 March	110–150	01 August	31 August
Kenya	Lower midland zone 3 (LM3)	Second	15 October	31 October	110–150	01 February	31 March
Kenya	Lower midland zone 4 (LM4)	First	01 March	31 March	90–120	01 July	15 August
Kenya	Lower midland zone 4 (LM4)	Second	01 October	31 October	90–120	01 January	28 February
Kenya	Lowerland zone 2 (L2) and (IL2)	First	01 April	15 April	90–120	15 July	15 August
Kenya	Lowerland zone 3 (L3)		15 April	30 April	90–120	15 July	15 August
Kenya	Lowerland zone 4 (L4)		15 March	15 April	90–120	15 July	15 August
Kenya	Upper Highland Zone 2 (UH2)	First	25 March	05 April	210–280	01 July	20 December
Kenya	Upper Highland Zone 2 (UH2)	Second	15 October	31 October	210–280	01 May	31 July
Kenya	Upper Highland Zone 1 (UH 1)	First	15 March	31 March	210–280	15 September	20 December
Kenya	Upper Highland Zone 1 (UH 1)	Second	15 October	31 October	210–280	01 May	31 July
Kenya	Upper midland zone 1(UM1)	First	15 February	31 March	135–160	01 August	15 September
Kenya	Upper midland zone 1(UM1)	Second	01 August	15 October	135–160	15 December	31 March
Kenya	Upper midland zone 2 (UM2)	First	15 March	15 April	135–160	01 August	30 September
Kenya	Upper midland zone 2 (UM2)	Second	01 August	15 October	135–160	15 December	31 March
Kenya	Upper midland zone 3 (UM3)	First	01 March	31 March	135–160	01 August	30 September
Kenya	Upper midland zone 3 (UM3)	Second	15 October	31 October	135–160	01 February	31 March
Kenya	Upper midland zone 4 (UM4)	First	01 March	15 April	135–160	15 August	30 September
Kenya	Upper midland zone 4 (UM4)	Second	15 October	21 October	135–160	28 February	31 March
Uganda	Busoga Farming System		25 February	15 August	110–120	15 June	15 December
Uganda	Eastern Highlands		15 March	30 April	120–180	15 August	30 October
Uganda	Eastern Savannah		15 March	15 August	110–120	05 July	20 December
Uganda	Karamoja Dry Zone		01 April	30 April	100–110	10 July	20 August
Uganda	Karamoja Wet Zone		20 March	20 April	100–120	30 June	20 August
Uganda	Lake Albert Crescent		20 March	31 August	110–120	10 July	31 December
Uganda	Lake Victoria Crescent		25 January	31 August	110–120	15 May	31 December
Uganda	Northern Farming System		15 March	20 July	110–120	05 July	20 December
Uganda	South Western Highlands		15 August	15 September	150–180	15 January	15 February
Uganda	West Nile Farming System		20 March	20 July	110–120	10 July	20 November
Uganda	Western Range Lands		15 August	15 September	110–120	05 December	15 January
Tanzania	Central Plateaux (Plains)		01 November	31 December	90–180	01 February	30 June
Tanzania	Coastal Plains		01 October	31 May	90–125	01 January	10 October
Tanzania	Eastern Plateaux and Mt. Blocks		01 December	15 June	90–125	01 March	10 November
Tanzania	High Plains and Plateaux		01 December	31 December	110–190	01 March	30 June
Tanzania	Inland Sediments		01 May	31 May	90–110	01 August	20 September
Tanzania	Ruaha Rift Zone - Alluvial Flats		01 December	31 December	90–110	01 March	15 April
Tanzania	Ufipa Plateau		01 January	31 January	180–190	01 July	15 August
Tanzania	Volcanoes and Rift Depressions		01 January	30 November	90–190	01 April	31 May
Tanzania	Western Highlands		01 January	31 January	90–190	01 April	15 August

References

1. Gao, C.; Kuklane, K.; Östergren, P.-O.; Kjellstrom, T. Occupational heat stress assessment and protective strategies in the context of climate change. *Int. J. Biometeorol.* **2018**, *62*, 359–371. [CrossRef] [PubMed]
2. Kjellstrom, T.; Lemke, B.; Otto, M.; Hyatt, O.; Briggs, D.; Freyberg, C. Climate Change and Increasing Heat Impacts on Labor Productivity. 2015. Available online: http://unfcc.int/files/science/workstreams/the_2013-2015_review/ (accessed on 12 February 2019).
3. Perkins-Kirkpatrick, S.; Gibson, P. Changes in regional heatwave characteristics as a function of increasing global temperature. *Sci. Rep.* **2017**, *7*, 12256. [CrossRef] [PubMed]
4. Jalloh, A.; Nelson, G.C.; Thomas, T.S.; Zougmoré, R.B.; Roy-Macauley, H. *West African Agriculture and Climate Change: A Comprehensive Analysis*; International Food Policy Research Institute: Washington, DC, USA, 2013.
5. Thamo, T.; Addai, D.; Pannell, D.J.; Robertson, M.J.; Thomas, D.T.; Young, J.M. Climate change impacts and farm-level adaptation: Economic analysis of a mixed cropping–livestock system. *Agric. Syst.* **2017**, *150*, 99–108. [CrossRef]
6. Kjellstrom, T.; Briggs, D.; Freyberg, C.; Lemke, B.; Otto, M.; Hyatt, O. Heat, human performance, and occupational health: A key issue for the assessment of global climate change impacts. *Annu. Rev. Public Health* **2016**, *37*, 97–112. [CrossRef] [PubMed]
7. Cubasch, U.; Wuebbles, D.; Chen, D.; Facchini, M.C.; Frame, D.; Mahowald, N.; Winther, J.-G. Introduction. In *Climate Change 2013: The Physical Science Basis—Contribution of Working Group I to the Fifth Assessment Report of the Intergovernmental Panel on Climate Change*; Stocker, T.F., Qin, D., Plattner, G.-K., Tignor, M., Allen, S.K., Boschung, J., Nauels, A., Xia, Y., Bex, V., Midgley, P.M., Eds.; Cambridge University Press: Cambridge, UK; New York, NY, USA, 2013; Volume 292, p. 1261.
8. Huang, C.; Barnett, A.G.; Wang, X.; Vaneckova, P.; FitzGerald, G.; Tong, S. Projecting future heat-related mortality under climate change scenarios: A systematic review. *Environ. Health Perspect.* **2011**, *119*, 1681. [CrossRef] [PubMed]
9. Jones, H.M. Climate Change and Increasing Risk of Extreme Heat. In *Human Health and Physical Activity during Heat Exposure*; Springer: Berlin/Heidelberg, Germany, 2018; pp. 1–13.
10. Krishnan, S. Assessment of Heat Stress and Its Health Impacts on Health of Workers from Different Occupational Sectors. Ph.D. Thesis, Sri Ramachandra University, Chennai, India, 2017.
11. Costello, A.; Abbas, M.; Allen, A.; Ball, S.; Bell, S.; Bellamy, R.; Friel, S.; Groce, N.; Johnson, A.; Kett, M. Managing the health effects of climate change: Lancet and University College London Institute for Global Health Commission. *Lancet* **2009**, *373*, 1693–1733. [CrossRef]
12. Adhikari, U.; Nejadhashemi, A.P.; Woznicki, S.A. Climate change and eastern Africa: A review of impact on major crops. *Food Energy Secur.* **2015**, *4*, 110–132. [CrossRef]
13. Descheemaeker, K.; Oosting, S.J.; Tui, S.H.-K.; Masikati, P.; Falconnier, G.N.; Giller, K.E. Climate change adaptation and mitigation in smallholder crop–livestock systems in sub-Saharan Africa: A call for integrated impact assessments. *Reg. Environ. Chang.* **2016**, *16*, 2331–2343. [CrossRef]
14. Pereira, L. Climate Change Impacts on Agriculture across Africa. Oxford Research Encyclopedia of Environmental Science, 2017. Available online: https://oxfordre.com/environmentalscience/view/10.1093/acrefore/9780199389414.001.0001/acrefore-9780199389414-e-292 (accessed on 22 September 2018).
15. Ramirez-Villegas, J.; Thornton, P.K. *Climate Change Impacts on African Crop Production*; CGIAR Research Program on Climate Change, Agriculture and Food Security (CCAFS): Copenhagen, Denmark, 2015; p. 27.
16. Pecl, G.T.; Araújo, M.B.; Bell, J.D.; Blanchard, J.; Bonebrake, T.C.; Chen, I.-C.; Clark, T.D.; Colwell, R.K.; Danielsen, F.; Evengård, B. Biodiversity redistribution under climate change: Impacts on ecosystems and human well-being. *Science* **2017**, *355*, eaai9214. [CrossRef]
17. Rojas-Downing, M.M.; Nejadhashemi, A.P.; Harrigan, T.; Woznicki, S.A. Climate change and livestock: Impacts, adaptation, and mitigation. *Clim. Risk Manag.* **2017**, *16*, 145–163. [CrossRef]
18. Weindl, I.; Lotze-Campen, H.; Popp, A.; Müller, C.; Havlík, P.; Herrero, M.; Schmitz, C.; Rolinski, S. Livestock in a changing climate: Production system transitions as an adaptation strategy for agriculture. *Environ. Res. Lett.* **2015**, *10*, 094021. [CrossRef]
19. Neumann, B.; Vafeidis, A.T.; Zimmermann, J.; Nicholls, R. Future coastal population growth and exposure to sea-level rise and coastal flooding-a global assessment. *PLoS ONE* **2015**, *10*, e0118571. [CrossRef] [PubMed]

20. Serdeczny, O.; Adams, S.; Baarsch, F.; Coumou, D.; Robinson, A.; Hare, W.; Schaeffer, M.; Perrette, M.; Reinhardt, J. Climate change impacts in Sub-Saharan Africa: From physical changes to their social repercussions. *Reg. Environ. Chang.* **2017**, *17*, 1585–1600. [CrossRef]
21. Kjellstrom, T.; Holmer, I.; Lemke, B. Workplace heat stress, health and productivity—An increasing challenge for low and middle-income countries during climate change. *Glob. Health Action* **2009**, *2*, 2047. [CrossRef] [PubMed]
22. Kjellstrom., T.; Lemke, B.; Otto, M.; Hyatt, O.; Briggs, D.; Freyberg, C. *Threats to Occupational Health, Labor Productivity and the Economy from Increasing Heat during Climate Change: An Emerging Global Health Risk and a Challenge to Sustainable Development and Social Equity Vol Technical Report 2014: 2*; Climate Change Health Impact & Prevention (ClimateCHIP): Mapua, New Zealand, 2014.
23. FAO. *FAOSTAT Statistical Database*; Food and Agriculture Organization of the United Nations: Rome, Italy, 2020; Available online: http://www.fao.org/faostat/en/ (accessed on 3 April 2020).
24. Olsson, L.; Chadee, D.D.; Hoegh-Guldberg, O.; Porter, J.R.; Pörtner, H.-O.; Smith, K.R.; Travasso, M.I.; Tschakert, P. Cross-chapter box on heat stress and heat waves. In *Climate Change 2014: Impacts, Adaptation, and Vulnerability. Contribution of WG II to the Fifth Assessment Report of the IPCC*; Field, C.B., Barros, V.R., Dokken, D.J., Mach, K.J., Mastrandrea, M.D., Bilir, T.E., Chatterjee, M., Ebi, K.L., Estrada, Y.O., Genova, R.C., et al., Eds.; Cambridge University Press: New York, NY, USA, 2014; pp. 109–111.
25. Coumou, D.; Robinson, A.; Rahmstorf, S. Global increase in record-breaking monthly-mean temperatures. *Clim. Chang.* **2013**, *118*, 771–782. [CrossRef]
26. Forzieri, G.; Cescatti, A.; e Silva, F.B.; Feyen, L. Increasing risk over time of weather-related hazards to the European population: A data-driven prognostic study. *Lancet Planet. Health* **2017**, *1*, e200–e208. [CrossRef]
27. Miralles, D.G.; Teuling, A.J.; Van Heerwaarden, C.C.; de Arellano, J.V.-G. Mega-heatwave temperatures due to combined soil desiccation and atmospheric heat accumulation. *Nature Geosci.* **2014**, *7*, 345. [CrossRef]
28. Lesk, C.; Rowhani, P.; Ramankutty, N. Influence of extreme weather disasters on global crop production. *Nature* **2016**, *529*, 84. [CrossRef]
29. Quiller, G.; Krenz, J.; Ebi, K.; Hess, J.J.; Fenske, R.A.; Sampson, P.D.; Pan, M.; Spector, J.T. Heat exposure and productivity in orchards: Implications for climate change research. *Arch. Environ. Occup. Health* **2017**, *72*, 313–316. [CrossRef]
30. Mbow, C.; Smith, P.; Skole, D.; Duguma, L.; Bustamante, M. Achieving mitigation and adaptation to climate change through sustainable agroforestry practices in Africa. *Curr. Opin. Environ. Sustain.* **2014**, *6*, 8–14. [CrossRef]
31. Wallace, P.J.; Mckinlay, B.J.; Coletta, N.A.; Vlaar, J.I.; Taber, M.J.; Wilson, P.M.; Cheung, S.S. Effects of motivational self-talk on endurance and cognitive performance in the heat. *Med. Sci. Sports Exerc.* **2017**, *49*, 191–199. [CrossRef] [PubMed]
32. Schmit, C.; Hausswirth, C.; Le Meur, Y.; Duffield, R. Cognitive functioning and heat strain: Performance responses and protective strategies. *Sports Med.* **2017**, *47*, 1289–1302. [CrossRef]
33. Zander, K.K.; Botzen, W.J.; Oppermann, E.; Kjellstrom, T.; Garnett, S.T. Heat stress causes substantial labour productivity loss in Australia. *Nat. Clim. Chang.* **2015**, *5*, 647. [CrossRef]
34. Kovats, R.S.; Hajat, S. Heat stress and public health: A critical review. *Annu. Rev. Public Health* **2008**, *29*, 41–55. [CrossRef]
35. Horie, S. Prevention of heat stress disorders in the workplace. *J. Jpn. Med Assoc.* **2012**, *141*, 289–293.
36. Shackleton, S.; Ziervogel, G.; Sallu, S.; Gill, T.; Tschakert, P. Why is socially-just climate change adaptation in sub-Saharan Africa so challenging? A review of barriers identified from empirical cases. *Wiley Interdiscip. Rev. Clim. Chang.* **2015**, *6*, 321–344. [CrossRef]
37. UNCTAD. *The Role of Smallholder Farmers in Sustainable Commodities Production and Trade*; United Nations Conference on Trade and Development (UNCTAD): Geneva, Switzerland, 2015; p. 17.
38. Rapsomanikis, G. *The Economic Lives of Smallholder Farmers—An Analysis Based on Household Data from Nine Countries*; United Nations Food and Agriculture Organization (FAO): Rome, Italy, 2016; p. 47.
39. Abraham, M.; Pingali, P. *Transforming smallholder agriculture to achieve the SDGs In The Role of Small Farms in Food and Nutrition Security*; Riesgo, L., Gomez-Y-Paloma, S., Louhichi, K., Eds.; Springer: New York, NY, USA, 2017; p. 41.
40. Mrema, G.; Baker, D.; Kahan, D. *Agricultural Mechanization in Africa: Time for Action*; Food and Agriculture Organization of the United Nations (FAO): Rome, Italy, 2008; p. 36.

41. Houmy, K.; Clarke, L.J.; Ashburner, J.E.; Kienzle, J. *Agricultural Mechanization in Sub-Saharan Africa—Guidelines for Preparing a Strategy*; Food and Agriculture Organization of the United Nations (FAO): Rome, Italy, 2013; p. 105.
42. Falola, T. *Africanizing Knowledge: African Studies across the Disciplines*; Routledge: London, UK, 2017.
43. Sims, B.G.; Kienzle, J. *Farm Power and Mechanization for Small Farms in Sub-Saharan Africa*; Food and Agriculture Organization of the United Nations (FAO): Rome, Italy, 2007; p. 92.
44. Lotter, D. Facing food insecurity in Africa: Why, after 30 years of work in organic agriculture, I am promoting the use of synthetic fertilizers and herbicides in small-scale staple crop production. *Agric. Hum. Values* **2015**, *32*, 111–118. [CrossRef]
45. Kassa, H.; Dondeyne, S.; Poesen, J.; Frankl, A.; Nyssen, J. Transition from forest-based to cereal-based agricultural systems: A review of the drivers of land use change and degradation in Southwest Ethiopia. *Land Degrad. Dev.* **2017**, *28*, 431–449. [CrossRef]
46. Fairhead, J.; Fraser, J.; Amanor, K.; Solomon, D.; Lehmann, J.; Leach, M. Indigenous Soil Enrichment for Food Security and Climate Change in Africa and Asia: A Review. In *Indigenous Knowledge-Enhancing Contribution to Natural Resource Management*; CABI: Wallingford, UK, 2017; pp. 99–115.
47. Willis, R.G. *The Fipa and Related Peoples of South-West Tanzania and North-East Zambia: East Central Africa*; Routledge: London, UK, 2017.
48. Knight, C.G. Ethnoscience and the African farmer: Rationale and strategy. In *Indigenous Knowledge Systems and Development*; University Press of America: Lanham, MD, USA, 1980; pp. 205–231.
49. Joy, P. The crisis of farming systems in Luapula Province, Zambia. *Nord. J. Afr. Stud.* **1993**, *2*, 118–141.
50. Murray, S.E. Conservation tillage. In *Sustainable Intensification of Crop Production*; Reddy, P.P., Ed.; Springer Nature: Singapore, 2016; pp. 27–40.
51. Njoh, A.J. Transportation infrastructure and economic development in sub-Saharan Africa. *Public Works Manag. Policy* **2000**, *4*, 286–296. [CrossRef]
52. USDAAF. *Heat Stress Control and Heat Casualty Management*; US Department of the Army and Air Force (USDAAF): Washington, DC, USA, 2003.
53. Liljegren, J.C.; Carhart, R.A.; Lawday, P.; Tschopp, S.; Sharp, R. Modeling the wet bulb globe temperature using standard meteorological measurements. *J. Occup. Environ. Hyg.* **2008**, *5*, 645–655. [CrossRef] [PubMed]
54. Yaglou, C.; Minaed, D. Control of heat casualties at military training centers. *Arch. Indust. Health* **1957**, *16*, 302–316.
55. Jones, C.; Giorgi, F.; Asrar, G. The Coordinated Regional Downscaling Experiment: CORDEX–an international downscaling link to CMIP5. *CLIVAR Exch.* **2011**, *16*, 34–40.
56. Arora, V.; Scinocca, J.; Boer, G.; Christian, J.; Denman, K.; Flato, G.; Kharin, V.; Lee, W.; Merryfield, W.J.G.R.L. Carbon emission limits required to satisfy future representative concentration pathways of greenhouse gases. *Geophys. Res. Lett.* **2011**, *38*, 1–6. [CrossRef]
57. Hajizadeh, R.; Farhang Dehghan, S.; Golbabaei, F.; Jafari, S.M.; Karajizadeh, M. Offering a model for estimating black globe temperature according to meteorological measurements. *Meteorol. Appl.* **2017**, *24*, 303–307. [CrossRef]
58. Stull, R. Wet-bulb temperature from relative humidity and air temperature. *J. Appl. Meteorol. Climatol.* **2011**, *50*, 2267–2269. [CrossRef]
59. Luo, M.; Liu, T.; Meng, F.; Duan, Y.; Frankl, A.; Bao, A.; De Maeyer, P. Comparing bias correction methods used in downscaling precipitation and temperature from regional climate models: A case study from the Kaidu River Basin in Western China. *Water* **2018**, *10*, 1046. [CrossRef]
60. Casanueva, A.; Kotlarski, S.; Herrera, S.; Fischer, A.M.; Kjellstrom, T.; Schwierz, C. Climate projections of a multivariate heat stress index: The role of downscaling and bias correction. *Geosci. Model Dev.* **2019**, *12*, 3419–3438. [CrossRef]
61. IPCC. Climate Change 2013: The Physical Science Basis. In *Contribution of Working Group I to the Fifth Assessment Report of the Intergovernmental Panel on Climate Change*; Stocker, T.F., Qin, D., Plattner, G.-K., Tignor, M., Allen, S.K., Boschung, J., Nauels, A., Xia, Y., Bex, V., Midgley, P.M., Eds.; Cambridge University Press: Cambridge, UK; New York, NY, USA, 2013; p. 1535.
62. Kohut, L. Stress tolerance of military personnel during exercise in hot, dry climates–Prevention and treatment. *Medicine* **2008**, *7*, 301–307.

63. Frimpong, K.; Eddie Van Etten, E.; Oosthuzien, J.; Nunfam, V.F. Heat exposure on farmers in northeast Ghana. *Int. J. Biometeorol.* **2017**, *61*, 397–406. [CrossRef]
64. Frimpong, K.; Odonkor, S.T.; Kuranchie, F.A.; Nunfam, V.F. Evaluation of heat stress impacts and adaptations: Perspectives from smallholder rural farmers in Bawku East of Northern Ghana. *Heliyon* **2020**, *6*, e03679. [CrossRef] [PubMed]
65. Sadiq, L.S.; Hashim, Z.; Osman, M. The Impact of Heat on Health and Productivity among Maize Farmers in a Tropical Climate Area. *J. Environ. Public Health* **2019**, *2019*. [CrossRef] [PubMed]
66. ICLARM. *The Context of Small-Scale Integrated Agriculture-Aquaculture Systems in Africa: A Case Study of Malawi*; Number 5878; International Center for Living Aquatic Resources Management (ICLARM), the WorldFish Center: Manilla, Philippines, 1991; Volume 18, p. 311.
67. FAO. *Small Family Farms Data Portrait: Basic Information on Document—Methodology and Data Description*; Food and Agriculture Organization of the United Nations Organization: Rome, Italy, 2017; p. 16. Available online: http://www.fao.org/fileadmin/user_upload/smallholders_dataportrait/docs/Data_portrait_variables_description_new2.pdf (accessed on 7 February 2020).
68. Field, C.B. *Climate change 2014–Impacts, Adaptation and Vulnerability: Regional Aspects*; Cambridge University Press: Cambridge, UK, 2014.
69. Sonwa, D.J.; Dieye, A.; El Mzouri, E.-H.; Majule, A.; Mugabe, F.T.; Omolo, N.; Wouapi, H.; Obando, J.; Brooks, N. Drivers of climate risk in African agriculture. *Clim. Dev.* **2017**, *9*, 383–398. [CrossRef]
70. Shiferaw, B.; Tesfaye, K.; Kassie, M.; Abate, T.; Prasanna, B.; Menkir, A. Managing vulnerability to drought and enhancing livelihood resilience in sub-Saharan Africa: Technological, institutional and policy options. *Weather Clim. Extrem.* **2014**, *3*, 67–79. [CrossRef]
71. Lankoski, J.; Ollikainen, M.; Uusitalo, P. *No-till Technology: Benefits to Farmers and the Environment*; Agrifood Research Finland: Helsinki, Finland, 2004; p. 36.
72. Mukasa, A.N.; Woldemichael, A.D.; Salami, A.O.; Simpasa, A.M. Africa's Agricultural Transformation: Identifying Priority Areas and Overcoming Challenges. *Afr. Econ. Brief* **2017**, *8*, 1–16.
73. Crowe, J.; Manuel Moya-Bonilla, J.; Román-Solano, B.; Robles-Ramírez, A. Heat exposure in sugarcane workers in Costa Rica during the non-harvest season. *Glob. Health Action* **2010**, *3*, 5619. [CrossRef]
74. Bodin, T.; García-Trabanino, R.; Weiss, I.; Jarquín, E.; Glaser, J.; Jakobsson, K.; Lucas, R.A.I.; Wesseling, C.; Hogstedt, C.; Wegman, D.H. Intervention to reduce heat stress and improve efficiency among sugarcane workers in El Salvador: Phase 1. *Occup. Environ. Med.* **2016**, *73*, 409–416. [CrossRef]
75. Watts, N.; Adger, W.N.; Agnolucci, P.; Blackstock, J.; Byass, P.; Cai, W.; Chaytor, S.; Colbourn, T.; Collins, M.; Cooper, A.J.T.L. Health and climate change: Policy responses to protect public health. *Lancet* **2015**, *386*, 1861–1914. [CrossRef]
76. Ngwenya, B.; Oosthuizen, J.; Cross, M.; Frimpong, K. Heat Stress and Adaptation Strategies of Outdoors Workers in the City of Bulawayo, Zimbabwe. *Community Med. Public Health Care* **2018**, *5*, 034. [CrossRef]
77. WHO. *Protecting Health in Europe from Climate Change—2017 Update*; The Regional Office for Europe of the World Health Organization (WHO): Copenhagen, Denmark, 2017; p. 85.
78. Wilhelmi, O.; Hayden, M. Reducing vulnerability to extreme heat through interdisciplinary research and stakeholder engagement. In *Extreme Weather, Health, and Communities*; Springer: Berlin/Heidelberg, Germany, 2016; pp. 165–186.
79. Conway, G. The science of climate change in Africa: Impacts and adaptation. *Grantham Inst. Clim. Chang. Discuss. Pap.* **2009**, *1*, 24.
80. Niang, I.; Ruppel, O.; Abdrabo, M.; Essel, A.; Lennard, C.; Padgham, J.; Urquhart, P. Contribution of Working Group II to the Fifth Assessment Report of the Intergovernmental Panel on Climate Change. In *Climate Change 2014: Impacts, Adaptation and Vulnerability*; Cambridge University Press: Cambridge, UK, 2014.
81. AGRA. *Africa Agriculture Status Report: Catalyzing Government Capacity to Drive Agricultural Transformation*; Alliance for a Green Revolution in Africa (AGRA): Nairobi, Kenya, 2018; p. 234.

MDPI
St. Alban-Anlage 66
4052 Basel
Switzerland
Tel. +41 61 683 77 34
Fax +41 61 302 89 18
www.mdpi.com

Atmosphere Editorial Office
E-mail: atmosphere@mdpi.com
www.mdpi.com/journal/atmosphere